VOLUME SIXTY ONE

Developments in Sedimentology

CARBONATES IN CONTINENTAL SETTINGS: FACIES, ENVIRONMENTS, AND PROCESSES

VOLUME SIXTY ONE

DEVELOPMENTS IN SEDIMENTOLOGY

CARBONATES IN CONTINENTAL SETTINGS: FACIES, ENVIRONMENTS, AND PROCESSES

Editors

A.M. Alonso-Zarza

Dpto Petrología y Geoquímica,
Fac. CC. Geológicas, IGE-CSIC, Universidad
Complutense de Madrid, Madrid, Spain

and

L.H. Tanner

Department of Biological Sciences,
Le Moyne College, Syracuse, NY, USA

ELSEVIER

Amsterdam • Boston • Heidelberg • London • New York • Oxford
Paris • San Diego • San Francisco • Singapore • Sydney • Tokyo

Elsevier
Radarweg 29, PO Box 211, 1000 AE Amsterdam, The Netherlands
Linacre House, Jordan Hill, Oxford OX2 8DP, UK

First edition 2010

British Library Cataloguing in Publication Data
A catalogue record for this book is available from the British Library

Library of Congress Cataloging-in-Publication Data
A catalog record for this book is available from the Library of Congress

ISBN: 978-0-444-53025-7
ISSN: 0070-4571

Printed and bound in Great Britain

10 11 12 13 14 10 9 8 7 6 5 4 3 2 1

CONTENTS

Contributors ix

Preface xi

1. Lacustrine Carbonates **1**
Elizabeth H. Gierlowski-Kordesch

1. Introduction 2
2. Lake Terminology 2
3. Mineralogy and Texture 3
4. Lake Types and Global Distribution 5
5. Controls on Lacustrine Carbonates 7
6. Depositional Processes 11
7. Preservation of Lacustrine Carbonates 14
8. Fauna/Flora 15
9. Carbon and Oxygen Isotope Analyses 20
10. Diagenesis 24
11. Facies Types 26
12. Carbonate Lake Models 61
13. Concluding Remarks 69

Acknowledgements 70
References 70

2. Palustrine Carbonates **103**
A.M. Alonso-Zarza and V.P. Wright

1. Introduction 103
2. The Sedimentary (Ancient) Record of Palustrine Carbonates 105
3. Wetlands: The Recent Analogues for Ancient Palustrine Environments? 115
4. Environmental Controls on Palustrine Deposition 119
5. Conclusions 125

Acknowledgements 126
References 126

3. Fluvial and Associated Carbonate Deposits **133**
C. Arenas-Abad, M. Vázquez-Urbez, G. Pardo-Tirapu and C. Sancho-Marcén

1. Introduction 133
2. Context of Formation 135

Contributors

A.M. Alonso-Zarza
Dpto Petrología y Geoquímica, Fac. CC. Geológicas, IGE-CSIC, Universidad Complutense de Madrid, 28040 Madrid, Spain

C. Arenas-Abad
Department of Earth Sciences, Group of Continental Sedimentary Basin Analysis, Stratigraphy, University of Zaragoza, 50009 Zaragoza, Spain

E.S. Bellosi
CONICET-División Icnología, Museo Argentino de Ciencias Naturales, Av. Angel Gallardo 470, 1405 Buenos Aires, Argentina

A. Borsato
School of Environmental and Life Sciences, The University of Newcastle, Callaghan, NSW 2308, Australia

S. Frisia
School of Environmental and Life Sciences, The University of Newcastle, Callaghan, NSW 2308, Australia

J.F. Genise
CONICET-Museo Paleontológico Egidio Feruglio, Av. Fontana 140, 9100 Trelew, Chubut, Argentina

Elizabeth H. Gierlowski-Kordesch
Department of Geological Sciences, Ohio University, Athens, OH 45701-2979, USA

Brian Jones
Department of Earth and Atmospheric Sciences, University of Alberta, Edmonton, Canada, AB T6G 2E3

R.N. Melchor
CONICET-Facultad de Ciencias Naturales, Universidad Nacional de La Pampa, Av. Uruguay 154, 6300 Santa Rosa, La Pampa, Argentina

G. Pardo-Tirapu
Department of Earth Sciences, Group of Continental Sedimentary Basin Analysis, Stratigraphy, University of Zaragoza, 50009 Zaragoza, Spain

Robin W. Renaut
Department of Geological Sciences, University of Saskatchewan, Saskatoon, Canada SK S7N 5E2

C. Sancho-Marcén
Department of Earth Sciences, Geomorphology, University of Zaragoza, C/Pedro Cerbuna 12, 50009 Zaragoza, Spain

M. Vázquez-Urbez
Department of Earth Sciences, Group of Continental Sedimentary Basin Analysis, Stratigraphy, University of Zaragoza, 50009 Zaragoza, Spain

M. Verde
SNI, ANII, Facultad de Ciencias, Universidad de la República, Iguá 4225, CP 11400, Montevideo, Uruguay

V.P. Wright
BG Group, 100 Thames Valley Park Drive, Reading RG6 1PT; School of Earth & Ocean Sciences, Cardiff University, Cardiff CF10 3YE, UK

Preface

Although sedimentary rocks constitute only a tiny fraction of the volume of Earth's crust, they cover roughly 75% of the continental area. Continental carbonates are a significant component of this continental cover. Today, these deposits occur in a wide variety of depositional (and diagenetic) environments, and presumably they formed in equally varied settings in ancient times as well. The modern environments where they are found include very fragile and sometimes visually spectacular settings, such as wetlands, caves, rivers and lakes. In many cases, these areas are considered to have special natural and historical significance, such as Mammoth Hot Springs in the United States, Pammukale in Turkey and the Guilin karst in China.

Unlike marine carbonates, the record of which extends back to the Archean, continental carbonates did not become prominent in the geological record until the Devonian, when plants widely colonised continental environments and promoted the release of ions through enhanced weathering. Ever since, these carbonates have served as important archives of ancient continental environments by preserving a variety of information about key factors in their formation, such as climate and resident ecosystems.

The study of both recent and ancient continental carbonates has involved earth scientists from varied disciplines, such as sedimentary petrology, sedimentology, geomorphology, geochemistry, pedology, hydrology, palaeontology and of sedimentary geology in general. This dictates that the proper understanding of these carbonates requires a multidisciplinary approach. Notably, the wide variety and types of continental carbonates, and in many cases the special characteristics of some of them, and of the environments in which they form, make many of these deposits vastly different from marine carbonates. All of this, in fact, may explain why continental carbonates have never been considered collectively in a single review, such as this. There are, of course, a number of books that include in their scope some aspects or types of continental carbonates, but this is the first collection dedicated solely to the full suite of continental carbonates.

The aim of this review, therefore, is to provide an up-to-date compilation of the current research on the formation of continental carbonate deposits, as well as their petrography, facies, diagenesis, geochemistry, palaeoenvironmental significance and economic aspects. It is intended to provide an intermediate step between the university-level sedimentology texts used by undergraduate students and the highly specialised literature of research journals on which working scientists depend. In this way it offers a bridge in understanding between the more basic levels of petrology, sedimentology, geochemistry and basin analysis and their applications for teaching at the graduate level or advanced research. However, adequate coverage of the broad range of topics within this subject area demands more space than can be accommodated by a single volume. For this reason, this review comprises two volumes: the first volume is dedicated to the formation of carbonate sediments in continental environments, covering the depositional settings, facies and sedimentological processes and the second volume of this set examines the geochemistry, diagenesis, sequence stratigraphy of these deposits, along with some of the practical applications.

The first four chapters (1–4) of this first volume are dedicated to the study of the environments in which carbonate sediments are deposited. The overview on lacustrine carbonates (Chapter 1) utilises an extensive database consisting of hundreds of basins distributed globally. After defining the basic terms of lacustrine environments, the author describes the mineralogy, texture and global distribution of lake carbonates. Particular emphasis is given to depositional processes, diagenesis, geochemistry and lake-facies models. Palustrine carbonates are discussed in Chapter 2. Ancient and modern deposits are analysed, and the influences of tectonism and climate on their formation are considered. Additionally, the authors consider that wetlands are a possible modern analogue for these deposits. Fluvial and associated carbonate deposits formed at ambient temperature are the focus of Chapter 3. This chapter includes detailed discussions on the facies associations, depositional models, controls on formation and the stable-isotope signals of these types of carbonates. Calcareous spring deposits are analysed in Chapter 4. This chapter provides a broad discussion of the terminology of these deposits, which are closely related to the fluvial deposits of the previous chapter. Furthermore, the role of temperature, the mechanisms of precipitation, the typical facies and the diagenesis of spring deposits are presented in detail.

The succeeding section of this volume (Chapters 5 and 6) focuses on carbonates formed by meteoric diagenetic processes (calcretes and karst) as well as by the diagenetic modifications of the different types of continental carbonate sediments. Calcretes are analysed in Chapter 5; their classification, micro- and macromorphology, and their types and mechanisms of formation are discussed. The final part of the chapter focuses on the significance of calcretes in Quaternary environments and ancient sedimentary basins. Chapter 6 discusses karst deposits and their significance as archives of environmental signals, such as climate, but also describes how gravitational, seismic and volcanic processes, among others, can be detected through the study of karst deposits. The mineralogy and morphological features, both macro- and microscale, are described in this chapter.

Trace fossils have become a very useful tool for interpreting ancient sedimentary basins. They were commonly used and applied in marine basins, but until recently were less studied and poorly understood in the continental realm. The final chapter of this volume presents the state of knowledge as well as new advances in the ichnology of continental carbonates. The case studies provided form the basis for a proposed ichnofacies model for continental carbonate environments.

Of course, a thorough understanding of carbonate sediments also requires study of their geochemistry and diagenesis. These topics are given thorough treatment in the second volume of this review, with specific chapters dedicated to the geochemical composition, carbonate diagenesis and silicification of carbonate sediments. Other chapters examine the applications of continental carbonates to palaeoclimate studies and sequence stratigraphy. The final two chapters of the volume describe the economic importance of continental carbonate deposits, and their value as natural heritage sites worldwide.

There are indeed several 'hot' topics of continental carbonates that do not have their own chapters, 'the dolomite problem' and the role of micro-organisms in the formation of carbonates, for example although these issues are covered in multiple chapters. We hope the information provided here is a good starting point, and that the reader finds this book both stimulating and informative, whether for teaching or for research. Overall, these chapters will, ideally, constitute a base for understanding that continental carbonates are an integral part of ancient and recent landscapes and contribute to the broader knowledge of continental basins.

This book in its final form would not have been possible without the work of the reviewers who dedicated their time to careful reviews and revisions. We were truly lucky to have had the help of the following colleagues: P. Anadón, C. Arenas, Ll. Cabrera, F. McDermott, D. Deocampo, S. Dunagan, B. Jones, D. Larsen, M. Pedley, R. Renaut, B.P. Singh, R. Sinha, N. Tabor and L. Tapanila. Our sincere thanks also go to our families and our departments: Departamento de Petrología y Geoquímica de la Universidad Complutense, IGE-CSIC, de Madrid, the Department of Biological Sciences at Le Moyne College. Lastly, producing this volume would not have been possible without the assistance of the book series editor, A.J. (Tom) van Loon, of Adam Mickiewicz University, Poznan.

A.M. Alonso-Zarza
L.H. Tanner
Editors

11.3. Microbial carbonates 53
11.4. Marginal carbonates 56
11.5. Open-water carbonates 59
12. Carbonate Lake Models 61
13. Concluding Remarks 69
Acknowledgements 70
References 70

1. Introduction

This chapter provides an overview of the type of carbonate facies and fossils found in saline and freshwater lake systems as well as of the distribution of these facies across perennial lake environments. The accumulation of carbonate sediments in lakes both modern and ancient has been studied for decades with respect to precipitation conditions, sedimentologic influences, petrology, isotopic composition, and basin case studies. Early last century, for example, Pollock (1918) discussed how the activity of blue-green algae produced marl in the freshwater lakes of Michigan (USA) and Bradley (1929a,b) studied the algal reefs, oolites, and varves of the Green River Formation in the western USA. Reconstruction of the water chemistries, climate, productivity, catchment drainage, and hydrology of lakes and their basins evolved in sophistication in the latter half of the last century and continues into the present century. This review contains examples from a data set compiled from the literature of modern unconsolidated to lithified, subaqueous lake carbonates of nearly 250 lakes and lake basins worldwide to illustrate the characteristics of carbonate deposits in lakes.

2. Lake Terminology

The term "perennial lake" refers to a continental depression with a continuous coverage by a water column of any depth (subaqueous conditions), whereas the term "ephemeral lake" describes a continental depression with periodic exposure of its sediments (subaqueous to subaerial conditions). Ephemeral lakes and their carbonate deposits are described in the chapter on palustrine carbonates; only perennial lakes and their carbonate deposits are described here.

The subaqueous zones of a lake extend from the shoreline to the open-water depocenter. Depending on the shape of the lake basin, a lake can have gentle or steep slopes from shallow to deeper water with relatively level areas in shallow zones leading to steep transitions into deeper water called "benches" or "terraces." The shallow water zone around the large margin is called the "littoral zone" while the "sublittoral zone" is the transition to deeper water. Light can still penetrate down to the bottom in the sublittoral zone. The open-water lake area of greatest depth with minimal to no sunlight penetration is termed the "profundal zone." The water column above the profundal zone within the upper photic zones of the water column is termed the "pelagic zone." Lake water can stratify mainly through salinity or temperature differences in the water column, essentially producing two independent bodies of water of different chemistries. The upper "epilimnion" is separated from the lower "hypolimnion" by a thermocline or "metalimnion" in freshwater lakes stratified by temperature, whereas the upper "mixolimnion" is separated by a chemocline from the lower "monimolimnion" in saline or meromictic (permanently stratified) lakes. Anoxia can develop in the hypolimnion or monimolimnion seasonally or permanently (Wetzel, 2001; Loeffler, 2004). Deep and narrow lakes stratify more easily (Valero Garcés et al., 1995). Interestingly, ecologists refer to the lake catchment area as the "paralimnion" (Binford et al., 1983).

3. Mineralogy and Texture

Most lakes contain some type of carbonate mineral (Kelts and Hsü, 1978; Kelts and Talbot, 1990). Hard water lakes contain waters with more than 1.0 milliequivalents l^{-1} of total dissolved alkaline-earth carbonates (mostly calcium and magnesium) and surface saturation with respect to carbonate at least during the summer months (Dean, 1981; Dean and Fouch, 1983). The types of minerals that can accumulate in lakes around the world encompass a wide spectrum, depending on source area, salinity of the waters, and hydrology of the lake basin. Table 1 lists the major carbonate and bicarbonate minerals reported from modern lakes, ancient lakes (extant but older than 100,000 years: Gorthner, 1994), as well as those from the geologic record ("extinct lakes"). These represent primary precipitates, detrital or clastic minerals, as well as minerals formed during early to later diagenetic processes. Most of the minerals in Table 1 accumulate in saline lake to playa sediments. Fresher lakes mostly contain calcite and aragonite as primary carbonates (endogenic). This review will focus on calcium carbonate to magnesium carbonate deposition under subaqueous lacustrine freshwater to saline conditions.

Most carbonate deposition is in the form of mud-sized material. This "mud" in both the freshwater and marine realm is referred to as carbonate mud or, if lithified, as mudstone, micrite, or calcilutite. This generally

Table 2 Lake classification based on a sequence-stratigraphic depositional model.

	Overfilled	Balanced-filled	Underfilled
	Fluvial–lacustrine facies association	**Fluctuating-profundal facies association**	**Evaporative facies association**
Climate	All types	All types	All types
Basin subsidence	Generally low	Low to high	Generally medium to high
Stratal stacking patterns	Progradational	Progradational and aggradational	Aggradational
	Parasequences: poorly defined	Parasequences: well defined	Parasequences: well and poorly defined
Lithology	Mudstone/ marlstone	Mudstone/marlstone	Mudstone/marlstone
	Siltstone	Siltstone	Siltstone
	Sandstone	Sandstone	Some sandstone
	Some carbonate	Carbonate	Kerogenite
	Coquina	Kerogenite	Evaporites
	Coal/coaly shale	Some evaporites/ coal	Carbonate
Shorelines	Stable	Unstable to stable	Mostly unstable
Surface water inflow	Maximum fluvial channel	Fluvial channel/ alluvial Sheetflooding	Alluvial sheetflooding
Hydrology	Open	Open/closed	Closed
Organic matter	Terrigenous and microbial	Microbial dominant	Hypersaline microbial
Hydrocarbons	Oil and gas	Mostly oil (low sulfur)	Mostly oil (moderate/high sulfur)

See Bohacs et al. (2000, 2003) for details. Carbonates can occur in all three types of lakes.

Lacustrine carbonates can be found in many different types of tectonic basins (Platt and Wright, 1991), classified into three major tectonic scenarios: (1) lithospheric divergence, (2) crustal convergence, and (3) transform regimes (Cohen, 2003). Lake basins formed in divergent regimes that contain carbonate deposits include the rift lakes of East Africa (e.g., Renaut et al., 1986; Haberyan and Hecky, 1987; Tiercelin et al., 1993; Casanova, 1994), the Triassic–Jurassic Newark Supergroup of eastern North America (e.g., De Wet et al., 2002), and the Early Cretaceous rift lakes of the South Atlantic Ocean (e.g., Abrahão and Warme, 1990; De Carvalho et al., 2000). Lake basins in convergent situations containing lacustrine carbonates include the Pennsylvanian rocks of the Appalachian Basin (e.g., Valero Garcés et al., 1994, 1997), the Miocene lakes of the central Tibetan Plateau (Wu et al.,

2008), and the western basins of the USA containing the Eocene Green River Formation (Buchheim and Eugster, 1998; Smith et al., 2008). Transform lake basins with carbonate deposits include the Dead Sea Basin (Bartov et al., 2002; Bookman et al., 2004), the Miocene Ridge Basin of California (Link and Osborne, 1978; Talbot, 1994; Link, 2003), and the Proterozoic Nonacho Basin of northern Canada (Asplar and Donaldson, 1985). Cratonic or sag basins also contain lakes with carbonate accumulation. A major example is Lake Eyre in Australia (Magee et al., 1995).

Carbonate deposits are also common in volcanic, impact, and glacial lake basins across the globe. Maar lakes, for example, Lake Keilambete, West Basin Lake, and East Basin Lake in Victoria (Australia) (Bowler, 1981; Last and DeDeckker, 1990; Last, 1992), impact craters, for example, Ries and Steinheim craters in Germany (Arp, 1995; Heizmann and Reiff, 2002), and volcanic craters, for example, Creede Formation in the Oligocene of Colorado (USA) (Larsen and Crossey, 1996; Rye et al., 2000) and Holocene Nar Gölü (Turkey) (Jones et al., 2005), can also accumulate carbonates. In addition, periglacial lakes can accumulate carbonate (Fairchild et al., 1994), including Quaternary to modern Lake Zürich (Zhao et al., 1984) and Lake Geneva (Moscariello, 1997; Davaud and Girardclos, 2001).

5. Controls on Lacustrine Carbonates

Three major factors control sedimentation patterns, and thus carbonate deposition, in lake basins: hydrology (including water output and input of surface water, rainfall, and groundwater), sediment input, and temperature changes (Tucker and Wright, 1990; Platt and Wright, 1991). Climate and tectonics influence these three factors in tandem; separating their effects on lake sedimentation is difficult. Water input is controlled by hydrology (tectonic situation) as well as climate.

5.1. Hydrology

Open or closed drainage and the amount of accommodation space in the basin for lake formation are key tectonic elements in basin hydrology (Bohacs et al., 2000, 2003), controlling most of the input and output of water. Climate as well as the tectonic configuration of a lake basin (including the resulting topography and bedrock characteristics) (e.g., Pietras et al., 2003) controls water input and hydrologic characteristics. Water enters a lake basin through rainfall, surface inflow, and groundwater (see Rosen, 1994; Winter, 2004). Surface inflow from rivers can bring in carbonates through bedload, suspended load, or in solution. For example, the Eocene lake carbonates of the Shulu Sag in north China (Jiang et al., 2007) comprise debris flows of matrix- to clast-supported rudstones, littoral to sublittoral pebbly carbonate arenites and calcisiltites, and profundal finely laminated calcilutites associated with alluvial input along its hinged western slope where the catchment area is composed of Cambro-Ordovician carbonate bedrock. A modern example is Kelly Lake in British Columbia (Canada), where a carbonate bench developed on the east side of the lake at the river input from a source area containing mostly Permian carbonate rocks (Renaut, 1990; Ferris et al., 1997). The west side of the lake contains siliciclastic sediments reflecting rocks exposed on the other side of Kelly Lake. It is clear that lakes that accumulate carbonates are not "sediment-starved," but receive an influx of carbonate (whether as traction bedload, suspended load, or as dissolved load) that is overwhelmingly larger than any siliciclastic input.

Groundwater input by springs and subaqueous seeps can promote carbonate precipitation along the fringes of alluvial fans (Nickel, 1985) and directly into lake waters (e.g., the Dead Sea in Israel; Gvirtzman, 2006; Yechieli, 2006). Subaqueous seeps along the margins of lakes can introduce large amounts of carbonates that can then be spread around the littoral zone by wave and current action (Schneider, 1977). For example, the exposed bedrock around Lake Tanganyika contains no carbonates, but the subaqueous seeps along the lake margin of this rift lake are calcium-rich and carbonate precipitates all along the seepage zones in the littoral to sublittoral zone (Cohen and Thouin, 1987; Barrat et al., 2000), attesting to subsurface carbonate-rich rocks. Lacustrine tufa mounds or bioherms, as in Mono Lake (California, USA) (Dunn, 1953; Scholl and Taft, 1964), Walker Lake (Nevada, USA) (Newton and Grossman, 1988), the Miocene "travertine" of the Nördlinger Ries crater in Germany (Arp, 1995), the Pleistocene saline lakes of the central Altiplano of Bolivia (Rouchy et al., 1996), the Great Salt Lake (USA) (Colman et al., 2002), and Fayetteville Green Lake (New York, USA) (Thompson et al., 1990), form at groundwater seep points, related to faulting or basin hydrology.

In addition to the influx of appropriate ions through surface water and groundwater, the atmosphere also directly contributes ions to lake waters (Stumm, 2004). Wet deposition, through rainfall, and dry deposition, through wind, can also introduce carbonate and calcium into a lake. Separating sedimentary carbonates introduced by rain from those from surface or groundwater input is presently impossible. General catchment studies provide evidence of a significant contribution by rainfall (Lindberg et al., 1986; Meybeck, 1995; Lesack and Melack, 1991; Stumm, 2004) to the salinity of lake and river waters. Wind deposits of carbonates are difficult to recognize in lake sediments. Modern examples from the desert (Capo and Chadwick, 1999) as well as the existence of carbonate dunes (Warren, 2006) suggest that this source of carbonate can be important for lakes. In more arid climates, dry deposition by wind can exceed wet deposition (Wetzel, 2001).

5.2. Sediment input

Lake sediments are largely the products of the erosion of the catchment area (both surface and subsurface rocks) (Hinderer and Einsele, 2001), so source area rocks control the lake sediment type (e.g., Calvo et al., 1989; Alonso-Zarza and Calvo, 2000; Gebhardt et al., 2000). Clearly widespread carbonate rocks in a source area will allow for thick carbonate accumulations in the associated lake basin (Jones and Bowser, 1978; Gierlowski-Kordesch, 1998), independent of climate (e.g., Kennedy and Smith, 1977; Glass and Wilkinson, 1980; Jiang et al., 2007). For example, Lake Ohrid in Albania and Macedonia (Stankovic, 1960; Roeloefs and Kilham, 1983; Wagner et al., 2009), Fayetteville Green Lake in New York State, USA (Brunskill, 1969; Thompson et al., 1990), and Sucker Lake in Michigan, USA (Treese and Wilkinson, 1982), all contain mostly carbonate sediments. Their provenance is dominated by Paleozoic to Mesozoic carbonate rocks. Thunmark (1937) showed in his study of more than 460 lakes in southern Sweden that carbonate sediments are preserved in lakes associated with carbonate bedrock or carbonate glacial deposits. Lake deposits of the Steinheim impact crater in Baden-Württemberg (Germany) are predominately carbonate since the impact occurred in Jurassic carbonate bedrock (Heizmann and Reiff, 2002).

Thick magnesite ($MgCO_3$) layers as well as hydromagnesite stromatolites are found in the Neogene to present-day lake basins of western Turkey. Source rocks for these lake systems include carbonate rocks as well as Mg-rich ultrabasic rocks, such as serpentinite (Braithwaite and Zedef, 1996; Zedef et al., 2000). The mineralogy of the source rocks clearly is reflected in the resultant basinal sediments.

On the opposite side of the spectrum, Lake Baikal (Siberia) contains no carbonates in its sediments (Fagel et al., 2007), its catchment area contains less than 20% carbonate rocks, and the cold lake waters inhibit accumulation. Climate clearly influences the general temperature of lake waters at the surface and at depth and it also affects the timing of carbonate precipitation through summer productivity of microbes and macrophytic plants (Kelts and Hsü, 1978; Dean, 1981), but it is not the primary reason for the occurrence of lacustrine carbonates.

Geochemical tracing of the precise source of carbonate lacustrine sediments can be derived from Sr isotopes. $^{87}Sr/^{86}Sr$ signals from the waters of the source area are directly comparable to the rocks/sediments of the receiving lake deposit, as demonstrated for modern Lac d'Annecy (French Alps) (Négrel et al., 1997; Brauer and Casanova, 2001), the Pleistocene Lisan Formation (Lake Lisan, precursor to the Dead Sea) (Stein et al., 1997), the Paleocene–Eocene Flagstaff Formation (Utah, USA) (Gierlowski-Kordesch et al., 2008), and the sedimentary record of the basins of the Green River Formation and other associated basins in the western USA (Carroll et al., 2008; Davis et al., 2008, 2009).

Once in a lake, transport of carbonates to sublittoral and profundal areas can occur through sediment gravity flows, as in turbidites and debris flows (Håkanson and Jansson, 1983; Johnson, 1984; Jiang et al., 2007; Sáez et al., 2007) as well as through the resuspension of sediments (Nuhfer et al., 1993; Shteinman et al., 1997). Movement of sedimentary material from the shallow zones to fill deeper zones is termed "sediment focusing" (Lehman, 1975; Johnson, 1984; Lamoureux, 1999) and is an important mechanism for the transport of sedimentary material into the profundal zone (e.g., Smoot and Benson, 1998; Colman, 2005). Of course, local geomorphic factors coupled with climate influence sedimentation rates and patterns as well, including basin configuration (e.g., a wide vs. a narrow littoral shelf) and current and circulation patterns (Sly, 1978; Johnson, 1980; Håkanson and Jansson, 1983; Imboden, 2004).

5.3. Climate

Carbonate lake deposits originate in all climates. Examples include modern lakes in both cold and hot climatic regions, such as in Antarctica (e.g., Wharton et al., 1982; Simmons et al., 1993; Wharton, 1994), in Australia (Last and DeDeckker, 1990), and in East Africa (e.g., Stoffers and Hecky, 1978; Casanova, 1986, 1994; Cohen and Thouin, 1987). An arid climate is not a key factor in carbonate precipitation/accumulation, not even for palustrine carbonates (Platt and Wright, 1992). Indeed, saline lakes occur at all latitudes (Carroll and Bohacs, 1999; Bohacs et al., 2003), so climate is not the ultimate controlling factor in evaporite or carbonate accumulation. For example, in the Late Quaternary Konya Basin in Turkey, the stable isotopic signal from carbonates in three lakes within three related sub-basins under presumably the same climatic regime exhibits contrasting patterns, attesting to the importance of localized hydrologic conditions in influencing carbonate deposition (Leng et al., 1999a). Groundwater patterns are key in understanding these processes (see Rosen, 1994; Winter, 2004).

Climate, however, clearly influences water and sediment input into a lake basin. Climatic effects, such as temperature and rainfall patterns, can influence the timing of carbonate precipitation, as in planktonic blooms in temperate lakes (e.g., Stabel, 1986; Thompson et al., 1997; Smith and Hollander, 1999), where calcitic or aragonitic laminite deposition occurs during summer months in lakes (e.g., Wright et al., 1980; Zhao et al., 1984; Hodell et al., 1998; Lajewski et al., 2003). Thus, temperature directly influences the nature and rate of biologic activity as well as erosion and sedimentation rates (Platt and Wright, 1991). Temperature also influences the solubility of carbonate and controls stratification processes affecting carbonate accumulation and preservation (Dean, 1981). Groundwater input sources can be less affected by short-term climate changes, depending on the lake system (Rosen, 1994). Longer term environmental changes promoting productivity and carbonate precipitation are linked to changes in humidity/aridity and temperature (Anderson and Dean, 1988; Mullins, 1998). For example, carbonate lamina thickness in modern sediments is correlated with winter rainfall patterns in Lake La Cruz (Spain) (Romero-Viana et al., 2008).

Thus, carbonate depositional patterns can be related to cyclic climatic processes through the analysis of varves and rhythmites in open-water lacustrine sediments (e.g., Glenn and Kelts, 1991). Sedimentation changes (see Section 11.1 for details) are related to climate and seasonal patterns as influenced by regional to hemispheric atmospheric oscillations (e.g., Mullins, 1998; Hammarlund et al., 2002; Kirby et al., 2002), as well as Milankovitch cyclicity. The recognition of such cyclicity patterns, however, requires understanding carbonate depositional processes as well as careful age control and determination of diagenetic alteration and preservation potential of all sedimentary features (Lamoureux, 1999; Gälman et al., 2006; Pietras and Carroll, 2006; Shanahan et al., 2008).

6. Depositional Processes

In the literature, carbonate deposition in lakes has been attributed to four main processes: (1) concentration through evaporation, (2) biogenic mediation, including high productivity of micro- and picoplankton, microfauna shell formation, and encrustations on any substrate, (3) water-borne clastic input, and (4) eolian supply.

6.1. Concentration through evaporation

Carbonate precipitation in freshwater and saline lakes is suggested to be mostly chemical in origin (Müller et al., 1972; Eugster and Kelts, 1983; Morgan and Stone, 1985; Given and Wilkinson, 1985; Last, 2001) with high temperatures contributing to evaporation and concomitant precipitation of carbonates. Lake-wide events of the precipitation of calcium carbonate, called "whitings," in the epilimnion of freshwater and saline lakes during summer and autumn are documented from many lakes (e.g., Strong and Eadie, 1978; Galat and Jacobsen, 1985; Thompson et al., 1997; Hodell et al., 1998; Lajewski et al., 2003; Romero-Viana et al., 2008) and interpreted as temperature-dependent chemical reactions. Despite chemical kinetics, supersaturation of water with respect to calcite is not correlated with calcite precipitation (Stabel, 1985). Stabel (1986) shows that supersaturation of waters in Lake Constance (Germany) with respect to calcite does not initiate precipitation of whitings, but the presence of algae in the epilimnion does. Whitings are now attributed to the temperature-dependent photosynthetic activities of microflora (e.g., Thompson et al., 1997). A direct connection between evaporation through concentration events and carbonate laminae production under subaqueous conditions is not clearly documented yet. Precipitation of carbonate by concentration through evaporation certainly occurs along shorelines under palustrine conditions as lakes contract in size (e.g., Lake Chilwa (McLachlan, 1979)).

6.2. Biogenic mediation

Primary precipitates or authigenic minerals comprise carbonates that precipitate directly from solution in lake and sediment pore waters, mostly through biomediation as influenced by physical conditions, such as temperature and isotopic water composition (Talbot, 1990; Talbot and Kelts, 1990; Leng and Marshall, 2004). Müller and Wagner (1978) describe the turbidity caused by the precipitation of magnesian calcite during phytoplankton blooms in Lake Balaton (Hungary) as "blondness," as translated from Hungarian. Rare whiting events also occur in the Dead Sea (Bloch et al., 1944) and are linked to enrichment in $\delta^{13}C$ and $\delta^{18}O$ in precipitated aragonite (Friedman, 1993), though Robbins and Blackwelder (1993) give evidence for a microbial origin. Interestingly, the Dead Sea exhibits diverse algal and bacterial populations in its water column (Oren, 1997) that bloom when conditions are optimal (Leng and Marshall, 2004). Andrews et al. (1993) demonstrate that cyanobacterial precipitation of primary carbonates reflects the environmental parameters for kinetic equilibrium so that stable isotopes of these primary precipitates, though biomediated, still represent proxies for environmental conditions within lake waters (e.g., regional rainfall and temperature).

Thus, calcitic and dolomitic carbonates precipitate through the help of microbiota as well as macrobiota (Kempe and Kazmierczak, 1990; Merz, 1992; Dean and Megard, 1993; Arp et al., 1999a,b; Wright, 1999). This can occur at groundwater input sites, forming bioherms (e.g., Kempe et al., 1991; Arp, 1995; Pache et al., 2001; Rosen et al., 2004; Sanz-Montero et al., 2008), ooids/oolites (Reitner et al., 1997; Davaud and Girardclos, 2001), and oncoids/oncolites in nearshore environments (Garber and Friedman, 1983; Popp and Wilkinson, 1983). Thompson and Ferris (1990) and Thompson et al. (1997) tie calcite precipitation during whiting events in Fayetteville Green Lake (New York, USA) to the same planktonic cyanobacteria that contribute to bioherm growth in the lake (Thompson et al., 1990).

Shelly invertebrate fauna, such as ostracodes, gastropods, and bivalves, precipitate carbonate into lake systems (Dean, 1981; Cohen, 2003). These animals can live in or on the substrate as well as in the water column, deriving their building materials from lake waters. Algal and macrophytic photosynthesis can also precipitate carbonate inside and around plants and algae, including charophytes and nearshore reeds and other macrophytic plants (Schneider et al., 1983; Robbins et al., 1991; Anadón et al., 2000, 2002a). Today macrophytic plants in the littoral zone of many lakes can become encrusted with large quantities of calcium carbonate that can contribute thick accumulations to the sedimentary record (Kelts and Hsü, 1978). Even cyanobacterial mats in the carbonate-rich Everglades (Florida, USA) mediate the production of carbonate deposits (Thiel et al., 1997) in shallow water.

6.3. Clastic input

A clastic origin (both extra- and intrabasinal) for carbonate grains can be recognized in open-water lake sediments through grain shape and size, mineralogy, or isotopic analysis. Detrital grains should represent the geochemical signal of source carbonate rocks (Kelts and Talbot, 1990; Jiang et al., 2007). Half of the carbonate sediment collected in traps in Lake Biel (Switzerland) is from the drainage area and groundwater input with the other half from phytoplanktonic activity (Wright et al., 1980). Thompson and Kelts (1974) identify dolomite grains within lacustrine chalk as detrital because of the freshwater conditions of Lake Zürich, with biogenic carbonate grains ranging in size from 8 to 30 μm and detrital grains from 1 to 20 μm. Müller (1969) does the same for dolomite identified in the sediments of Lake Balaton (Hungary). Drummond et al. (1996) attribute dolomite in Cretaceous lacustrine carbonates to diagenesis.

However, primary dolomite has been proposed for many lake successions in the geologic record (e.g., Janaway and Parnell, 1989; Anadón and Utrilla, 1993; Arenas et al., 1999). Recent evidence of biomediated precipitation of dolomite grains in the laboratory and in lakes (Wright, 1999; Calvo et al., 2003; Sánchez-Román et al., 2008) may alter the perception of dolomite as an allochthonous mineral in freshwater lake sediments or a later diagenetic mineral in lacustrine rock successions. For example, Sanz-Montero et al. (2008) have identified the role of organotemplates in the precipitation of primary dolomite in association with microbial mat growth of dolomite/silica stromatolites in the Miocene Intermediate Unit of the Duero Basin (Spain). In addition, Bréhéret et al. (2008) attribute dolomite precipitation to microbial activities and degradation in the Holocene sediments of the Sarliève lake in the Limagne rift (France). The amount of detrital dolomite in comparison to authigenic biomediated dolomite in the units of this paleolake deposit varies from 0% to 100% (Fourmont et al., 2009).

6.4. Eolian supply

Contributions of eolian carbonate grains to laminae in profundal lake sediments are not well documented. Small grain sizes of eolian particles ($<20\,\mu m$ in size) are not easy to distinguish from the sedimentary matrix material, whether carbonate or siliciclastic (see Lamoureux and Gilbert, 2004). For example, Anderson (1993) and Dean et al. (2002) attribute changes in varve thickness in the sediments of Elk Lake (Minnesota, USA) mostly to variations in detrital input, while Dean (1997) interprets the siliciclastic components of the varves specifically to wind activity because of the lack of permanent fluvial input. Halfman et al. (1994) interpret some inconsistent ^{14}C dates from some carbonate fractions in the sediments of Lake Turkana as contamination by eolian contributions from the region east of the lake. A powerful technique to differentiate the origins of carbonate within laminae or in bulk sediment samples involves analysis of stable isotopes of carbon and oxygen. This leads to the recognition of diagenetic alteration and preservation parameters and aids in the reconstruction of climate and other physical and biologic parameters affecting primary carbonate sedimentation in lakes (Stuiver, 1970).

7. Preservation of Lacustrine Carbonates

Unfortunately, precipitation or deposition of carbonates does not guarantee preservation, which is dependent on internal lake processes, such as circulation patterns, seasonality and temperature ranges, stratification patterns, lake basin geometry, and general water chemistry (e.g., Dean, 1981), in addition to wave and current erosion, bioturbation by benthos (e.g., Cohen, 1984), and cryoturbation during low winter temperatures (Last and Vance, 1997). Recognizing original textures is important in identifying changes in the modes of preservation of carbonate layers. Prokaryotic and eukaryotic algae form calcium carbonate crystals of many sizes and shapes during their life (Freytet and Verrecchia, 1998). Crystals range from 1 to $120\,\mu m$ or more in length and shapes include rhombic dodecahedrons, platelets, and euhedral crystals with planar or curved faces. These biomediated crystals can also be diagenetically transformed quickly during the degradation of associated microbes. As carbonate crystals form in surface waters from photosynthetic activities of microbes, they slowly sink and can begin to dissolve, as evidenced by the rounding of crystals and surface etching features (Kelts and Hsü, 1978).

The solubility of carbonates can be affected at great depths where temperatures are low and decay of organic matter influences pH and redox conditions. In the thermally stratified, East African rift lakes, precipitated carbonates preferentially dissolve in anoxic waters of the hypolimnion

(Degens et al., 1972; Degens and Stoffers, 1976). Hodell et al. (1998) link the chroococcoid cyanobacteria *Synechococcus* with the calcite crystals associated with annual whiting events in Lake Ontario (Canada/USA), though preservation of calcite in the sediments did not occur every year over the duration of their study. Holocene North American freshwater lake sediments with abundant organic matter do not preserve carbonate in profundal sediments, whereas those with a lower organic matter content do preserve carbonate (Dean and Gorham, 1976). For example, Knowlton Lake in Ontario (Canada) contains marl benches, but the profundal sediments are dominated by gyttja (a mixture of clay and/or silt with fine-grained reworked organic material, commonly as peat fragments) with minor carbonate (Gilbert and Lesak, 1981).

The temperature range of lake waters and resulting stratification patterns, as influenced by climate, also affect carbonate deposition. In the Holocene sediments of Lake Juusa in Estonia, Punning et al. (2005) connect the relationship between the accumulation of calcite and organic matter at depths through time as a proxy for both productivity and stratification stability. Such colder climate lakes may not preserve carbonate in profundal sediments where organic accumulation is rapid in the summer (promoting anoxia at depth). Ice cover during long winters favors high concentrations of carbon dioxide due to the degradation of organics leading to the dissolution of carbonates.

Organic matter, on the other hand, clearly can be preserved in association with carbonate-dominated profundal lake deposits (Katz, 1988), as demonstrated by the Tertiary oil-shale basins of northeastern Spain (Anadón et al., 1989; de Las Heras et al., 2003), the Green River oil shales (Horsfield et al., 1994; Katz, 1994), and the finely laminated, Las Hoyas deposit of central Spain (Fregenal-Martínez and Meléndez, 1994). The key here is the significant input of carbonate that encloses and quickly preserves organic matter. Peat associated with carbonates occurs in modern to Quaternary lakes (Treese and Wilkinson, 1982; Zanchetta et al., 2007) while coal is associated with carbonates in Cenozoic lakes (Cabrera and Sáez, 1987; Flores and M'Gonigle, 1991; Anadón et al., 1998; Cabrera et al., 2002; Mandic et al., 2009), Mesozoic lakes (Gierlowski-Kordesch et al., 1991), and Paleozoic lakes (Valero Garcés et al., 1994, 1997).

8. Fauna/Flora

There are many kinds of animals and plants preserved in lake successions. They can be found in all lake environments. This is not meant to be an exhaustive list – only references to major faunal components that preserve well as fossils in carbonates are summarized. For a more complete overview on the fauna and flora in lakes, see DeDeckker

8.2. Diatoms

This photosynthesizing phytoplankton consists of algae (class Bacillariophyceae) that house themselves within an amorphous silica structure, called a "frustule" (Stoermer and Smol, 1999; Battarbee et al., 2001; Starratt, 2007a). The oldest freshwater diatoms are documented from the Late Cretaceous of México (Chacón-Baca et al., 2002) in association with volcanic activity, finally dominating lakes by the beginning of the Miocene (Kidder and Gierlowski-Kordesch, 2005; Starratt, 2007b) when silica from grasslands could continuously replenish supplies to lake and river environments. Both freshwater and saline diatoms have been studied to reconstruct salinity, lake levels, productivity, paleoecology, and climate (e.g., Roeloefs and Kilham, 1983; Fritz et al., 1999; Leng and Barker, 2006; Ekdahl, 2007; Fritz, 2007). Diatoms are preserved within and associated with carbonate deposits in lakes from the Miocene to today. Examples include modern Lake Geneva (Davaud and Girardclos, 2001; Anadón et al., 2006) (Figure 1B) and Lake Tanganyika (Cocquyt, 1991), the Holocene of Lake Tilo (Ethiopia) (Lamb et al., 2000), the Miocene carbonates of the Bicorb (Anadón et al., 1998) and Cenajo (Bellanca et al., 1989) Basins (Spain), Miocene–Pliocene laminated marls of the Denzili Basin (Turkey) (Alçiçek et al., 2007), Miocene to Quaternary diatomites/carbonates of Chile (Sáez et al., 1999, 2007), and the Holocene paleolakes of the western Sahara and Sahel (Africa) (Gasse, 2002).

8.3. Ostracodes

Ostracodes are microcrustaceans encased in a calcitic bivalved shell, living a benthic as well as pelagic life style in continental environments (De Deckker and Forester, 1988; De Deckker, 2002; Horne et al., 2002). Found in a wide range of aqueous surface environments (as well as leaf litter and groundwater aquifers), the high preservation potential of ostracode shells makes them excellent indicators for paleoenvironmental conditions (Holmes, 2001; Frogley et al., 2002; Park and Smith, 2003). Ostracode shells can accumulate as coquinas (Figure 1C), surviving mechanical compaction and diagenesis (Figure 1D) as far back as the Paleozoic. Though small in size (average 1 mm in length), the shell chemistry of ostracodes, including carbon and oxygen isotopes and trace metals, can reveal salinity, water chemistry, sedimentation rate (mode of preservation), eutrophication, diagenesis, nutrient levels, and temperature (Carbonel et al., 1988; Palacios-Fest et al., 1993; De Deckker, 2002; Holmes and Chivas, 2002). Changes in these parameters through time as recorded in lake sediments give information on the evolution of lake systems and provide a stratigraphic framework for lacustrine successions (e.g., Colman et al., 1994; Schwalb et al., 1994, 1995; Filippi et al., 1999; Ricketts et al., 2001; Dettman et al., 2005; Anadón et al., 2006, 2008).

8.4. Molluscs (bivalves and gastropods)

The phylum Mollusca is well represented in lacustrine deposits through most of the Phanerozoic as bivalves and gastropods (Cohen, 2003; Park and Gierlowski-Kordesch, 2007). These animals mainly inhabit littoral, sublittoral, and even profundal environments that are well oxygenated. These molluscs that live in or on the substrate (see Dillon, 2000; Cohen, 2003) take their building materials from lake waters. Their shells can even be found *in situ* or can accumulate as coquina layers from reworking (Figure 1E). Molluscs are used as stratigraphic (La Rocque, 1960) as well as paleoecologic indicators, especially for gauging lake-level fluctuations (e.g., Miller et al., 1985; Digerfeldt et al., 1997; Yang et al., 2001).

Stable isotope analyses of molluscan skeletal carbonates, when carefully measured and interpreted in the context of vital vs. kinetic effects (physiological influence vs. isotopic equilibrium with lake waters – see Leng and Marshall, 2004), are powerful tools in reconstructing paleoenvironments (Fritz and Poplawski, 1974; Miller and Tevesz, 2001). Paleoclimatic and paleoenvironmental parameters, such as temperature (Abell, 1982; Filippi et al., 1997; Abell and Hoelzmann, 2000), seasonal changes (Dettman and Lohmann, 1993), productivity (Leng et al., 1999b), and trends in climate-induced water body chemistry and depth (Tevesz et al., 1997; Zanchetta et al., 1999; Udomkan et al., 2003; Anadón et al., 2008), can be deduced. Diagenetic effects should be considered in the use of stable isotope ratios in the geologic record of lake deposits.

8.5. Insects

Insects use lake waters as breeding grounds and are preserved in both siliciclastic and carbonate lacustrine sediments of the Phanerozoic (Dayvault et al., 1995; Grimaldi and Engel, 2005). As a representative example, insect-produced structures preserved within carbonate bioherms are identified as caddisfly larval/pupal cases from lake successions since the Early Cretaceous. Examples from lacustrine bioherms include the Cretaceous Jinju Formation (Korea) (Paik, 2005), the Eocene Green River Formation in the Fossil Basin and the Greater Green River Basin of Wyoming (USA) (Loewen et al., 1999; Leggitt and Cushman, 2001; Leggitt and Loewen, 2002; Leggitt et al., 2007) (Figure 1F), the Oligo-Miocene lakes of the Limagne Graben in the Massif Central (France) (Bertrand-Sarfati et al., 1994), and the modern Big Soda Lake in Nevada (USA) (Rosen et al., 2004). In all these examples, accumulations of caddisfly cases alternate with carbonate layers, showing that bioherms were used as a substrate for colonization or discarded cases between calcification events. In addition, chironomid or midge larvae may also be preserved within the microbial laminae of bioherms (Bertrand-Sarfati et al., 1994), and brine-shrimp egg cases are replaced by aragonite in the growth cavities of modern stromatolites in the Great Salt Lake of Utah (USA) (Pedone and Folk, 1996).

9. Carbon and Oxygen Isotope Analyses

9.1. Overview

Isotopic analyses aid in the identification of the mechanisms causing the accumulation of carbonates in lake sediments, leading to the reconstruction of the history of a lake body. For example, lower lake levels can facilitate precipitation of calcite because the reduced lake volume results in a higher concentration of dissolved carbonates (e.g., Ricketts and Anderson, 1998), but preservation of carbonate resulting from higher than average productivity of phytoplankton can also contribute preservable carbonate to profundal sediments (e.g., Thompson et al., 1997). Schelske and Hodell (1991) and Hodell et al. (1998) show that biomediated calcite precipitation in Lake Ontario (USA) is dependent on lake water temperature, as are evaporation rates. In addition, detrital or clastic carbonate input needs to be considered as well (e.g., Filippi et al., 1999; Brauer and Casanova, 2001). Distinguishing these carbonate sources archived in both saline and freshwater lacustrine carbonates through stable isotope analyses requires a deep understanding of the local lake dynamics and the biology of the lake system.

Stiller (1977), in a groundbreaking study, analyzes the carbon and oxygen stable isotopic composition of the sediments, suspended-load river input, and the carbonate bedrock of the catchment area surrounding Lake Kinneret (Israel) to deduce differences in allochthonous and autochthonous carbonate samples. Bright et al. (2006) identify the detrital carbonate fraction in the sediments of Bear Lake (USA) in a similar way, as does Griffiths et al. (2002) for the marls of Quaternary Lake Kopais (Greece). In addition, the isotopic values of rainfall, groundwater, and lake waters in the studied extant lake and those in nearby catchment areas also are measured for more precise understanding of isotopic trends in lake carbonates (e.g., Lamb et al., 2000; Schwalb and Dean, 2002; Anderson and Leng, 2004; Diefendorf and Patterson, 2005).

Applying stable isotopic analysis to Lake Kinneret (Israel) sediments, Stiller and Kaufman (1985) connect productivity and climatic changes to specific calcitic intervals in core material. Warm phases with moderate productivity and cooler phases with enhanced productivity both preserve carbonate sediments in the profundal zone. The first application of isotope analysis to Cenozoic lake deposits in the Ebro Basin (Spain) also identifies the importance of river input as well as processes within the lake waters, such as evaporation and productivity, on the isotopic signatures of lake carbonates (Oberhänsli and Allen, 1987).

Authigenic carbonate attributed to increased productivity in hydrologically open (freshwater) lakes would record maxima in $\delta^{13}C$ and minima in $\delta^{18}O$, namely "inverse covariance" (Schelske and Hodell, 1991). Productivity in hydrologically closed lakes is documented by maxima in $\delta^{13}C$ and maxima in $\delta^{18}O$ in carbonates, termed "covariance" (Bein, 1986; Robbins and Blackwelder, 1993; Talbot and Kelts, 1990; Li and Ku, 1997). Enriched oxygen isotope ratios in carbonate sediments are attributed to aridity and increased evaporation in any lake, that is, temperature (maxima in $\delta^{18}O$) (Kelts and Talbot, 1990; Drummond et al., 1996; Ricketts and Johnson, 1996a,b; Dean and Schwalb, 2000; Yuan et al., 2006). However, covariance of carbon/oxygen isotopic ratios is measured in freshwater marls of temperate regions as well (Drummond et al., 1995); enriched $\delta^{18}O$ in the summer is attributed to isotopically heavy rainfall in comparison to a lighter winter precipitation as the moisture source changes seasonally. Temperature is not the only control on the composition of oxygen isotope values.

Clastic (both water- and wind-transported) carbonate grains can also affect trends in the stable isotopic analysis of bulk carbonate sediment. Detrital influence in carbonate composition of lake sediments of Quaternary age is identified through the recognition of stable isotope trends within and among samples. On the other hand, there are no established criteria to separate carbonate grains of an eolian vs. a water-transported origin (Filippi et al., 1997). In pre-Quaternary lake sediments, differentiation of detrital vs. authigenic carbonate mineral origin is difficult, especially considering the diagenetic alteration of sedimentary carbonates (Talbot and Kelts, 1986, 1990; Poyato-Ariza et al., 1998).

Thus, it is recognized that the stable isotopes of carbon and oxygen are affected by various geochemical and physical processes within lake systems, including the influence of groundwater (e.g., Valero Garcés et al., 1995; Leng et al., 1999a; Schwalb and Dean, 2002). Since multiple sources contribute stable isotopes into a lake system and multiple mechanisms promote carbonate precipitation, interpretation of the stable isotope ratios of oxygen and carbon from carbonates is challenging indeed. Detailed studies of the climate, hydrology, and provenance of a lake are necessary for a reliable reconstruction and interpretation of the isotope trends archived in its sedimentary carbonate (Leng and Marshall, 2004).

9.2. Carbon isotope analysis

The $\delta^{13}C$ of lake waters is influenced by the geochemical composition of input waters, atmospheric CO_2 exchange, gas mixing through bacterial methanogenesis in the degradation of organic matter, and pelagial photosynthesis (productivity) (McKenzie, 1985; Kelts and Talbot, 1990; Talbot and Kelts, 1990; Leng and Marshall, 2004). Temperature clearly is an important parameter in controlling gas exchange and promoting photosynthesis. In examining the Holocene sediments of two of the Finger Lakes in New York (USA), Anderson et al. (1997) and Mullins (1998) theorize that biomediated calcite precipitation can be limited by a shortened season of productivity in cooler summers, and that such patterns are recorded in the carbonate sedimentary record. For the Miocene Los Monegros carbonates of the Ebro Basin (Spain), Arenas et al. (1997) interpret the light carbon and oxygen isotope composition of the massive and bioturbated facies as indicative of a high productivity associated with high lake levels, inferring a wet climate; conversely, the enriched values of the laminated and stromatolitic facies are thought to indicate a dry climate with low lake levels.

Stratification in a lake as well as ^{13}C-enriched groundwater in karstic regions can influence the mixture of ^{12}C and ^{13}C in lake waters (Leng and Marshall, 2004). Bein (1986) interprets the precipitation of the late Cenozoic lacustrine marls of the Hula Basin (Jordan) in terms of productivity and changes in stratification patterns induced by hydrologic changes with increases in lake volume. Meromictic stratification and its effect on the growth of photosynthetic microbial communities in the water column (^{12}C-enriched microbes) in Holocene Medicine Lake (South Dakota, USA) is interpreted to increase aragonitic ^{13}C in a fresh mixolimnion with depleted $\delta^{18}O$ values, producing an inverse covariance (Valero Garcés et al., 1995).

Variations in soil productivity in the surrounding catchment area can be recognized through soil water input depleted in ^{13}C linked to isotopic carbon ratios in authigenic lacustrine carbonates (Talbot, 1990; Anderson and Leng, 2004). For the closed Oligocene lake in the As Pontes Basin (Spain), Sáez and Cabrera (2002) reconstruct the hydrologic regime of the basin lakes as well as possible vegetation and soil development through time using sedimentology, stable isotope analyses, fossil data, mineralogy, and diagenesis of the sedimentary basin fill. For example, the early diagenetic carbonates of the shallow lacustrine to palustrine facies exhibit an extremely wide range of mostly negative $\delta^{13}C$ values indicating oscillating water levels linked with vegetated areas containing possible varying proportions of C3 ($\delta^{13}C \sim -12$ to -14‰) and C4 ($\delta^{13}C \sim -25$ to -28‰) plants and an input of depleted $\delta^{13}C$ from dissolved CO_2 in soil water. Bacterial sulfate reduction of organic matter is postulated as the main mechanism for early diagenetic calcite precipitation. In contrast, the deeper water carbonates of higher lake levels exhibit mostly primary carbonate laminae with more enriched $\delta^{13}C$ values punctuated by short excursions to more depleted $\delta^{13}C$. This pattern is attributed to hydrologically closed conditions (and high productivity) interspersed with flood events diluting the mixolimnion of a meromictic lake.

9.3. Oxygen isotope analysis

The $\delta^{18}O$ of lake waters is influenced by the isotopic composition of waters supplied to the lake, including precipitation, surface inflow, and groundwater inflow. Temperature and thus evaporation also control the output of the lighter oxygen (^{16}O), affecting the isotope ratio. Changes in temperature, rainfall sources (especially through seasonal changes), riverine influx, and even groundwater input are surmised to be preserved as oxygen isotope signals in the carbonates precipitating in lake waters (Teranes et al., 1999; Lamb et al., 2000; Schwalb and Dean, 2002; Leng and Marshall, 2004; Yansa et al., 2007). Smith and Hollander (1999) use $\delta^{18}O$ of sedimentary carbonates in Lake Mendota in Wisconsin (USA) to surmise changes in rainfall sources through the last 100 years of cyclonic activity. Kirby et al. (2002) link $\delta^{18}O$ signals in the carbonate sediments of Fayetteville Green Lake (New York, USA) to circumpolar vortex variability affecting rainfall patterns. Oxygen isotope records of carbonates in sediments of Lake Turkana are interpreted as proxies for a drier climate, taking into account the measurement of projected rainfall isotope contributions as well as variable river input through time (Muchane, 1996; Johnson et al., 1991; Ricketts and Johnson, 1996b; Ricketts and Anderson, 1998).

Clearly, the assessment of $\delta^{18}O$ signals in carbonates is not solely dependent on any one parameter, whether rainfall patterns, temperature, or altitude (Carroll et al., 2008; Davis et al., 2008, 2009). A baseline on isotopic water composition must be established from the sampling of all input waters (or inferences about isotopic values for all possible lake inputs in the geologic record). Then, the proportional effects of these inputs are inferred, so that past environmental and climatic changes can be deduced more accurately. Sampling of ancient lake waters and groundwater is not possible but isotopic values for rainfall, water, and sediment can be deduced (e.g., Valero Garcés et al., 1997). For example, one way to reconstruct the isotope ratio of paleowater in the geologic record is from isotope measurements of turtle bones and even mammal teeth enamel (Tütken et al., 2006).

Application of these rigorous isotope techniques, using carbon, oxygen, and strontium isotopes along with tectonic and climatic data, allows for a complete reconstruction of a lacustrine catchment area through time. The evolution of the Green River lake system in the western USA, including drainage patterns and water flow directions through time, is the first example on a regional scale applying these techniques. For example, a dramatic decrease in $\delta^{18}O$ across an unconformity within the Greater Green River Basin, denoting a change from a balanced-fill to an overfilled characterization of a lake system and its post-sedimentary processes through time (e.g., Bellanca et al., 1989; Camoin et al., 1997; Utrilla et al., 1998).

11. Facies Types

The carbonate database for this review contains information on nearly 250 lakes and lake basins. Though far from complete (not all studies describe the petrology and facies distribution of the carbonate sediments across an entire lake or lake basin), the data set is large enough for recognizing patterns of accumulation. A partial list of lakes and lake basins from the carbonate database is provided in Table 3. Descriptions and photos illustrating each facies type are found in Figures 2–6.

Lacustrine carbonates that form under saline to freshwater aqueous conditions can be categorized into five general facies types. These are: (1) laminated carbonates, (2) massive carbonates, (3) microbial carbonates, (4) marginal carbonates, and (5) open-water carbonates. This classification is greatly simplified from that of Freytet and Verrecchia (2002). Table 4 lists the environments and elements for each general facies type. Microbial carbonates, macrophytic fossils and debris, and invertebrate and vertebrate fossil remains can be components of all the facies types.

11.1. Laminated carbonates

11.1.1. Definition

This lacustrine facies is composed of regular to irregular (thickness of laminae), continuous to discontinuous (lateral extent of laminae lenses) lamination composed of varying percentages of siliciclastic and carbonate material deposited in the profundal zone. Sedimentation occurs as a result of settling from suspension in surface waters or interflows (river input plumes) (Sturm and Matter, 1978) interlaminated with grains and clasts supplied by turbidites or underflows. The term "varves" refers to regular laminae that are deposited on an annual basis within a glacial lake (De Geer, 1912) while the term "non-glacial varves" refers to similar lamination in a non-glacial lake. The term "rhythmites" refers to regular lamination in which the time frame for deposition is not known (Glenn and Kelts, 1991). The term "laminites" simply describes regular to non-regular lamination.

Laminated sediment forms where the lake bottom (1) receives variable sediment input and (2) retains conditions for the preservation of sediment with minimal disturbance or erosion (*cf.* Glenn and Kelts, 1991). Variable sediment input commonly is attributed to seasonal changes in the mineralogy or grain size of sedimentary particles (Anderson and Dean, 1988; Gruszka, 2007) as well as in the remains of organisms

Table 3 Over 100 lakes and lake basins containing carbonate deposits (only a partial list from carbonate database).

Land	Basin/lake	Deposit	Age	References
Africa				
Congo/ Rwanda	Lake Kivu		Cz–H	Degens et al. (1972), Stoffers and Hecky (1978), Haberyan and Hecky (1987)
Congo, etc.	Lake Tanganyika		Cz–Pl–H	Stoffers and Hecky (1978), Cohen and Thouin (1987), Tiercelin et al. (1993, 1994), Cohen et al. (1997)
Congo	Congo	Marnes Noires Fm.	Mz–K	Harris et al. (1994)
Ethiopia	Lake Tilo		Cz–H	Lamb et al. (2000)
Ghana	Lake Bosumtwi		Cz–Pl–H	Talbot et al. (1984), Talbot and Kelts (1986), Turner et al. (1996), Shanahan et al. (2008)
Kenya	Turkana	Koobi Fora Fm.	Cz–P–Ple	Johnson (1974), Abell et al. (1982), Abell and McClory (1986), Cerling et al. (1988)
Kenya	Lake Turkana		Cz–H	Yuretich (1979), Cohen (1984), Halfman and Johnson (1988), Halfman et al. (1994)
Kenya	Lake Bogoria		Cz–H	Vincens et al. (1986), Renaut et al. (1986)
Tanzania	Ruhuhu	Ruhuhu Fm.	Pz–P	Kaaya and Kreuser (1994), Kreuser (1995)
Antarctica/Australia/New Zealand				
Antarctica	Lake Fryxell		Cz–H	Wharton et al. (1982), Wharton (1994)
Antarctica	Lake Vanda		Cz–H	Wharton (1994)
Australia	Lake Walyungup		Cz–H	Coshell et al. (1998)
Australia	Lake Clifton		Cz–H	Moore and Burne (1994)
Australia	West/East Basin Lakes		Cz–H	Last and DeDeckker (1990), Last (1992)
Australia	Lake Keilambete		Cz–H	Bowler (1981), Chivas et al. (1993)

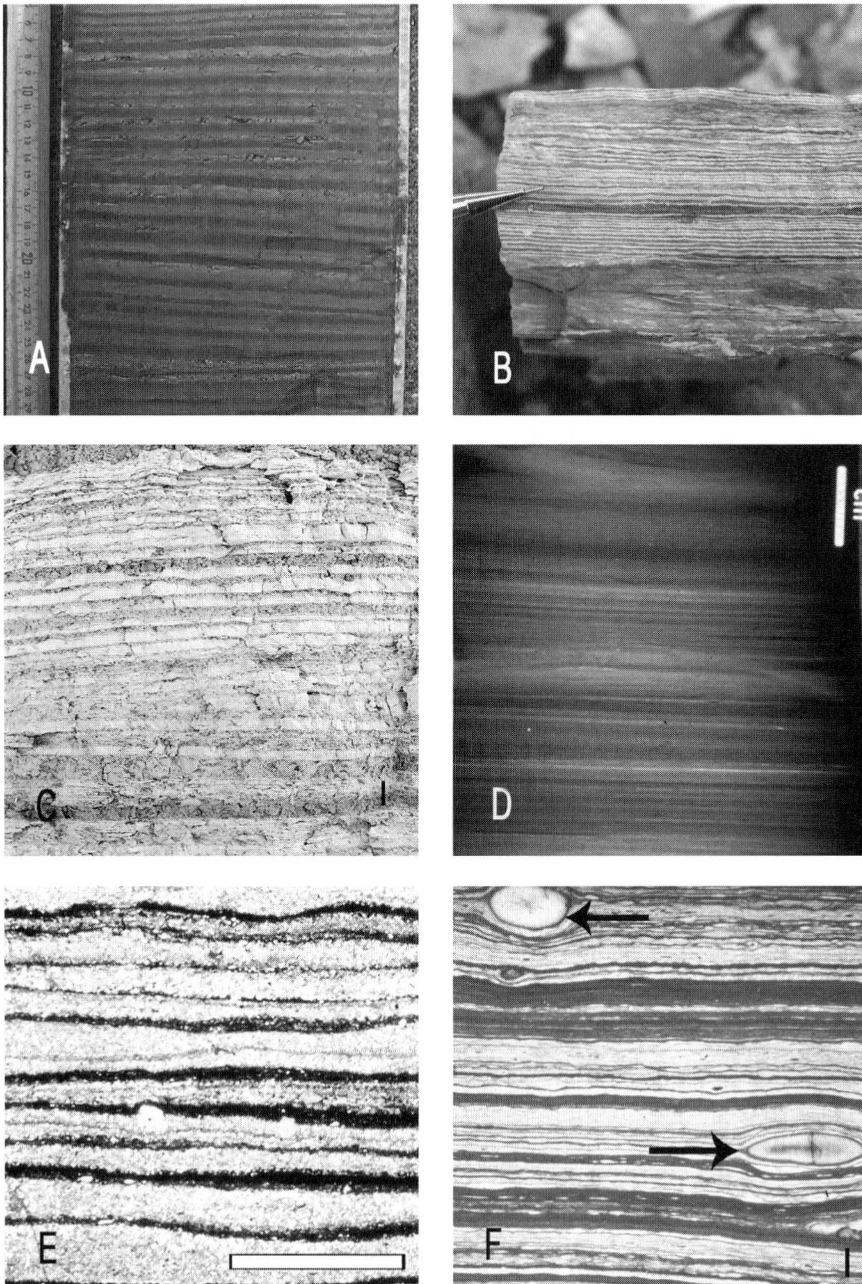
A
B
C
cm
D
E
F

(e.g., Goth et al., 1988; Zolitschka, 1989; Clausing and Boy, 2000), or to variable underflows or turbidites (Lambert and Hsü, 1979). Most lamination is described as couplets of laminae produced by alternating mechanisms, whether from allochthonous or autochthonous lake processes (Glenn and Kelts, 1991). Detailed microscopic and SEM studies of varves identify sublamination or laminae sets with more than two components, as in the varves from Lake Kassjön, a high-latitude lake in Sweden (Dean et al., 1999), and Lake Rõuge in Estonia (Veski et al., 2004). Three component varves can record sedimentation changes from spring snowmelt, summer biologic productivity, and winter ice cover. The ideal varves for the maar lakes of the Eifel region of Germany (Zolitschka, 1989) include five components: (1) an early spring lamina composed of chrysophyte cysts, (2) a spring lamina consisting of diatoms, (3) a summer lamina comprised of diatoms and calcite crystals, (4) a fall lamina with organics and vivianite (hydrated iron phosphate), and (5) a winter lamina containing laminated clay. The thickness of the carbonate laminae does not correlate well with summer temperatures in general, but may be related to rainfall patterns and concomitant changes in water input (Romero-Viana et al., 2008).

The location within a lake and the overall hydrologic regime, in addition to dating sediments and assessing lithologic changes through time, are important for assessing whether the lamination pattern represents true annual sedimentation events (Anderson et al., 1985; Lamoureux and Gilbert, 2004; Gälman et al., 2006; Chutko and Lamoureux, 2008). For example, sediments of some proglacial lakes in high-latitude regions appear to be varved, but time analysis research shows no correlation to any annual signal. Caution is advisable in the assessment of timing of sedimentation patterns, especially in the geologic record, because erosion or dissolution may affect the preservation of laminae.

Figure 2 Laminated carbonates. (A) Core with glacio-lacustrine varves from Lake Superior (USA/Canada), deposited prior to 8 ka. The laminae alternate between light carbonate-rich and dark terrigenous types. Meter stick at left for scale. Photo by Tom Johnson. (B) Carbonate/clay rhythmites of facies association 3 of the Upper Unit of the Miocene Bicorb Basin (Spain), interpreted as deposits of a deep anoxic lake. Pen at left for scale. Photo by Pere Anadón. (C) Micritic aragonite/siliciclastic lamination in the profundal facies of the Pleistocene Lisan Formation (Israel). Scale bar 10 cm. Photo courtesy of Yehouda Enzel and Yuval Bartov. (D) Core with deep basinal laminites (micrite/clay/evaporites) from the Holocene sediments of saline Freefight Lake, Saskatchewan (Canada). Scale bar 1 cm. Photo by William Last. (E) Thin section (regular light) of carbonate rhythmites (micrite/sparite) with some clay/organics laminae from the Cretaceous Fort Apache Formation in the Bisbee Basin, Arizona (USA). Scale bar 1 cm. Photo by Michael Soreghan. (F) Laminated limestone containing fish coprolites (arrows) from the Eocene Green River Formation in the Fossil Basin, Wyoming (USA). Scale bar at lower right 1 cm. Photo by Hans Eugster, courtesy of Walt Dean.

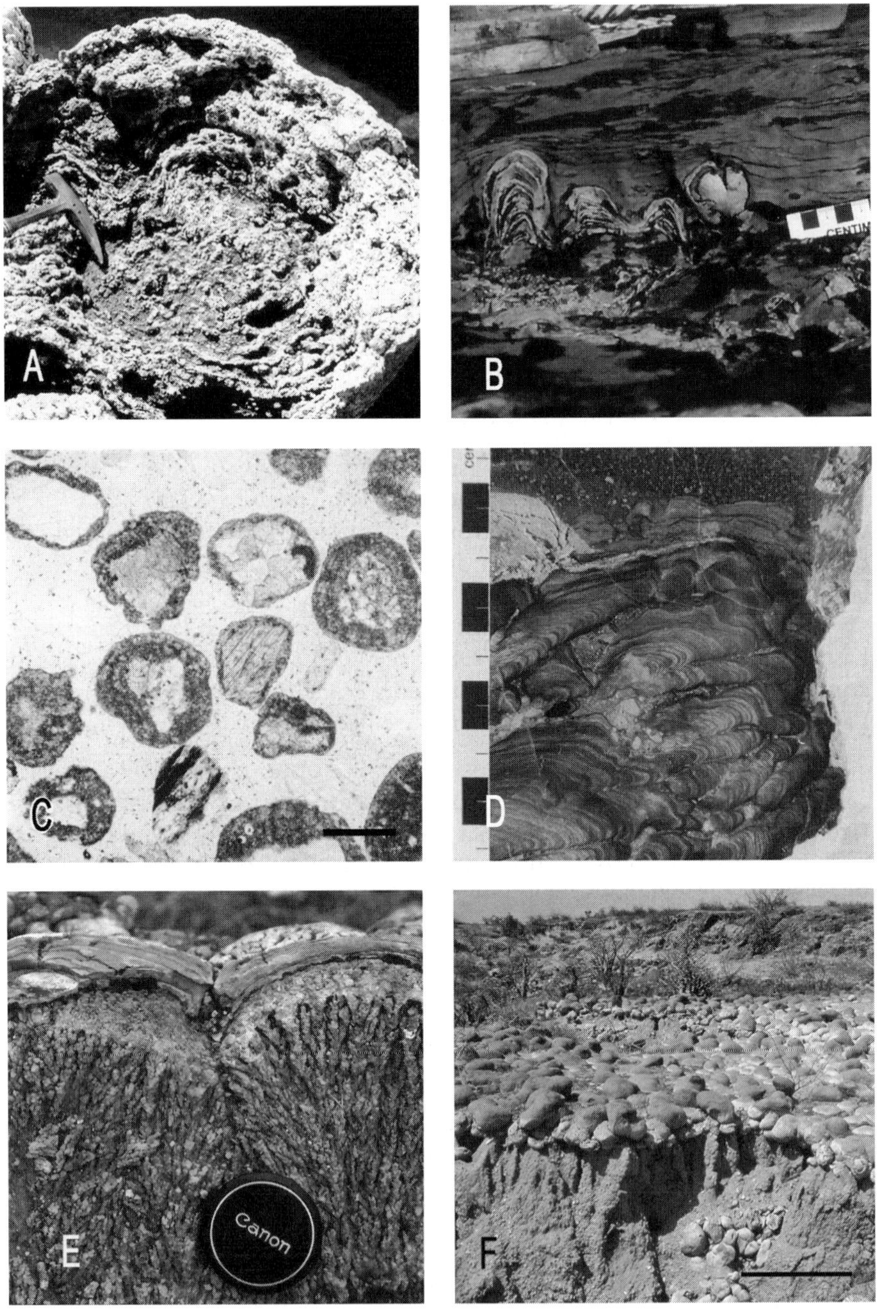
A
B
C
D
E
Canon
F

1990). This is due to reduced bioturbation of a limited benthos since dissolved salts can exclude most benthos in some lakes (e.g., Gierlowski-Kordesch and Rust, 1994), an extremely stable stratification of highly saline bottom brines, or the cohesiveness of the sediment from microbial activity (Finkelstein et al., 1999).

11.1.3. Examples

Examples of carbonate laminae within siliciclastic lacustrine sediments are numerous. Lake Superior (USA/Canada) contains glacio-lacustrine varves with laminae alternating between light carbonate-rich and dark terrigenous ones (Figure 2A). Lake Zürich (Switzerland) (see Figure 7) contains non-glacial varves in the profundal areas; these varves contain a lower dark layer of organics containing algal filaments, clay, and iron oxide grading upward into a middle layer of diatom frustules, organic matter, and clay, with the upper layer comprising a transition from calcite polyhedra (30 μm in length) mixed with diatom frustules to pure calcite at the top of the varve (Kelts and Hsü, 1978).

Nar Gölü, a volcanic crater lake in Turkey (Jones et al., 2005), contains Holocene varves comprising couplets of white carbonate laminae alternating with dark laminae containing organic matter and diatoms. The carbonate laminae are composed of fining-upward sequences of calcite polyhedral grains, 10–40 μm in length, or simply aragonite crystals, less than 5 μm in size. Changes in Mg concentrations in the lake water over time or different precipitation mechanisms are responsible for the generation of these carbonates.

Holocene lamination from cores in Lake Turkana (Kenya) (Halfman and Johnson, 1988) features alternating light/dark couplets with the light laminae containing higher concentrations of carbonate that is comprised of

Figure 4 Microbial carbonates. (A) Cross-sectional slice near the base of a modern bioherm from Big Soda Lake, Nevada (USA). Hammer head at left for scale. Photo by Michael Rosen. (B) Dolomitized stromatolites preserved with mudrocks of the Devonian lacustrine rocks of the northern Orcadian Basin (Scotland). Centimeter scale on outcrop. (C) Thin section (regular light) of modern ooids from the epoxy-impregnated, nearshore sandbar sediments of Lake Geneva (Switzerland), demonstrating various types of cortex shapes and nucleus grains. Scale bar 0.3 mm. Photo by Stephanie Girardclos. (D) Stromatolites topped by massive oolite layer from the Eocene Green River Formation in the Piceance Basin (USA). Centimeter scale on left. Slab provided by H. Paul Buchheim. (E) Tufa mounds (bioherms) exposed along a relict shoreline (Holocene) along the edge of Pyramid Lake, Nevada (USA). Lens cap for scale. (F) Early Holocene oncolites (10–50 cm in diameter) preserved in the Galana Boi Formation, shoreline of Lake Turkana (Kenya), interpreted as algal colonies defining a higher lake level. Scale bar 30 cm. Photo by Robin Renaut.

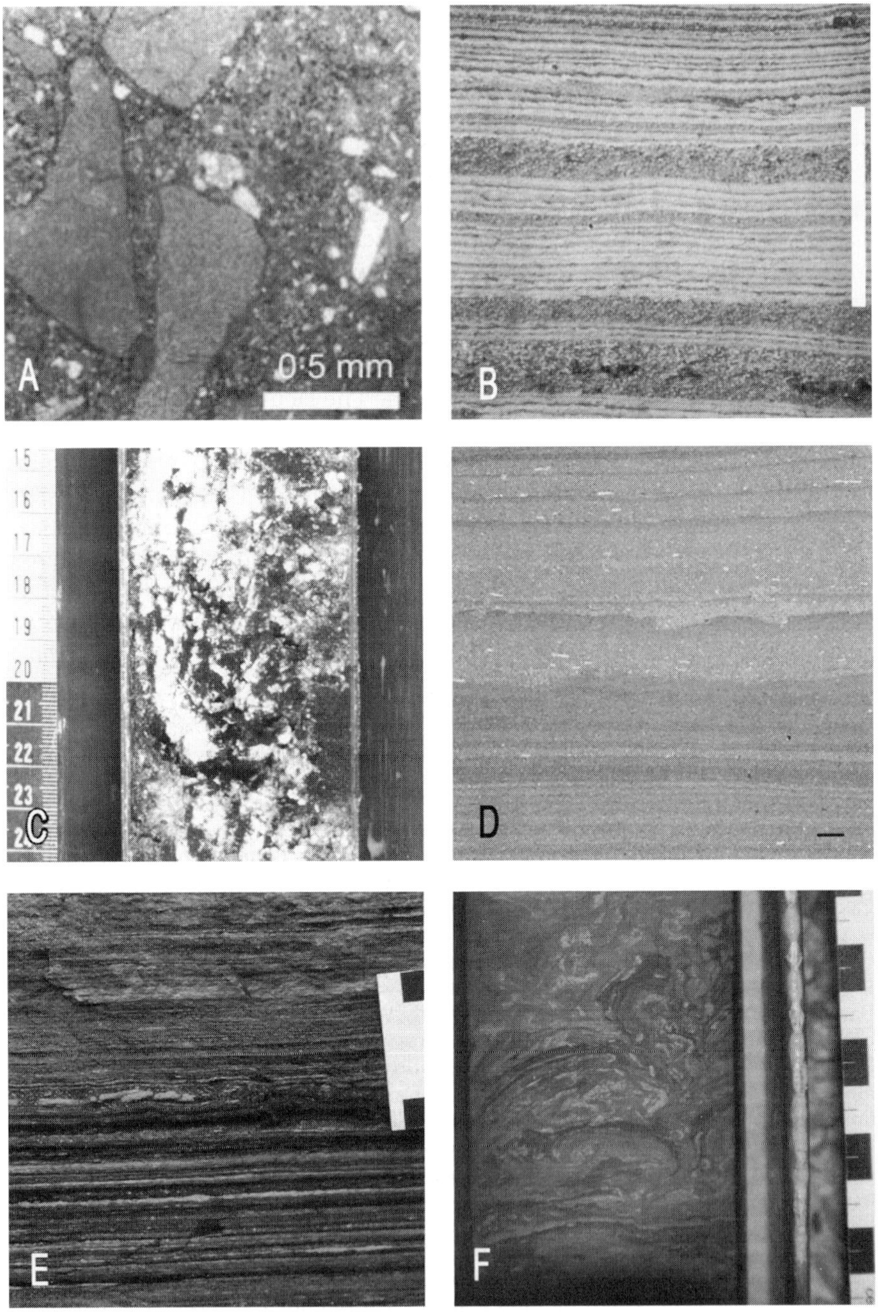
A
0·5 mm
B
15
16
17
18
19
20
21
22
23
C
D
E
F

alternating dark siliciclastic and light aragonite sublaminae (Figure 2C). These are interpreted as varves from winter and summer accumulation, respectively. This is also true for the late Holocene profundal sediments of the Dead Sea (Bookman et al., 2004). The deep basinal laminites (Figure 2D) from the modern sediments of saline Freefight Lake (Saskatchewan, Canada) include salts (sulfates), clays (illite, smectite, and kaolinite), and carbonates (calcium-rich dolomite, aragonite, magnesite, and hydromagnesite) (Last, 1993). The carbonate/siliciclastic laminae of saline 83 Mile Lake and Goodenough Lake of the Cariboo Plateau of British Columbia, Canada, exhibit carbonate/clay alternations. Carbonates, including calcite, Mg-calcite, aragonite, and protodolomite, comprise grains that are 0.5–2 μm wide, anhedral aggregates precipitated from the water column in warmer, drier summer months accompanying biologic activity (Renaut and Long, 1989).

Mostly siliciclastic-free carbonate lamination can be composed of alternating micrite and sparite laminae. For example, the carbonate couplets

Figure 6 Open lake carbonates. (A) Thin section (regular light) of matrix-supported "carbonate arenite" or grainstone from the Eocene Shulu Sag of the Hebei Province (North China). Grains are composed of Cambro-Ordovician source rocks of dolomudstones and siliciclastics. The matrix consists of calcisiltite and carbonate and siliciclastic mudstone. This facies is interpreted as a turbidite deposit; it is associated with finer-grained carbonate facies. Scale bar 0.5 mm. Photo by Zaixing Jiang. (B) Core with carbonate sediments from Fayetteville Green Lake, New York (USA). The laminites consist of light-colored laminae mostly composed of low-Mg calcite crystals settled from surface waters and dark-colored laminae consisting of clay, organic matter, and low-Mg calcite. The thick coarser layers are interpreted as carbonate turbidites. Scale bar 5 cm. Photo by Stuart Ludlam, courtesy of Walt Dean. (C) Core containing carbonate breccia facies (total thickness: 40 cm) from mid-Holocene sediments of Lake Chungará within the Lauca Basin of Chile. Angular carbonate clasts (centimeter-scale) float within a brown diatomite matrix. Facies is interpreted as a debris-flow sediments deposited in the profundal zone with materials from the marginal lake zone. Centimeter scale at left. Photo by Alberto Sáez. (D) Profundal facies of the Early Cretaceous Las Hoyas Lake in the Las Hoyas sub-basin of the Serranía de Cuenca Basin (Spain). Fine lamination at the bottom alternates between dark and light micrites. Upper layers exhibit thick coarser-grained beds interpreted as turbidites or underflow deposits, as evidenced by oriented clasts and scouring. Scale bar 1 mm. Photo by Nieves Meléndez. (E) Sublittoral to profundal facies of the Jurassic East Berlin Formation in the Hartford Basin of Connecticut (USA). Laminites of the third black shale alternate between carbonate and mudrock. The transported clasts of one carbonate lamina show that carbonate layers were cemented early, and then brecciated and transported before the next sedimentation event. Early cementation of carbonate lamination also occurs in the shallow salt lakes of Saskatchewan. Scale on right shows two 1-cm-long blocks. (F) Core with contorted laminites from the Jurassic Shuttle Meadow Formation in the Hartford Basin, Connecticut (USA), interpreted as profundal deposits on a slope affected by seismicity. Centimeter scale at right. Core material supplied by Paul Olsen.

Table 4 Overview of carbonate facies, sediment lithologies, and locations in subaqueous environments of perennial lakes.

Facies	Location/depth	Types of deposit
Laminated	Profundal	Varves Laminites Rhythmites
Massive	Sublittoral Profundal	Wackestones Mudstones Micrites Marls/marlstones
Microbial	Littoral Sublittoral Profundal	Ooids/oolites Pisoids/pisolites Oncoids/oncolites Bioherms Stromatolites/thrombolites
Marginal	Littoral Sublittoral	Grainstones Packstones Shell coquinas
Open water	Sublittoral Profundal	Bioclasts Grainstones/siltstones Rudstones Turbidites Other sediment gravity flows (debris flows, hyperpycnal flows)

in the laminated carbonate facies of the Lower Cretaceous Apache Canyon Formation in the Bisbee Basin of Arizona (Soreghan, 1998) (see Figure 9) comprise laminae of dolomitic micrite with minor organic and clay elements alternating with calcitic microspar (Figure 2E). Also present in this laminated carbonate facies are stacked dolomite laminae up to 5 mm thick composed of fining-upward crystals. The carbonate fine-scale laminites of the Upper Chattian (Oligocene) "Calcaires d'eau douce et dolomie" of the Molasse Basin of Switzerland (Platt, 1992) (Figure 10) are composed of alternating dark gray micrite with charophytes and ostracode fragments and light gray fine sparite laminae that are discontinuous.

Carbonate lamination is commonly preserved in association with organic matter, attesting to anoxic conditions and high sedimentation rates within the paleolake environment. Rhythmically laminated limestones of the Lower Cretaceous Las Hoyas fossil site in the Serranía de Cuenca Basin in Spain (Fregenal-Martínez and Meléndez, 1994) (Figure 11) comprise

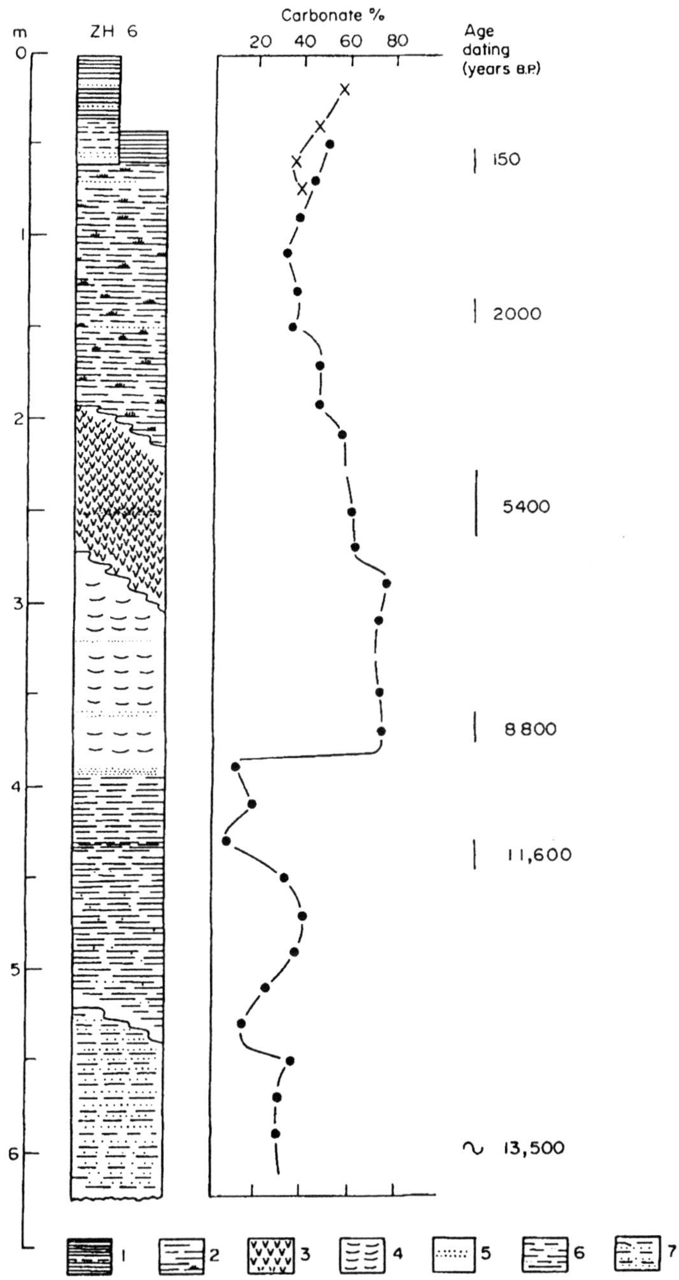

Figure 7 Core log from Lake Zürich, Switzerland (ZH-6). Ages are based on pollen analysis and varve counting in concert with an absolute age scale. Legend: (1) non-glacial black varves, (2) spotted, greenish-gray marls with faint lamination, (3) impure gray lacustrine chalk ($CaCO_3$: 50–60%), (4) light gray lacustrine chalk ($CaCO_3$: 60–80%), (5) intercalated sandy layers, (6) bluish-gray, laminated plastic muds, and (7) reddish-gray laminated muds with interbeds of thin, yellow sand layers (from Thompson and Kelts, 1974 with copyright permission from Wiley–Blackwell).

alternating with laminae composed of organic detritus dispersed in quartz and carbonate silt. Carbonate-rich laminites of Devonian lakes of the Orcadian Basin (Scotland) (Rogers and Astin, 1991) comprise interlaminations of light micrite and organic matter, both containing minor clastic components of quartz, mica, feldspar, and clay, up to silt-sized mineral grains. Couplets of carbonate/organic material average in thickness from 0.25 to 0.5 mm with continuous parallel lamination. Finally, the brown laminites of the modern sediments of Fayetteville Green Lake (NY, USA) (Hilfinger et al., 2001) (Figure 12) consist mostly of calcite (up to 90%) with organic matter (up to 15%) and clay with minor quartz (up to 15%).

11.2. Massive carbonates

11.2.1. Definition

This facies is composed of structureless fine-grained carbonates that can contain preserved fauna and flora. Massive limestones occur as mudstones, wackestones, packstones, grainstones, and marlstones, or they can contain fine carbonate siltstone textures. Modern examples include fossiliferous marl, carbonate silt to sand, and carbonate mud. The lack of sedimentary structures is interpreted as the result of bioturbation in this nearshore to basinal facies. Oxygenated waters with low sedimentation rates allow animals to churn the sediment in search of organic matter. Water agitation allows the settlement of organic matter that contributes food to the ecosystem. Alternate interpretations can include diagenetic recrystallization of various textures or homogenization from evaporite growth.

11.2.2. Examples

Holocene examples of this facies include the massive marls of Bear Lake of Idaho/Utah (USA) (Dean et al., 2006) and the thick massive to indistinctly laminated marl succession of Malham Tarn (UK) (Nuñez et al., 2002). Both lakes contain carbonate and siliciclastic rocks in their catchment area.

Examples from the geologic record include massive carbonate units from the Cenozoic and Paleozoic. The burrowed to massive limestone associated with pelecypod–ostracode wackestones and packstones of the lower Green River Formation in the Uinta Basin (Utah, USA) is interpreted as a nearshore facies (Wiggins and Harris, 1994). The massive charophyte packstones and grainstones of the Lower Unit of the Bicorb Basin (Spain) are inferred to be bioturbated, sublittoral charophyte "meadows" (Anadón et al., 2000). The massive limestone facies of the Miocene Alcubierre Formation in the north-central portion of the Ebro Basin (Arenas et al., 1997) exhibits faint bioturbation features (Figure 3A) as ostracode and gastropod wackestones from productive, shallow lacustrine zones. The Lower Cretaceous La Huérguina Formation of the Uña

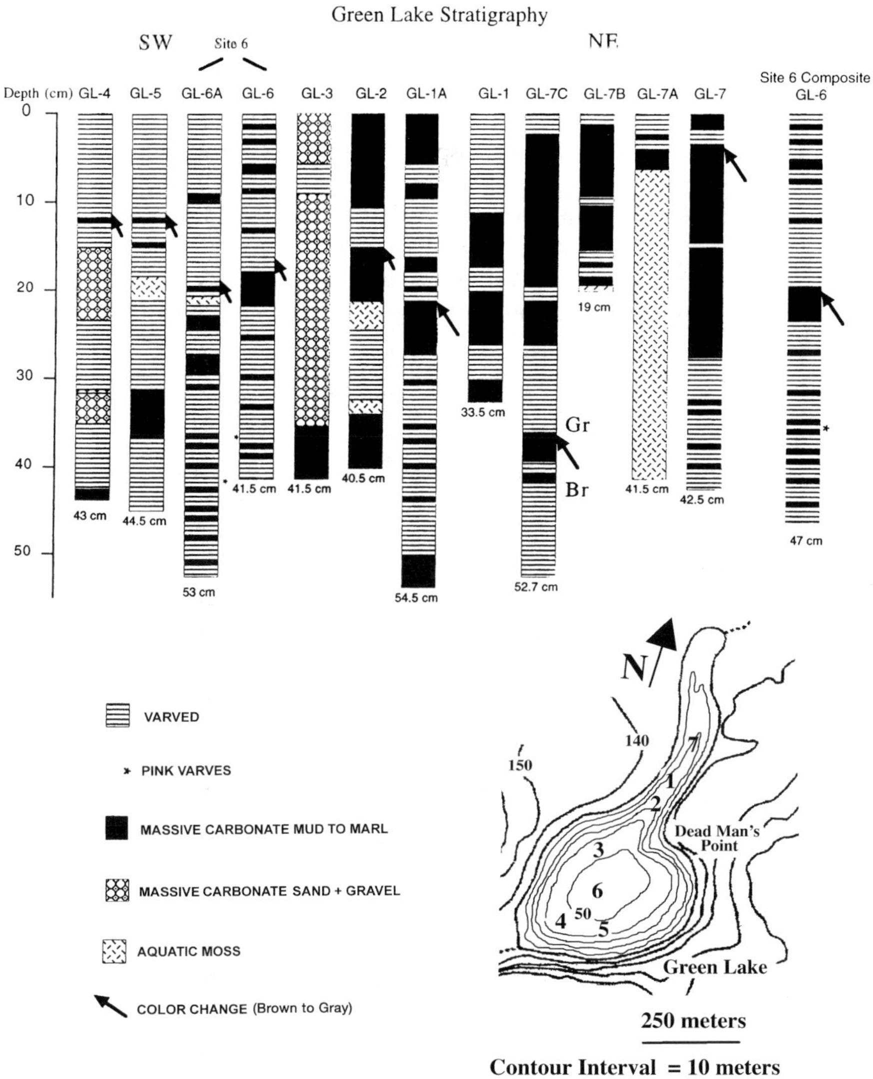

Figure 12 Schematic core descriptions from Fayetteville Green Lake of New York (USA). Note that 12 cores were recovered from seven sites (see map at lower right). Arrows mark the boundary between the older brown sediment (Br) and the overlying gray sediment (Gr). Average organic matter content is 10%. The carbonate content of brown sediments ranges from 75% to 90% and in the gray sediments average ~60% with 20–60% siliciclastic grains. The change in mineralogy is attributed to either deforestation or a new pattern for the polar front jet stream (from Hilfinger et al., 2001 with copyright permission from Springer).

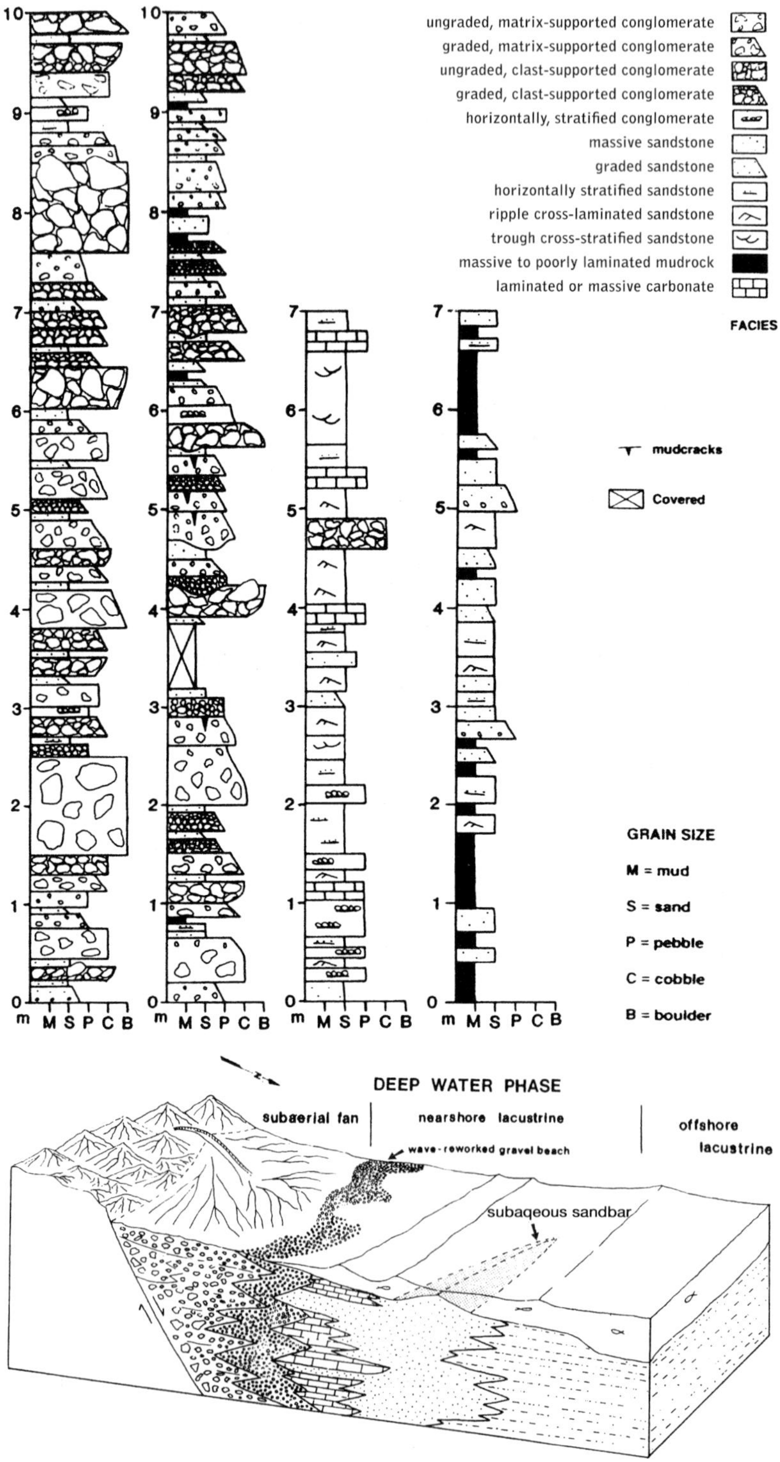
ungraded, matrix-supported conglomerate
graded, matrix-supported conglomerate
ungraded, clast-supported conglomerate
graded, clast-supported conglomerate
horizontally, stratified conglomerate
massive sandstone
graded sandstone
horizontally stratified sandstone
ripple cross-laminated sandstone
trough cross-stratified sandstone
massive to poorly laminated mudrock
laminated or massive carbonate
FACIES
mudcracks
Covered
GRAIN SIZE
M = mud
S = sand
P = pebble
C = cobble
B = boulder
m M S P C B
DEEP WATER PHASE
subaerial fan
nearshore lacustrine
offshore lacustrine
wave-reworked gravel beach
subaqeous sandbar

sub-basin of the Serranía de Cuenca Basin in central Spain is composed of lacustrine/palustrine carbonates associated with an alluvial plain to deltaic environment (Gierlowski-Kordesch et al., 1991). Massive wackestone beds associated with lignite represent shallow deltaic sediments containing charophytes (Figure 3B) as well as ostracodes, gastropods, and carbonized plant material (Figure 3C).

The limestones of the Miocene Horse Camp Formation in Nevada (USA) (Horton and Schmitt, 1996) (Figure 13) are laminated to massive, but are restricted to nearshore areas during high lake levels. Groundwater input in the extensional basin may have produced carbonates in shallow water only. Bioturbation and reworking by currents are evidenced by ooids and pisoids, abundant pelecypod and charophyte fragments suspended in structureless micrites, and reworked beach gravels. In the floodplain lakes of the Pennsylvanian Upper Freeport Formation (Pennsylvania, USA) (Valero Garcés et al., 1994) (Figure 14), massive micrites are present in shallow (Figure 3D) and deep water (Figure 3E). While the structureless texture is due to bioturbation here, early diagenesis can also destroy original sedimentary textures. The Jurassic Scots Bay Formation of Nova Scotia (Canada) (Birney De Wet and Hubert, 1989) (Figure 15) contains mudstone and wackestones (Figure 3F) with silicified logs, gastropod molds, microspar, and stylolites. Peloids and burrows survive in patches not associated with the hydrothermal fluid zones.

11.3. Microbial carbonates

11.3.1. Definition

Microbial carbonates, found within the other four lake facies as well, are characterized by microbially induced sedimentary structures (*cf.* Noffke et al., 2001) and carbonate biostructures, interpreted as primary structures where algal and bacterial films and mats can determine sedimentation patterns and develop on erosional and deformational surfaces. These carbonates include coated grains (ooids/oolites, pisoids/pisolites, and oncoids/oncolites), stromatolites/thrombolites (microbialites), tufa mounds, bioherms, algal biolithites, microreefs, and microbial/algal crusts (e.g., Monty and Mas,

Figure 13 Sedimentologic sections (above) of four facies associations of the Miocene Horse Camp Formation in Nevada (USA). The first column (from left to right) comprises the subaerial fan-delta facies association. The second column represents the subaqueous fan-delta facies association. The third column contains the nearshore lacustrine facies association. The fourth column is a schematic section of an open (offshore) lacustrine facies association. The paleoenvironmental reconstruction (below) shows the spatial relationships of facies associations. The limestone facies are restricted to the lake margin only during the deep water phase, perhaps delineating groundwater input areas (from Horton and Schmitt, 1996 with copyright permission from Wiley–Blackwell).

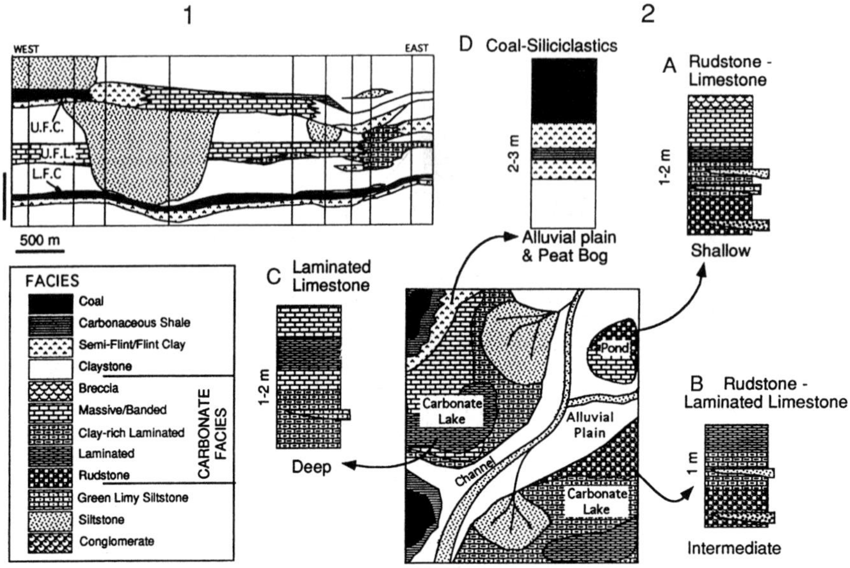

Figure 14 Sedimentologic data on the Pennsylvanian Upper Freeport Limestone of the northern Appalachian Basin in Pennsylvania (USA). (1) Subsurface cross-section showing stratigraphic relationships of the Upper Freeport Limestone (UFL) (top is the datum), Upper Freeport Coal (UFC), and the Lower Freeport Coal (LFC) in southwestern Pennsylvania near the town of Indiana. Vertical lines denote drill holes. (2) Schematic depositional model with observed facies associations showing carbonate lacustrine facies associated with floodbasin lakes of an anastomosed river system (from Valero Garcés et al., 1997 with copyright permission from Elsevier).

1981; Schneider et al., 1983, 1984; Arenas et al., 1993; Bertrand-Sarfati et al., 1966, 1994; Lindqvist, 1994).

These microbial elements can be found anywhere in the photic zone of a lake as *in situ* growths, encrustations on fossils, sedimentary grains, and plant material as well as transported crusts, intraclasts, or microbial laminae in the shallow and deeper zones of a lake. An environment of calcium-rich waters, especially flows from springs or seeps (Scholl, 1960; Scholl and Taft, 1964; Pache et al., 2001), and a reduced input of siliciclastic grains are envisioned as ideal conditions for the development of these microbial structures. These microbial structures commonly can be found in association with shelled organisms/fossils, such as molluscs and ostracodes, except in highly saline conditions.

11.3.2. Examples

Points of input of calcium-rich waters into lake waters, whether through surface or groundwater flow, without the masking effects of siliciclastic input, promote microbial precipitation of carbonates in both modern and

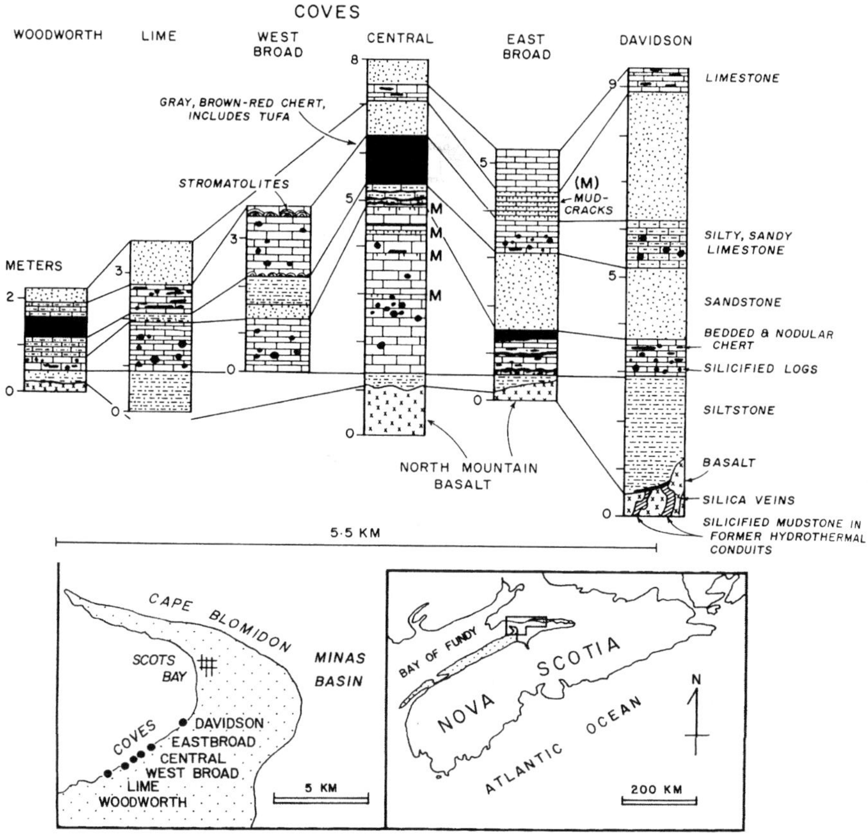

Figure 15 Measured sections from coves along the Bay of Fundy in Nova Scotia (Canada) of the Scots Bay Formation of the Jurassic Fundy Basin of the Newark Supergroup. This carbonate lake deposit contains stromatolites, oncolites, and a diverse invertebrate fauna with littoral wackestones and packstones and profundal lime mudstone. Silicified logs and carbonates are interpreted as hydrothermal spring and seep deposits (from Birney De Wet and Hubert, 1989 with copyright permission from Wiley–Blackwell).

ancient lakes (Dean and Eggleston, 1975; Halley, 1976; Schneider et al., 1983; Casanova and Thouin, 1990; Horton and Schmitt, 1996 (Figure 13); Pache et al., 2001). In modern Big Soda Lake of Nevada (USA) (Rosen et al., 2004), bioherms (Figure 4A) with incorporated caddisfly cases are observed in association with groundwater input levels. The presence of dolomitized stromatolites within the siliciclastic mudrocks of the Devonian Orcadian Basin of Scotland (Figure 4B) shows that groundwater seeps are critical for formation despite siliciclastic input.

Interestingly, even ooid formation, generally associated with microbial carbonates, as demonstrated from the Great Salt Lake (Reitner et al., 1997)

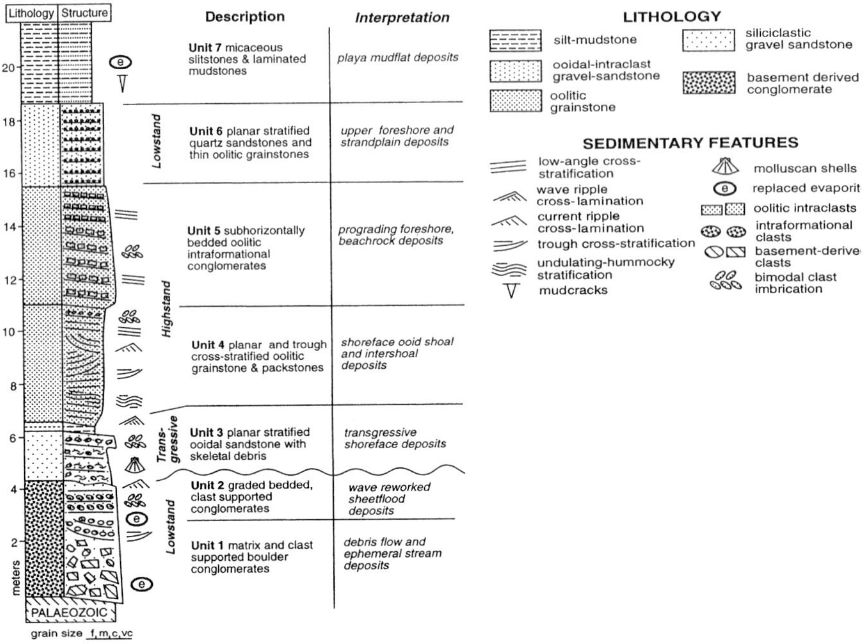

Figure 16 A composite sedimentary section of the Clevendon Succession of the Mercia Mudstone Group (UK) illustrates a high-energy lacustrine shoreline with a Carboniferous limestone source area. Rock types include oolitic intraformational conglomerates, rudstones, grainstones, and packstones exhibiting traction to grain flow deposition (from Milroy and Wright, 2000 with copyright permission from Wiley–Blackwell).

many of the nuclei, which are surrounded by concentric laminae or radial fabric.

There are also modern ooid banks and oncoids making up to 90% of coastal and shallow sublittoral sands in Lake Geneva (Switzerland) (Moscariello, 1997; Davaud and Girardclos, 2001). The origin of the ooids and oncoids is interpreted as biogenic through the presence of cyanobacterial filaments as biofilm coatings on nuclei of quartz and feldspar grains, lithic fragments, and shell pieces (Figure 4C). Very small oncoids resemble ooids in structure.

Carbonate grains, intraclasts, and extraclasts also represent transport and reworking of sediment in shallow water. Microbial crusts suspended within grainstones (Figure 5C) in the Pennsylvanian Pittsburgh Limestone in the northern Appalachian Basin (Ohio, USA) indicate reworking in the littoral zone of floodbasin lakes in an anastomosed river system. Reworking of alluvial-fan carbonate gravels (rudstones) by currents and waves along the Eocene shoreface of a lake in the Shulu Sag (Hebei Province, China) (Jiang et al., 2007) (Figure 17) causes the accumulation of clasts of various carbonate lithologies sourced from the catchment area (Figure 5D). Planar laminated calcisiltites (Figure 5E) of the Oligocene Mequinenza Formation in the southeastern Ebro Basin of Spain (Cabrera et al., 2002) are interbedded with allochthonous, sub-bituminous coal layers (Figure 18) and probably indicate tractive movement of carbonate grains within a shallow sublittoral zone. Limestone "megabreccias," both stratified and non-stratified, are present along the littoral zone of the open lacustrine facies of the Upper Unit of the Rubielos de Mora Basin (Spain) (Anadón et al., 1991). Up to 3.5 m in thickness, these breccias are polymodal and clast- to matrix-supported; they are interpreted as subaqueous debris-flow deposits associated with a fault margin.

Traction load movement of carbonate grains is sometimes difficult to surmise because of early diagenetic alteration into micrite. As an example, trough cross-stratification in micritic deltaic deposits is outlined subtly by plant debris (Figure 5F) in the Lower Cretaceous Uña sub-basin in central Spain (Gierlowski-Kordesch et al., 1991).

11.5. Open-water carbonates

11.5.1. Definition

This facies represents carbonate sediments that have been deposited under a current or density flow regime in the open waters of a lake, from the sublittoral to profundal zone. Grain sizes can range from silt to cobble. Sediment gravity flows from currents and overloading (turbidity currents and debris flows) can move sediment distally (e.g., Gruszka, 2007). Seismic activity (Rodríguez-Pascua et al., 2003) and even large seiches (Schnellman et al., 2006) can contort sediment layering through shearing, liquefaction to water escape, fluidization, plastic deformation, "*in situ*" brecciation, and even resuspension (Ringrose, 1989; Monecke et al., 2006; Agnon et al., 2006; Gruszka and Van Loon, 2007; Moretti and Sabato, 2007).

11.5.2. Examples

An excellent example of calcareous turbidites and subaqueous debris-flow deposits is in the lowermost basinal carbonates (Eocene) of the Shulu Sag half-graben in Hebei Province (China) (Jiang et al., 2007) (Figure 17). Lower Paleozoic carbonate successions in the nearby source area contributed to nine carbonate-dominated facies in the Shulu Lake. Pebbly carbonate arenites and matrix- to clast-supported carbonate rudstones are interpreted as debris flows with the latter facies sorted by longshore currents in the sublittoral zone. Carbonate arenites with rare granule to pebble-sized lithoclasts are from a few centimeters to 1 m thick and are interpreted as possible turbidites (Figure 6A) or sublacustrine wave- or current-reworked

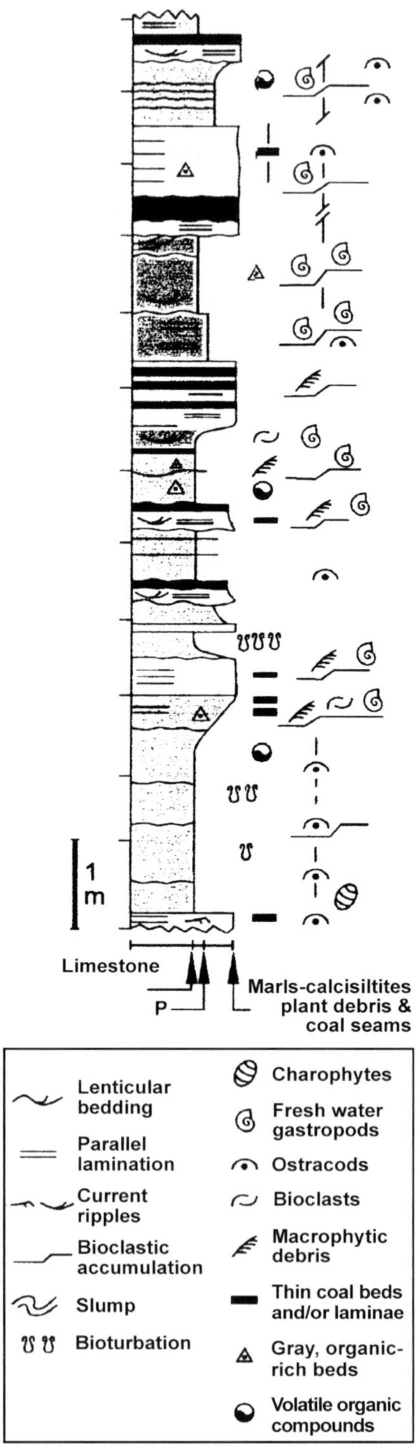

1 m
Limestone
P
Marls-calcisiltites plant debris & coal seams
Lenticular bedding
Parallel lamination
Current ripples
Bioclastic accumulation
Slump
Bioturbation
Charophytes
Fresh water gastropods
Ostracods
Bioclasts
Macrophytic debris
Thin coal beds and/or laminae
Gray, organic-rich beds
Volatile organic compounds

(profundal) conditions, producing "dominantly carbonate lakes." Lakes associated with catchment areas containing both siliciclastic and carbonate bedrock (perhaps 40–60% of carbonate) will contain carbonate facies distributed according to the mode of water input and current regime as well as location and volume of the carbonate bedrock in relation to the lake body, producing "partially carbonate lakes." Lakes associated with mostly siliciclastic bedrock comprising few calcium-rich minerals and less than 30–40% carbonate-rich bedrock may only develop carbonate laminae or rare beds in offshore sediments, producing "sparsely carbonate lakes."

Tucker and Wright (1990) and Platt and Wright (1991) present facies models describing lake margins for both partially and dominantly carbonate lakes, based on littoral, sublittoral, to profundal or pelagial carbonate deposits, dependent on the slope of the lake margin and on energy conditions. The four types of carbonate marginal lake successions are: (1) low-energy bench, (2) high-energy bench, (3) low-energy ramp, and (4) high-energy ramp (see Figure 19 and Table 5). These terms essentially define the steep (bench margins) or gentle (ramp margins) slopes of the littoral region of the carbonate lakes. The gradient of a littoral slope influences carbonate sedimentation patterns. Also, wave action in a lake is dependent on the orientation of the lake and prevalent weather patterns that determine fetch (Sly, 1978). However, the key factors in the formation of these types of lake margins are surface and groundwater flow into a lake from a catchment area or aquifer dominated by carbonate rocks.

Low-energy bench lakes contain margins with a steep littoral to sublittoral zone with limited wave and current action where carbonate accumulation produces a shelf or bench prograding into a lake of bioclastic accumulation and bio-induced precipitation associated with calcium-rich surface water input. Typical facies are massive carbonates with microbial carbonates and bedload-transported carbonates in the littoral zone. Because of the steep slope margin, sublittoral to profundal zones can contain bedload-transported carbonates with slumps, turbidites, and grain flow deposits containing resedimented carbonate produced in the littoral zone. High-energy bench lakes with steep slope margins commonly receive sediments from shallow water areas as well as microbial carbonates. Classic Gilbert-type delta features can develop with topset beds in the highly productive photic zone with material directed downslope and along shore through current reworking, grain flow, and turbidite currents, as well as

Figure 18 Sedimentologic log of the carbonate coal-bearing association of the Mina Pilar section of the Oligocene Mequinenza Formation of the Ebro Basin (Spain). Rock types include pure limestones, marly limestones and marlstones, calcisiltites, carbonate litharenites, and siliciclastics. The source area contains limestone-dominated Mesozoic rocks (from Cabrera et al., 2002 with copyright permission from Elsevier).

Table 6 Examples of the four types of facies models for carbonate marginal-lake environments from the literature.

Low-energy bench margin (steep gradient)	High-energy bench margin (steep gradient)	Low-energy ramp margin (gentle gradient)	High-energy ramp margin (gentle gradient)
Fayetteville Green Lake (USA), Dean and Eggleston (1975)	Shoofly Oolite, Pliocene (USA), Swirydczuk et al. (1979, 1980)	Flagstaff Formation, Paleocene–Eocene (USA), Wells (1983)	Lake Tanganyika (East Africa), Cohen and Thouin (1987)
Lake Littlefield (USA), Murphy and Wilkinson (1980)	La Huerguina Fm., Uña sub-basin, Lower Cretaceous (Spain), Gierlowski-Kordesch et al. (1991)	La Huerguina Fm., Las Hoyas sub-basin, Lower Cretaceous (Spain), Fregenal-Martínez and Meléndez (1994, 2000)	Lagoa Feia Formation, Lower Cretaceous (Brazil), De Carvalho et al. (2000)
Sucker Lake (USA), Treese and Wilkinson (1982)		Green River Formation, Fossil Basin Eocene (USA), Buchheim (1994)	Great Salt Lake (USA), Eardley (1966)
Kelly Lake (Canada), Ferris et al. (1997)		Draney Limestone, Lower Cretaceous (USA), Brown and Wilkinson (1981)	Green River Formation in Uinta Basin (Eocene) (USA), Ryder et al. (1976), Wiggins and Harris (1994)
Peterson Limestone, Lower Cretaceous (USA), Glass and Wilkinson (1980)			
Rambla del Saltador, Las Minas Basin Oligocene (Spain), Calvo et al. (2000)			
Calcaires d'eau douce et dolomie, Freshwater Molasse (Switzerland), Platt (1992)			

with micritic and bioclastic limestones and conglomerates containing clasts of Paleozoic carbonate from the catchment area (Flores and M'Gonigle, 1991). No carbonates are present at the eastern side of the basin. Water-column circulation patterns, water input velocities from rivers and alluvial fans, and the width and gradient of the lake slope probably control the extent and depth of carbonate influence on sedimentation patterns in lakes.

Distribution patterns of carbonates in lakes that have calcium-rich surface water inflow, as illustrated by the four types of carbonate lake margins (Figure 19), are clearly different from lakes that receive calcium-rich input mostly from groundwater. Calcium-rich waters can enter lake waters mostly through subaqueous seeps along the littoral to sublittoral zone, especially in rift lakes where groundwater aquifers of the surrounding rift blocks intersect the lake waters. Examples include modern Lake Tanganyika (Cohen and Thouin, 1987; Casanova and Thouin, 1990; Barrat et al., 2000), the Cretaceous Lagoa Feia Formation (Brazil) (Abrahão and Warme, 1990; De Carvalho et al., 2000), and the Triassic Bundsandstein (Germany) (Paul and Peryt, 2000). In these partially carbonate lakes, banks of carbonate sediment and fossil shells or oolites as well as bioherms and stromatolitic microreefs (Moore and Burne, 1994) concentrate in the littoral to sublittoral zone. The Great Salt Lake in Utah is rimmed by ooid sand bank and algal reefs along its shoreline (Eardley, 1966; Cohenour, 1966; Halley, 1976; Gwynn and Murphy, 1980). Shoreline springs and seeps are identified definitively along the eastern part of the lake below the Wasatch Mountains (Baskin, 1998).

Movement of carbonate downslope to the profundal zone in these nearshore groundwater-influenced accumulations would depend on the width and gradient of the lake slope and on local current patterns. Slumps, debris-flow deposits, and turbidites of carbonate clasts and grains, fossil shells, ooids/oolites, microbial crusts, and other microbial material produced in the shallow photic zone can be transported downslope through circulation currents, earthquake activity, hydrodynamic energy from river inflow, or subaqueous alluvial-fan deposition. Carbonate can also precipitate in association with lineaments and fault planes on the bottom of lakes where groundwater seeps can occur, as evidenced by carbonate bioherms above normal faults across the shallow floor of the Great Salt Lake (Colman et al., 2002).

If the extent of calcium-rich rocks in a lake catchment area, including in the subsurface, is limited relative to the volume of siliciclastic bedrock, then carbonates accumulate in less significant quantities in sparsely carbonate lakes. In these examples, carbonate occurs only as rare laminae, thin layers, or disseminated in low total percentages within siliciclastic muds and silts, mostly from photosynthetic activity in the water column. Examples include: (1) the Holocene sediments of Lake Ontario (McFadden et al., 2005) where calcite comprises less than 30% of the bulk minerals, (2) the

the tectonic situation (hydrology), and the climate of the depositional lake system. Also, internal lake dynamics are responsible for accumulation patterns and preservation or dissolution of carbonates. More detailed facies analyses of freshwater carbonate facies in more lake basins will contribute to the refining of carbonate accumulation models. Biomediated precipitation of carbonate on all levels may be very important indeed within lake systems, as it is in the marine realm. Future work will involve quantifying the extent of microbiologic influence on precipitation and accumulation of freshwater to saline lacustrine carbonates under subaqueous conditions. Also, more precise mineralogic and stable isotope analyses of micrites will deepen knowledge about the evolution of sedimentation patterns and paleoenvironmental conditions within a lake throughout its lifetime. Provenance studies, inferred hydrologic regime, tectonic situation, detailed facies studies, and stable isotopic analyses in addition to paleoclimatic interpretations are crucial in discerning carbonate accumulation patterns in continental settings.

ACKNOWLEDGEMENTS

This work was dependent on the help of the entire limnogeologic community as I begged and borrowed photos of lacustrine carbonates from all my colleagues around the world and asked many of them questions about their favorite lacustrine deposits. I alone am responsible for any errors in this work. My kind contributors of photos and information include: Pere Anadón, Lluís Cabrera, Nieves Meléndez, Walt Dean, Stephanie Girardclos, Eric Davaud, Antje Schwalb, Robin Renaut, David Finkelstein, Paul Buchheim, Zaixing Jiang, Yehouda Enzel, Yuval Bartov, Hülya Alçíçek, Alberto Sáez, Concha "Sandy" Arenas, Blas Valero, Mike Talbot, Tom Johnson, Michael Rosen, Michael Soreghan, Leroy Leggitt, Paul Buchheim, Bill Last, Elena and Victor Kuzevanov, and Mike Smith. Nieves Meléndez and Robin Renaut kindly read this manuscript and made comments before submission. Ana Alonso-Zarza and Pere Anadón reviewed the second draft of this contribution and helped immensely in helping the author focus its theme and presentation. Special thanks go to Sila Plá Pueyo, Kevin Kallini, and Chris Cassle for a student's perspective. In addition, my family deserves thanks for their patience these many months. ExxonMobil Upstream Research Company partially supported the collection of the lacustrine carbonate database.

REFERENCES

Abell, P.I., 1982. Palaeoclimates at Lake Turkana, Kenya, from oxygen isotope ratios of gastropod shells. Nature 297, 321–323.

Abell, P.I., Awramik, S.M., Osborne, R.H., Tomellini, S., 1982. Plio-Pleistocene lacustrine stromatolites from Lake Turkana, Kenya: morphology, stratigraphy, and stable isotopes. Sedimentary Geology 32, 1–26.

Abell, P.I., Hoelzmann, P., 2000. Holocene palaeoclimates in northwestern Sudan: stable isotope studies on molluscs. Global and Planetary Change 26, 1–12.

Abell, P.I., McClory, J.P., 1986. Sedimentary carbonates as isotopic marker horizons at Lake Turkana, Kenya. In: Frostick, L.E., Renaut, R.W., Reid, I., Tiercelin, J.-J. (Eds),

Sedimentation in the African Rifts, Geological Society (London) Special Publication 25, pp. 153–158.
Abrahão, D., Warme J.E., 1990. Lacustrine and associated deposits in a rifted continental margin – Lower Cretaceous Lagoa Feia Formation, Campos Basin, offshore Brazil. In: Katz, B.J. (Ed.), Lacustrine Basin Exploration: Case Studies and Modern Analogs, AAPG Memoir 50, pp. 287–305.
Agnon, A., Migowski, C., Marco, S., 2006. Intraclast breccias in laminated sequences reviewed: recorders of paleo-earthquakes. In: Enzel, Y., Agnon, A., Stein, M. (Eds), New Frontiers in Dead Sea Paleoenvironmental Research, Geological Society of America Special Paper 401, pp. 195–214.
Alçiçek, H., Varol, B., Ozkul, M., 2007. Sedimentary facies, depositional environments and palaeogeographic evolution of the Neogene Denizli Basin, SW Anatolia, Turkey. Sedimentary Geology 202, 596–637.
Alonso-Zarza, A.M., Calvo, J.P., 2000. Palustrine sedimentation in an episodically subsiding basin: the Miocene of the northern Teruel Graben (Spain). Palaeogeography, Palaeoclimatology, Palaeoecology 160, 1–21.
Alonso-Zarza, A.M., Wright, V.P., 2010. Palustrine carbonates. In: Alonso-Zarza, A.M., Tanner, L.H. (Eds), Carbonates in Continental Settings: Facies, Environments and Processes. Developments in Sedimentology (Elsevier, Amsterdam) 61, pp. 103–132.
Anadón, P., Burjachs, F., Martín, M., Rodríguez-Lázaro, J., Robles, F., Utrilla, R., Vázquez, A., 2002b. Paleoenvironmental evolution of the Pliocene Villarroya Lake, northern Spain. A multidisciplinary approach. Sedimentary Geology 148, 9–27.
Anadón, P., Cabrera, Ll., Juliá, R., 1988. Anoxic-oxic cyclical lacustrine sedimentation in the Miocene Rubielos de Mora Basin, Spain. In: Fleet, A.J., Kelts, K., Talbot, M.R. (Eds), Lacustrine Petroleum Source Rocks, Geological Society (London) Special Publication 40, pp. 353–367.
Anadón, P., Cabrera, Ll., Juliá, R., Marzo, M., 1991. Sequential arrangement and asymmetrical fill in the Miocene Rubielos de Mora Basin (northeast Spain). In: Anadón, P., Cabrera, Ll., Kelts, K. (Eds), Lacustrine Facies Analysis, International Association of Sedimentologists Special Publication 13, pp. 257–275.
Anadón, P., Cabrera, Ll., Juliá, R., Roca, E., Rosell, L., 1989. Lacustrine oil-shale basins in Tertiary grabens from NE Spain (western European rift system). Palaeogeography, Palaeoclimatology, Palaeoecology 70, 7–28.
Anadón, P., Moscariello, A., Rodríquez-Lázaro, J., Filippi, M.L., 2006. Holocene environmental changes of Lake Geneva (Lac Léman) from stable isotopes ($\delta^{13}C$, $\delta^{18}O$) and trace element records of ostracod and gastropod carbonates. Journal of Paleolimnology 35, 593–616.
Anadón, P., Robles, F., Roca, E., Utrilla, R., Vázquez, A., 1998. Lacustrine sedimentation in the diapir-controlled Miocene Bicorb Basin, eastern Spain. Palaeogeography, Palaeoclimatology, Palaeoecology 140, 217–243.
Anadón, P., Rosell, L., Talbot, M.R., 1992. Carbonate replacement of lacustrine gypsum deposits in two Neogene continental basins, eastern Spain. Sedimentary Geology 78, 201–216.
Anadón, P., Utrilla, R., 1993. Sedimentology and isotope geochemistry of lacustrine carbonates of the Oligocene Campins Basin, north-east Spain. Sedimentology 40, 699–720.
Anadón, P., Utrilla, R., Vázquez, A., 2000. Use of charophyte carbonates as proxy indicators of subtle hydrological and chemical changes in marl lakes: example from the Miocene Bicorb Basin, eastern Spain. Sedimentary Geology 133, 325–347.
Anadón, P., Utrilla, R., Vázquez, A., 2002a. Mineralogy and Sr–Mg geochemistry of charophyte carbonates: a new tool for paleolimnological research. Earth and Planetary Science Letters 197, 205–214.

Bell, C.M., 1989. Saline lake carbonates within an Upper Jurassic–Lower Cretaceous continental redbed sequence in the Atacama region of northern Chile. Sedimentology 36, 651–663.
Bellanca, A., Calvo, J.P., Censi, P., Elizaga, E., Neri, R., 1989. Evolution of lacustrine diatomite carbonate cycles of Miocene age, southeastern Spain: petrology and isotope geochemistry. Journal of Sedimentary Petrology 59, 45–52.
Bellanca, A., Calvo, J.P., Censi, P., Neri, R., Pozo, M., 1992. Recognition of lake-level changes in Miocene lacustrine units, Madrid Basin, Spain. Evidence from facies analysis, isotope geochemistry and clay mineralogy. Sedimentary Geology 76, 135–153.
Bellanca, A., Calvo, J.P., Neri, R., Mirto, E., 1995. Lake margin carbonate deposits of Las Minas Basin, Upper Miocene, southeastern Spain. A sedimentological and geochemical approach to the study of lacustrine and palustrine paleoenvironments. Mineralogica et Petrografica Acta 38, 113–128.
Benammi, M., Chaimanee, Y., Jaeger, J.-J., Suteethorn, V., Durcrocq, S., 2001. Eocene Krabi basin (southern Thailand): paleontology and magnetostratigraphy. Geological Society of America Bulletin 113, 265–273.
Benson, L., 1994. Carbonate deposition, Pyramid Lake Subbasin, Nevada: 1. Sequence of formation and elevational distribution of carbonate deposits (tufas). Palaeogeography, Palaeoclimatology, Palaeoecology 109, 55–87.
Benson, L.V., 1978. Fluctuation in the level of pluvial Lake Lahontan during the last 40,000 years. Quaternary Research 9, 300–318.
Beraldi-Campesi, H., Cevallos-Ferriz, S.R.S., Centeno-Garcia, E., Arenas-Abad, C., Fernández, L.P., 2006. Sedimentology and paleoecology of an Eocene–Oligocene alluvial–lacustrine arid system, southern México. Sedimentary Geology 191, 227–254.
Bertrand-Sarfati, J., Freytet, P., Plaziat, J.-C., 1966. Les calcaires concrétionnés de la limite Oligocène-Miocène des environs de Saint-Pourçain-sur-Sioule (Limagne d'Allier): rôle des Algues dans leur édification; analogie avec les stromatolites et rapports avec la sédimentation. Bulletin de la Société Géologique de France 8, 652–662.
Bertrand-Sarfati, J., Freytet, P., Plaziat, J.-C., 1994. Microstructures in Tertiary nonmarine stromatolites (France), comparisons with Proterozoic. In: Bertrand-Sarfati, J., Monty, C. (Eds), Phanerozoic Stromatolites II. Kluwer Academic Publishers, Dordrecht, pp. 155–191.
Binford, M.W., Deevey, E.S., Crisman, T.L., 1983. Paleolimnology: an historical perspective on lacustrine ecosystems. Annual Review of Ecology and Systematics 14, 255–286.
Binkley, K.L., Wilkinson, B.H., Owen, R.M., 1980. Vadose beachrock cementation along a southeastern Michigan marl lake. Journal of Sedimentary Petrology 50, 953–962.
Birney De Wet, C.C., Hubert, J.F., 1989. The Scots Bay Formation, Nova Scotia, Canada, a Jurassic carbonate lake with silica-rich hydrothermal springs. Sedimentology 36, 857–873.
Bloch, M.R., Litman, H.Z., Elazari-Volcani, B., 1944. Occasional whiteness of the Dead Sea. Nature 154, 402.
Bohacs, K.M., Carroll, A.R., Neal, J.E., 2003. Lessons from large lake systems – thresholds, nonlinearity, and strange attractors. In: Chan, M.A., Archer, A.W. (Eds), Extreme Depositional Environments: Mega End Members in Geologic Time. Geological Society of America Special Paper 370, pp. 75–90.
Bohacs, K.M., Grabowski, G., Jr., Carroll, A.R., 2007. Lithofacies architecture and variations in expression of sequence stratigraphy within representative intervals of the Green River Formation, Greater Green River Basin, Wyoming and Colorado. The Mountain Geologist 44, 39–60.

Bohacs, K.M., Carroll, A.R., Neal, J.E., Mankiewicz, P.J., 2000. Lake-basin type, source potential, and hydrocarbon character: an integrated sequence–stratigraphic–geochemical framework. In: Gierlowski-Kordesch, E.H., Kelts, K.R. (Eds), Lake Basins through Space and Time, AAPG Studies in Geology 46, AAPG, Tulsa, OK, pp. 3–33.
Bookman, R., Bartov, Y., Enzel, Y., Stein M., 2006. Quaternary lake levels in the Dead Sea basin: two centuries of research. In: Enzel, Y., Agnon, A., Stein, M. (Eds), New Frontiers in Dead Sea Paleoenvironmental Research. Geological Society of America Special Paper 401, pp. 155–170.
Bookman, R., Enzel, Y., Agnon, A., Stein, M., 2004. Late Holocene lake levels of the Dead Sea. Geological Society of America Bulletin 116, 555–571.
Bowler, J.M., 1981. Australian salt lakes: a palaeohydrologic approach. Hydrobiologia 82, 431–444.
Bradbury, J.P., Dean, W.E. (Eds), 1993. Elk Lake, Minnesota: evidence for rapid climate change in the north-central United States. Geological Society of America Special Paper 276, 336 pp.
Bradley, W.H., 1929a. Algae reefs and oolites in the Green River Formation. U.S. Geological Survey Professional Paper 154-G, pp. 203–233.
Bradley, W.H., 1929b. The varves and climate of the Green River epoch. U.S. Geological Survey Professional Paper 158-E, pp. 87–110.
Braithwaite, C.J.R., Zedef, V., 1996. Hydromagnesite stromatolites and sediments in an alkaline lake, Salda Gölü, Turkey. Journal of Sedimentary Research 66, 991–1002.
Brauer, A., Casanova, J., 2001. Chronology and depositional processes of the laminated sediment record from Lac d'Annecy, French Alps. Journal of Paleolimnology 25, 163–177.
Breckenridge, A., Johnson, T.C., Beske-Diehl, S., Mothersill, J.S., 2004. The timing of regional Lateglacial events and post-glacial sedimentation rates from Lake Superior. Quaternary Science Reviews 23, 2355–2367.
Bréhéret, J.-G., Fourmont, A., Macaire, J.-J., Négrel, P., 2008. Microbially mediated carbonates in the Holocene deposits from Sarliève, a small ancient lake of the French Massif Central, testify to the evolution of a restricted environment. Sedimentology 55, 557–578.
Brenner, M., Whitmore, T.J., Song, X.L., Long, R.H., Hodell, D.A., Curtis, J.H., 2000. In: Gierlowski-Kordesch, E.H., Kelts, K.R. (Eds), Lake Basins through Space and Time, AAPG Studies in Geology 46, AAPG, Tulsa, OK, pp. 619–624.
Bright, J., Kaufman, D.S., Forester, R.M., Dean, W.E., 2006. A continuous 250,000 yr record of oxygen and carbon isotopes in ostracode and bulk-sediment carbonate in Bear Lake, Utah-Idaho. Quaternary Science Reviews 25, 2258–2270.
Brown, R.E., Wilkinson, B.H., 1981. The Draney Limestone: Early Cretaceous lacustrine carbonate deposition in western Wyoming. Contributions to Geology, University of Wyoming 20, pp. 23–31.
Brunskill, G.J., 1969. Fayetteville Green Lake, New York. II. Precipitation and sedimentation of calcite in a meromictic lake with laminated sediments. Limnology and Oceanography 14, 830–847.
Buchheim, H.P., 1994. Paleoenvironments, lithofacies, and varves of the Fossil Butte Member of the Eocene Green River Formation, southwestern Wyoming. Contributions to Geology, University of Wyoming 30, pp. 3–14.
Buchheim, H.P, Eugster, H.P., 1998. Eocene Fossil Lake: the Green River Formation of Fossil Basin, southwestern Wyoming. In: Pitman, J.K., Carroll, A.R. (Eds), Modern and Ancient Lake Systems. Utah Geological Association Guidebook 26, C&M Press, Denver, CO, pp. 191–207.
Burne, R.V., Bauld, J., De Deckker, P., 1980. Saline lake charophytes and their geological significance. Journal of Sedimentary Petrology 50, 281–293.

Bustillo, M.A., Arribas, M.E., Bustillo, M., 2002. Dolomitization and silicification in low-energy lacustrine carbonates (Paleogene, Madrid Basin, Spain). Sedimentary Geology 151, 107–126.
Cabaleri, N.G., Armella, C., 2005. Influence of a biohermal belt on the lacustrine sedimentation of the Cañadón Asfalto Formation (Upper Jurassic, Chubut province, southern Argentina). Geologica Acta 3, 205–214.
Cabrera, Ll., Cabrera, M., Gorchs, R., de las Heras, F.X.C., 2002. Lacustrine basin dynamics and organosulphur compound origin in a carbonate-rich lacustrine system (Late Oligocene Mequinenza Formation, SE Ebro basin, NE Spain). Sedimentary Geology 148, 289–317.
Cabrera, Ll., Sáez, A., 1987. Coal deposition in carbonate-rich lacustrine systems: the Calaf and Mequinenza sequences (Oligocene, eastern Ebro Basin, NE Spain). Journal of the Geological Society London 144, 451–461.
Calvo, J.P., Alonso-Zarza, A.M., García del Cura, M.A., 1989. Models of Miocene marginal lacustrine sedimentation in response to varied depositional regimes and source areas in the Madrid Basin (central Spain). Palaeogeography, Palaeoclimatology, Palaeoecology 70, 199–214.
Calvo, J.P., Alonso Zarza, A.M., García del Cura, M.A., Ordóñez, S., Rodríguez-Aranda, J.P., Sanz Montero, M.E., 1996. Sedimentary evolution of lake systems through the Miocene of the Madrid Basin: paleoclimate and paleohydrological constraints. In: Friend, P.F., Dabrio, C.J. (Eds), Tertiary Basins of Spain: The Stratigraphic Record of Crustal Kinematics. Cambridge University Press, Cambridge, pp. 272–277.
Calvo, J.P., Gómez-Gras, D., Alonso-Zarza, A.M., Jiménez, S., 2000. Architecture of a bench-type carbonate lake margin and its relation to fluvially dominated deltas, Las Minas Basin, Upper Miocene, Spain. Journal of Sedimentary Research 70, 240–254.
Calvo, J.P., McKenzie, J.A., Vasconcelos, C., 2003. Microbially-mediated lacustrine dolomite formation: evidence and current research trends. In: Valero-Garcés, B.L. (Ed.), Limnogeology in Spain: A Tribute to Kerry Kelts. Consejo Superior de Investigaciones Científicas, Madrid, pp. 229–251.
Calvo, J.P., Ordóñez, S., García del Cura, M.A., Hoyos, M., Alonso-Zarza, A.M., 1994. Madrid Basin (Neogene), Spain. In: Gierlowski-Kordesch, E., Kelts, K. (Eds), Global Geological Record of Lake Basins. Cambridge University Press, Cambridge, pp. 303–305.
Camoin, G., Casanova, J., Rouchy, J.-M., Blanc-Valleron, M.-M., Deconinck, J.-F., 1997. Environmental controls on perennial and ephemeral carbonate lakes: the central palaeo-Andean Basin of Bolivia during Late Cretaceous to early Tertiary times. Sedimentary Geology 113, 1–26.
Cañaveras, J.C., Sánchez-Moral, S., Calvo, J.P., Hoyos, M., Ordóñez, S., 1996. Dedolomites associated with karstification. An example of early dedolomitization in lacustrine sequences from the Tertiary Madrid Basin, central Spain. Carbonates and Evaporites 11, 85–103.
Cañaveras, J.C., Sánchez-Moral, S., Sanz-Rubio, E., Hoyos, M., 1998. Meteoric calcitization of magnesite in Miocene lacustrine deposits (Calatayud basin, NE Spain). Sedimentary Geology 119, 183–194.
Capo, R.C., Chadwick, O.A., 1999. Sources of calcium and strontium in desert soil and calcrete. Earth and Planetary Science Letters 170, 61–72.
Carbonel, P., Colin, J.-P., Danielopol, D.L., Loeffler, H., Neustrueva, I., 1988. Paleoecology of limnic ostracodes: a review of some major topics. Palaeogeography, Palaeoclimatology, Palaeoecology 62, 413–461.
Carozzi, A.V., 1964. Complex ooids from Triassic lake deposit, Virginia. American Journal of Science 262, 231–241.

Carroll, A.R., Bohacs, K.M., 1999. Stratigraphic classification of ancient lakes: balancing tectonic and climatic controls. Geology 27, 99–102.
Carroll, A.R., Doebbert, A.C., Booth, A.L., Chamberlain, C.P., Rhodes-Carson, M.K., Smith, M.E., Johnson, C.M., Beard, B.L., 2008. Capture of high altitude precipitation by a low altitude Eocene lake. Western US Geology 36, 791–794.
Casanova, J., 1986. East African rift stromatolites. In: Frostick, L.E., Renaut, R.W., Reid, I., Tiercelin, J.-J. (Eds), Sedimentation in the African Rifts, Geological Society (London) Special Publication 25, pp. 201–210.
Casanova, J., 1994. Stromatolites from the East African rift: a synopsis. In: Bertrand-Sarfati, J., Monty, C. (Eds), Phanerozoic Stromatolites II. Kluwer Academic Publishers, Dordrecht, pp. 193–226.
Casanova, J., Thouin, C., 1990. Biosédimentologie des stromatolites du lac Tanganyika (Burundi): implications hydrologiques. Bulletin de la Société Géologique de France 8, 647–656.
Cerling, T.E., Bowman, J.R., O'Neil, J.R., 1988. An isotopic study of a fluvial–lacustrine sequence: the Plio-Pleistocene Koobi Fora sequence, East Africa. Palaeogeography, Palaeoclimatology, Palaeoecology 63, 335–356.
Chacón-Baca, E., Beraldi-Campesi, H., Cevallos-Ferriz, S.R.S., Knoll, A.H., Golubic, S., 2002. 70 Ma nonmarine diatoms from México. Geology 30, 279–281.
Chivas, A.R., De Deckker, P., Cali, J.A., Chapman, A., Kiss, E., Shelley, J.M.G., 1993. Coupled stable-isotope and trace-metal measurements of lacustrine carbonates as paleoclimatic indicators. In: Swart, P., Lohmann, K.C., McKenzie, J., and Savin, S. (Eds), Climate Change in Continental Isotopic Records. American Geophysical Union Geophysical Monograph 78, pp. 113–121.
Chutko, K.J., Lamoureux, S.F., 2008. Identification of coherent links between interannual sedimentary structures and daily meteorological observations in Arctic proglacial lacustrine varves: potentials and limitations. Canadian Journal of Earth Sciences 45, 1–13.
Clausing, A., Boy, J.A., 2000. Lamination and primary production in fossil lakes: relationship to palaeoclimate in the Carboniferous–Permian transition. In: Hart, M.B. (Ed.), Climates: Past and Present. Geological Society (London) Special Publication 181, pp. 5–16.
Clemmensen, L.B., 1978. Lacustrine facies and stromatolites from the middle Triassic of East Greenland. Journal of Sedimentary Petrology 48, 1111–1128.
Cocquyt, C., 1991. Epilithic diatoms from thrombolitic reefs of Lake Tanganyika. Belgian Journal of Botany 124, 102–108.
Cohen, A.S., 1984. Effect of zoobenthic standing crop on laminae preservation in tropical lake sediment, Lake Turkana, East Africa. Journal of Paleontology 58, 499–510.
Cohen, A.S., 2003. Paleolimnology. Oxford University Press, Oxford, 500 pp.
Cohen, A.S., Talbot, M.R., Awarnik, S.M., Dettman, D.L., Abell, P., 1997. Lake level and paleoenvironmental history of Lake Tanganyika, Africa, as inferred from late Holocene and modern stromatolites. Geological Society of America Bulletin 109, 444–460.
Cohen, A.S., Thouin, C., 1987. Nearshore carbonate deposits in Lake Tanganyika. Geology 15, 414–418.
Cohenour, R.E., 1966. Industrial development and potential of Great Salt Lake with notes on engineering geology and operational problems. In: Stokes, W.L. (Ed.), The Great Salt Lake. Guidebook to the Geology of Utah 20. Utah Geological Society, Salt Lake City, pp. 153–163.
Colman, S.M., 2005. Acoustic stratigraphy of Bear Lake, Utah-Idaho – Late Quaternary sedimentation in a simple half-graben. Sedimentary Geology 185, 113–125.
Colman, S.M., Forester, R.M., Reynolds, R.L., Sweetkind, D.S., King, J.W., Gangemi, P., Jones, G.A., Keigwin, L.D., Foster, D.S., 1994. Lake-level history of Lake Michigan for

the past 12,000 years: the record from deep lacustrine sediments. Journal of Great Lakes Research 20, 73–92.

Colman, S.M., Kelts, K.R., Dinter, D.A., 2002. Depositional history and neotectonics in Great Salt Lake, Utah, from high-resolution seismic stratigraphy. Sedimentary Geology 148, 61–78.

Coshell, L., Rosen, M.R., McNamara, K.J., 1998. Hydromagnesite replacement of biomineralized aragonite in a new location of Holocene stromatolites, Lake Walyyungup, Western Australia. Sedimentology 45, 1005–1018.

Crausbay, S.D., Russell, J.M., Schnurrenberger, D.W., 2006. A ca. 800-year lithologic record of drought from sub-annually laminated lake sediment, East Java. Journal of Paleolimnology 35, 641–659.

Dam, G., Stemmerik, L., 1994. East Greenland lacustrine complexes. In: Gierlowski-Kordesch, E., Kelts, K. (Eds), Global Geological Record of Lake Basins. Cambridge University Press, Cambridge, pp. 19–27.

Davaud, E., Girardclos, S., 2001. Recent freshwater ooids and oncoids from western Lake Geneva (Switzerland): indications of a common organically-mediated origin. Journal of Sedimentary Research 71, 423–429.

Davis, R.L., Wilkinson, B.H., 1983. Sedimentology and petrology of freshwater lacustrine carbonate – Mid-Tertiary Camp Davis Formation, northwestern Wyoming. University of Wyoming Contributions to Geology 22, 45–55.

Davis, S.J., Mulch, A., Carroll, A.R., Horton, T.W., Chamberlain, C.P., 2009. Paleogene landscape evolution of the central North American Cordillera: developing topography and hydrology in the Laramide foreland. Geological Society of America Bulletin 120, 100–133.

Davis, S.J., Wiegand, B.A., Carroll, A.R., Chamberlain, C.P., 2008. The effect of drainage reorganization on paleoaltimetry studies: an example from the Paleogene Laramide foreland. Earth and Planetary Science Letters 275, 258–268.

Dayvault, R.D., Codington, L.A., Kohls, D., Hawes, W.D., Ott, P.M., 1995. Fossil insects and spiders from three locations in the Green River Formation of the Piceance Creek Basin, Colorado. In: Averett, W.R. (Ed.), The Green River Formation in Piceance Creek and Eastern Uinta Basins. Grand Junction Geological Society, Grand Junction, pp. 97–115.

de Las Heras, F.X.C., Anadón, P., Cabrera, Ll., 2003. Biomarker record variations in lacustrine coals and oil shales: contributions from Tertiary basins in NE Spain. In: Valero-Garcés, B.L. (Ed.), Limnogeology in Spain: A Tribute to Kerry Kelts. Consejo Superior de Investigaciones Científicas, Madrid, pp. 187–228.

De Carvalho, M.D., Praça, U.M., da Silva-Telles, Jr., A.C., Jahnert, R.J., Dias, J.L., 2000. Bioclastic carbonate lacustrine facies models in the Campos Basin (Lower Cretaceous), Brazil. In: Gierlowski-Kordesch, E.H., Kelts, K.R. (Eds), Lake Basins through Space and Time, AAPG Studies in Geology 46, AAPG, Tulsa, OK, pp. 245–256.

Dean, J.M., Kemp, A.E.S., Bull, D., Pike, J., Patterson, G., Zolitschka, B., 1999. Taking varves to bits: scanning electron microscopy in the study of laminated sediments and varves. Journal of Paleolimnology 22, 121–136.

Dean, W.E., 1981. Carbonate minerals and organic matter in sediments of modern north temperate hard-water lakes. In: Ethridge, F.G., Flores, R.M. (Eds), Recent and Ancient Nonmarine Depositional Environments: Models for Exploration. SEPM Special Publication 31, pp. 213–231.

Dean, W.E., 1997. Rates, timing, and cyclicity of Holocene eolian activity in north-central United States: evidence from varved lake sediments. Geology 25, 331–334.

Dean, W.E., Anderson, R., Bradbury, J.P., Anderson, D., 2002. A 1500-year record of climatic and environmental change in Elk Lake, Minnesota. I: Varve thickness and gray-scale density. Journal of Paleolimnology 27, 287–299.

Dean, W.E., Eggleston, J.R., 1975. Comparative anatomy of marine and freshwater algal reefs, Bermuda and central New York. Geological Society of America Bulletin 86, 665–676.
Dean, W.E., Forester, R.M., Bright, J., Anderson, R.Y., 2007. Influence of the diversion of Bear River into Bear Lake (Utah and Idaho) on the environment of deposition of carbonate minerals. Limnology and Oceanography 52, 1094–1111.
Dean, W.E., Fouch, T.D., 1983. Lacustrine environment. In: Scholle, P.A., Bebout, D.G., Moore, C.H. (Eds), Carbonate Depositional Environments, AAPG Memoir 33, pp. 97–130.
Dean, W.E., Gorham, E., 1976. Major chemical and mineral components of profundal surface sediments in Minnesota lakes. Limnology and Oceanography 21, 259–284.
Dean, W.E., Megard, R.O., 1993. Environment of deposition of $CaCO_3$ in Elk Lake, Minnesota. In: Bradbury, J.P., Dean, W.E. (Eds), Elk Lake, Minnesota: Evidence for Rapid Climate Change in the North-Central United States. Geological Society of America Special Paper 276, pp. 97–113.
Dean, W.E., Rosenbaum, J., Skipp, G., Colman, S., Forester, R., Liu, A., Simmons, K., Bischoff, J., 2006. Unusual Holocene and late Pleistocene carbonate sedimentation in Bear Lake, Utah and Idaho, U.S.A.. Sedimentary Geology 185, 93–112.
Dean, W.E., Schwalb, A., 2000. Holocene environmental and climatic change in the northern Great Plains as recorded in the geochemistry of sediments in Pickerel Lake. Quaternary International 67, 5–20.
DeDeckker, P., 1988a. Biological and sedimentary facies of Australian salt lakes. Palaeogeography, Palaeoclimatology, Palaeoecology 62, 237–270.
DeDeckker, P., 1988b. Large Australian lakes during the last 20 million years: sites for petroleum source rock or metal ore deposition, or both? In: Fleet, A.J., Kelts, K. Talbot, M.R. (Eds), Lacustrine Petroleum Source Rocks. Geological Society (London) Special Publication 40, pp. 45–58.
De Deckker, P., 2002. Ostracod palaeoecology. In: Holmes, J.A., Chivas, A.R. (Eds), The Ostracoda: Applications in Quaternary Research. Geophysical Monograph 131. American Geophysical Union, Washington, pp. 121–134.
De Deckker, P., Forester, R.M., 1988. The use of ostracods to reconstruct continental palaeoenvironmental records. In: De Deckker, P., Colin, J.-P., and Peypouquet, J.-P. (Eds), Ostracoda in the Earth Sciences. Elsevier, Amsterdam, pp. 175–199.
De Geer, G., 1912. A geochronology of the last 12,000 years. 11th International Geological Congress (Stockholm, 1910), Comptes Rendus, pp. 241–253.
Degens, E.T., Stoffers, P., 1976. Stratified waters as a key to the past. Nature 263, 22–27.
Degens, E.T., von Herzen, R.P., Wong, H.K., Deuser, W.G., Jannasch, H.W., 1972. Lake Kivu: structure, chemistry, and biology of an East African rift lake. Geologische Rundschau 62, 245–277.
Del Papa, C.E., 1999. Sedimentation on a ramp-type lake margin: Paleocene–Eocene Maiz Gordo Formation, northwestern Argentina. Journal of South American Earth Sciences 12, 389–400.
Détriché, S., Bréhéret, J.-G., Zarki, J., Karrat, L.H., Macaire, J.-J., Fontugne, M., 2008. Lake Holocene palaeohydrology of Lake Afourgagh (Middle Atlas, Morocco) from deposit geometry and facies. Bulletin de la Société Géologique de France 179, 41–50.
Dettman, D.L., Lohmann, K.C., 1993. Seasonal change in Paleogene surface water $\delta^{18}O$: freshwater bivalves of western North America. In: Swart, P., Lohmann, K.C., McKenzie, J., and Savin, S. (Eds), Climate Change in Continental Isotopic Records. American Geophysical Union Geophysical Monograph 78, pp. 153–163.
Dettman, D.L., Palacios-Fest, M.R., Nkotagu, H.H., Cohen, A.S., 2005. Paleolimnological investigations of anthropogenic environmental change in Lake Tanganyika: VII.

Fritz, S.C., Cumming, B.F., Gasse, F., Laird, K.R., 1999. Diatoms as indicators of hydrologic and climatic change in saline lakes. In: Stoermer, E.F., Smol, J.P. (Eds), The Diatoms: Applications for the Environmental and Earth Sciences. Cambridge University Press, Cambridge, pp. 41–72.
Frogley, M.R., Griffiths, H.I., Martens, K., 2002. Modern and fossil ostracods from ancient lakes. In: Holmes, J.A., Chivas, A.R. (Eds), The Ostracoda: Applications in Quaternary Research. Geophysical Monograph 131. American Geophysical Union, Washington, pp. 167–184.
Galat, D.L., Jacobsen, R.L., 1985. Recurrent aragonite precipitation in saline-alkaline Pyramid Lake, Nevada. Archiv für Hydrobiologie 105, 137–159.
Gälman, V., Petterson, G., Renberg, I., 2006. A comparison of sediment varves (1950–2003 AD) in two adjacent lakes in northern Sweden. Journal of Paleolimnology 35, 837–853.
Garber, R.A., Friedman, G.M., 1983. Coated grains along the Dead Sea shore. In: Peryt, T.M. (Ed.), Coated Grains. Springer-Verlag, Berlin, pp. 142–153.
García, A., 1994. Charophyta: their use in paleolimnology. Journal of Paleolimnology 10, 43–52.
García, A., Chivas, A.R., 2006. Diversity and ecology of extant and Quaternary Australia charophytes (Charales). Cryptogramie, Algologie 27, 323–340.
Gasse, F., 2002. Diatom-inferred salinity and carbonate oxygen isotopes in Holocene waterbodies of the western Sahara and Sahel (Africa). Quaternary Science Reviews 21, 737–767.
Gebhardt, U., Merkel, T., Szabados, A., 2000. Karbonatsedimentation in siliziklastischen fluviatilen Abfolgen. Freiberger Forschungshefte C 490, 133–168.
Gibling, M.R., 1988. Cenozoic lacustrine basins of Southeast Asia, their tectonic setting, depositional environment, and hydrocarbon potential. In: Fleet, A.J., Kelts, K., Talbot, M.R. (Eds), Lacustrine Petroleum Source Rocks. Geological Society (London) Special Publication 40, pp. 341–351.
Gibling, M.R., Tantisukrit, C., Uttamo, W., Thanasuthipitak, T., Haraluck, M., 1985. Oil shale sedimentology and geochemistry in Cenozoic Mae Sot Basin, Thailand. AAPG Bulletin 69, 767–780.
Gierlowski-Kordesch, E.H., 1998. Carbonate deposition in an ephemeral siliciclastic alluvial system: Jurassic Shuttle Meadow Formation, Newark Supergroup, Hartford Basin, U.S.A. Palaeogeography, Palaeoclimatology, Palaeoecology 140, 161–184.
Gierlowski-Kordesch, E.H., Gómez-Fernández, J.C., Meléndez, N., 1991. Carbonate and coal deposition in an alluvial lacustrine setting: Lower Cretaceous (Weald) in the Iberian Range (east-central Spain). In: Anadón, P., Cabrera, Ll., Kelts, K. (Eds), Lacustrine Facies Analysis. International Association of Sedimentologists Special Publication 13, pp. 109–125.
Gierlowski-Kordesch, E.H., Jacobson, A.D., Blum, J.D., Valero Garcés, B.L., 2008. Watershed reconstruction of a Paleocene–Eocene lake basin using Sr isotopes in carbonate rocks. Geological Society of America Bulletin 120, 85–95.
Gierlowski-Kordesch, E.H., Rust, B.R., 1994. The Jurassic East Berlin Formation, Hartford Basin, Newark Supergroup (Connecticut and Massachusetts): a saline lake–playa–alluvial plain system. In: Renaut, R.W., Last, W.M. (Eds), Sedimentology and Geochemistry of Modern and Ancient Saline Lakes. SEPM Special Publication 50, pp. 249–265.
Gilbert, R., Lesak, S.M., 1981. Factors affecting marl deposition in Knowlton Lake, southeastern Ontario. Journal of Great Lakes Research 7, 286–289.
Giralt, S., Julià, R., Klerkx, J., 2001. Microbial biscuits of vaterite in Lake Issyk-kul (Republic of Kyrgyzstan). Journal of Sedimentary Research 71, 430–435.

Giralt, S., Klerkx, J., Riera, S., Julià, R., Lignier, V., Beck, C., De Batist, M., Kalugin, I., 2002. Recent environmental evolution of Lake Issyk-kul. In: Klerkx, J., Imanackunov, B. (Eds), Lake Issyk-kul: Its Natural Environment. Kluwer Academic Publishers, Dordrecht, pp. 125–145.
Girardclos, S., 1993. Enrégistrement des événements de haute énergie et formation d'ooïdes dans des sédiments littoraux du Léman (Petit-lac). Diploma Thesis, Department of Geology and Paleontology, Université de Genève, 112 pp.
Given, R.K., Wilkinson, B.H., 1985. Kinetic control of morphology, composition, and mineralogy of abiotic sedimentary carbonates. Journal of Sedimentary Petrology 55, 109–119.
Glass, S.W., Wilkinson, B.H., 1980. The Peterson Limestone – Early Cretaceous lacustrine carbonate deposition in western Wyoming and southeastern Idaho. Sedimentary Geology 27, 143–160.
Glenn, C.R., Kelts, K., 1991. Sedimentary rhythms in lake deposits. In: Einsele, G., Ricken, W., and Seilacher, A. (Eds), Cycles and Events in Stratigraphy. Springer-Verlag, New York, pp. 188–221.
Gorthner, A., 1994. What is an ancient lake? Archiv für Hydrobiologie Beiheft/Ergebnisse der Limnologie 44, 97–100.
Goth, K., De Leeuw, J.W., Püttmann, W., Tegelaar, E.W., 1988. Origin of Messel oil shale kerogen. Nature 336, 759–761.
Griffiths, S.J., Street-Perrott, F.A., Holmes, J.A., Leng, M.J., Tzedakis, C., 2002. Chemical and isotopic composition of modern water bodies in the Lake Kopais Basin, central Greece: analogues for the interpretation of the lacustrine sedimentary sequence. Sedimentary Geology 148, 79–103.
Grimaldi, D.A., Engel, M.S., 2005. Evolution of the Insects. Cambridge University Press, Cambridge, 755 pp.
Grosjean, M., 1994. Paleohydrology of the Laguna Lejia (north Chilean Altiplano) and climatic implications for late glacial times. Palaeogeography, Palaeoclimatology, Palaeoecology 109, 89–100.
Gruszka, B., 2007. The Pleistocene glaciolacustrine sediments in the Belchatów mine (central Poland): endogenic and exogenic controls. Sedimentary Geology 193, 149–166.
Gruszka, B., Van Loon, A.J., 2007. Pleistocene glaciolacustrine breccias of seismic origin in an active graben (central Poland). Sedimentary Geology 193, 93–104.
Gvirtzman, H., 2006. Groundwater hydrology and palaeohydrology of the Dead Sea rift valley. In: Enzel, Y., Agnon, A., Stein, M. (Eds), New Frontiers in Dead Sea Paleoenvironmental Research. Geological Society of America Special Paper 401, pp. 95–111.
Gwynn, J.W., Murphy, P.J., 1980. Recent sediments of the Great Salt Lake basin. In: Gwynn, J.W. (Ed.), Great Salt Lake: A Scientific, Historical, and Economic Overview. Utah Geological and Mineral Survey Bulletin 116, Artistic Printing, Salt Lake City, UT, pp. 83–96.
Haberyan, K.A., Hecky, R.E., 1987. The Late Pleistocene and Holocene stratigraphy and paleolimnology of Lakes Kivu and Tanganyika. Palaeogeography, Palaeoclimatology, Palaeoecology 61, 169–197.
Hakala, A., 2004. Meromixis as part of lake evolution – observations and a revised classification of true meromictic lakes in Finland. Boreal Environment Research 9, 37–53.
Håkanson, L., Jansson, M., 1983. Principles of Lake Sedimentology. Springer-Verlag, Berlin, 316 pp.
Halfman, J.D., Johnson, T.C., 1988. High-resolution record of cyclic climatic changes during the past 4 ka from Lake Turkana, Kenya. Geology 16, 496–500.

Halfman, J.D., Johnson, T.C., Finney, B.P., 1994. New AMS dates, stratigraphic correlations and decadal climatic cycles for the past 4 ka at Lake Turkana, Kenya. Palaeogeography, Palaeoclimatology, Palaeoecology 111, 83–98.
Halley, R.B., 1976. Textural variation within Great Salt Lake algal mounds. In: Walter, M.R. (Ed.), Stromatolites. Elsevier, Amsterdam, pp. 435–445.
Hammarlund, D., Barnekow, L., Birks, H.J.B., Burchardt, B., Edwards, T.W.D., 2002. Holocene changes in atmospheric circulation recorded in the oxygen isotope stratigraphy of lacustrine carbonates from northern Sweden. The Holocene 12, 339–351.
Harris, N.B., Sorriaux, P., Toomey, D.F., 1994. Geology of the Lower Cretaceous Viodo Carbonate, Congo Basin: a lacustrine carbonate in the South Atlantic rift. In: Lomando, A.J., Schreiber, B.C., Harris, P.M. (Eds), Lacustrine Reservoirs and Depositional Systems. SEPM Core Workshop 19, pp. 143–172.
Hazan, N., Stein, M., Agnon, A., Marco, S., Nadel, D., Negendank, J.F.W., Schwab, M.J., Neev, D., 2005. The late Quaternary limnological history of Lake Kinneret (Sea of Galilee), Israel. Quaternary Research 63, 60–77.
Heizmann, E.P.J., Reiff, W., 2002. Der Steinheimer Meteorkrater. Verlag Dr. Friedrich Pfeil, Munich, 160 pp.
Hilfinger, M.F., IV, Mullins, H.T., Burnett, A., Kirby, M.E., 2001. A 2500 year sediment record from Fayetteville Green Lake, New York: evidence for anthropogenic impacts and historic isotopic shift. Journal of Paleolimnology 26, 293–305.
Hinderer, M., Einsele, G., 2001. The world's large lake basins as denudation–accumulation systems and implications for their lifetimes. Journal of Paleolimnology 26, 355–372.
Hodell, D.A., Schelske, C.L., Fahnenstill, G.L., Robbings, L.L., 1998. Biologically-induced calcite and its isotopic composition in Lake Ontario. Limnology and Oceanography 43, 187–199.
Holmes, J.A., 2001. Ostracoda. In: Smol, J.P., Birks, H. J. B., and Last, W.M. (Eds), Tracing Environmental Change using Lake Sediments, Volume 4: Zoological Indicators. Kluwer Academic Publishers, Dordrecht, pp. 125–151.
Holmes, J.A., Chivas, A.R., 2002. Ostracod shell chemistry – overview. In: Holmes, J.A., Chivas, A.R. (Eds), The Ostracoda: Applications in Quaternary Research. Geophysical Monograph 131. American Geophysical Union, Washington, pp. 185–204.
Horne, D.J., Cohen, A., Martens, K., 2002. Taxonomy, morphology, and biology of Quaternary and living Ostracoda. In: Holmes, J.A., Chivas, A.R. (Eds), The Ostracoda: Applications in Quaternary Research. Geophysical Monograph 131. American Geophysical Union, Washington, pp. 5–36.
Horsfield, B., Curry, D.J., Bohacs, K., Littke, R., Rullkötter, J., Schenk, H.J., Radke, M., Schaefer, R.G., Carroll, A.R., Isaksen, G., Witte, .G., 1994. Organic geochemistry of freshwater and alkaline lacustrine sediments in the Green River Formation of the Washakie Basin, Wyoming, U.S.A.. Organic Geochemistry 22, 415–440.
Horton, B.K., Schmitt, J.G., 1996. Sedimentology of a lacustrine fan-delta system, Miocene Horse Camp Formation, Nevada, U.S.A.. Sedimentology 43, 133–155.
Icole, M., Masse, J.-P., Perinet, G., Taieb, M., 1990. Pleistocene lacustrine stromatolites, composed of calcium carbonate, fluorite, and dolomite, from Lake Natron, Tanzania: depositional and diagenetic processes and their paleoenvironmental significance. Sedimentary Geology 69, 139–155.
Imboden, D.M., 2004. The motion of lake waters. In: O'Sullivan, P.E., Reynolds, C.S. (Eds), The Lakes Handbook, Volume 1 – Limnology and Limnetic Ecology. Blackwell Publishing, Malden, pp. 115–152.
Janaway, T.M., Parnell, J., 1989. Carbonate production within the Orcadian Basin, northern Scotland: a petrographic and geochemical study. Palaeogeography, Palaeoclimatology, Palaeoecology 70, 89–105.

Jiang, Z., Chen, D., Qiu, L., Liang, H., Ma, J., 2007. Source-controlled carbonates in a small Eocene half-graben basin (Shulu Sag) in central Hebei Province, North China. Sedimentology 54, 265–292.
Johnson, G.D., 1974. Cainozoic lacustrine stromatolites from hominid-bearing sediments east of Lake Rudolf, Kenya. Nature 247, 520–523.
Johnson, T.C., 1980. Sediment redistribution by waves in lakes, reservoirs, and embayments. In: Stefan, H. (Ed.), Proceedings of the Symposium on Surface Water Impoundments. American Society of Civil Engineering, NY, pp. 1307–1317.
Johnson, T.C., 1984. Sedimentation in large lakes. Annual Review of Earth and Planetary Sciences 12, 179–204.
Johnson, T.C., Halfman, J.D., Showers, W.J., 1991. Paleoclimate of the past 4000 years at Lake Turkana, Kenya, based on isotopic composition of authigenic calcite. Palaeogeography, Palaeoclimatology, Palaeoecology 85, 189–198.
Jones, B.F., Bowser, C.J., 1978. The mineralogy and related chemistry of lake sediments. In: Lerman, A. (Ed.), Lakes: Chemistry, Geology, Physics. Springer, Berlin, pp. 179–235.
Jones, F.G., Wilkinson, B.H., 1978. Structure and growth of lacustrine pisoliths from recent Michigan marl lakes. Journal of Sedimentary Petrology 48, 1103–1110.
Jones, M.D., Leng, M.J., Roberts, C.N., Türkes, M.R., 2005. A coupled calibration and modeling approach to the understanding of dry-land lake oxygen isotope records. Journal of Paleolimnology 34, 391–411.
Kaaya, C.Z., Kreuser, T., 1994. Permian lakes in East Africa (Tanzania). In: Gierlowski-Kordesch, E., Kelts, K. (Eds), Global Geological Record of Lake Basins. Cambridge University Press, Cambridge, pp. 83–87.
Kalff, J., 2002. Limnology: Inland Water Ecosystems. Prentice Hall, Upper Saddle River, 592 pp.
Katz, B.J., 1988. Clastic and carbonate lacustrine systems: an organic geochemical comparison (Green River Formation and East African lake sediments). In: Fleet, A.J., Kelts, K., Talbot, M.R. (Eds), Lacustrine Petroleum Source Rocks. Geological Society (London) Special Publication 40, pp. 81–90.
Katz, B.J., 1994. The Green River shale: an Eocene carbonate lacustrine source rock. In: Katz, B.J. (Ed.), Petroleum Source Rocks. Springer-Verlag, Berlin, pp. 309–324.
Kelts, K., Chen, K.Z., Lister, G., Yu, J.Q., Gao, Z.H., Niessen, F., Bonani, G., 1989. Geological fingerprints of climate history: a cooperative study of Qinghai Lake, China. Eclogae Geologicae Helvetiae 82, 167–182.
Kelts, K., Hsü, K.J., 1978. Freshwater carbonate sedimentation. In: Lerman, A. (Ed.), Lakes: Chemistry, Geology, Physics. Springer, Berlin, pp. 295–323.
Kelts, K., Talbot, M., 1990. Lacustrine carbonates as geochemical archives of environmental change and biotic/abiotic interactions. In: Tilzer, M.M., Serruya, C. (Eds), Large Lakes: Ecological Structure and Function. Science Technical Publications, Madison, pp. 290–317.
Kempe, S., Kazmierczak, J., 1990. Calcium carbonate supersaturation and the formation of *in situ* calcified stromatolites. In: Ittekkot, V., Kempe, S., Michaelis, W., and Spitzy, A. (Eds), Facets of Modern Biogeochemistry. Springer-Verlag, Berlin, pp. 255–278.
Kempe, S., Kazmierczak, J., Landmann, G., Konuk, T., Reimer, A., Lipp, A., 1991. Largest known microbialites discovered in Lake Van, Turkey. Nature 349, 605–608.
Kennedy, S.K., Smith, N.D., 1977. The relationship between carbonate mineralogy and grain size in two alpine lakes. Journal of Sedimentary Petrology 47, 411–418.
Kidder, D.L., Gierlowski-Kordesch, E.H., 2005. Impact of grassland radiation on the nonmarine silica cycle and Miocene diatomite. Palaios 20, 198–206.

Kirby, M.E., Lund, S.P., Anderson, M.A., Bird, B.W., 2007. Insolation forcing of Holocene climate change in southern California: a sediment study from Lake Elsinore. Journal of Paleolimnology 38, 395–417.

Kirby, M.E., Patterson, W.P., Mullins, H.T., Burnett, A.W., 2002. Post-Younger Dryas climate interval linked to circumpolar vortex variability: isotopic evidence from Fayetteville Green Lake, New York. Climate Dynamics 19, 321–330.

Kreuser, T., 1995. Tectonic and climatic controls of lacustrine sedimentation in pre-rift and rift settings in the Permian-Triassic of East Africa. Journal of Paleolimnology 13, 3–19.

La Rocque, A., 1960. Molluscan faunas of the Flagstaff Formation of central Utah. Geological Society of America Memoir 78, 100 pp.

Laabs, B.J.C., Kaufman, D.S., 2003. Quaternary highstands in Bear Lake Valley, Utah and Idaho. Geological Society of America Bulletin 115, 463–478.

Lajewski, C.K., Mullins, H.T., Patterson, W.P., Callinan, C.W., 2003. Historic calcite record from the Finger Lakes, New York: impact of acid rain on a buffered terrane. Geological Society of America Bulletin 115, 373–384.

Lamb, A.L., Leng, M.J., Lamb, H.F., Mohammed, M.U., 2000. A 9000-year oxygen and carbon isotope record of hydrological change in a small Ethiopian crater lake. The Holocene 10, 167–177.

Lamb, A.L., Leng, M.J., Lamb, H.F., Telford, R.J., Mohammed, M.U., 2002. Climatic and non-climatic effects on the $\delta^{18}O$ and $\delta^{13}C$ compositions of Lake Awassa, Ethiopia, during the last 6.5 ka. Quaternary Science Reviews 21, 2199–2211.

Lambert, A., Hsü, K.J., 1979. Non-annual cycles of varve-like sedimentation in Walensee, Switzerland. Sedimentology 26, 453–461.

Lamoureux, S., 1999. Spatial and interannual variations in the sedimentation patterns, recorded in nonglacial varved sediments from the Canadian High Arctic. Journal of Paleolimnology 21, 73–84.

Lamoureux, S.F., Gilbert, R., 2004. Physical and chemical properties and proxies of high latitude lake sediments. In: Pienitz, R., Douglas, M.S.V., Smol, J.P. (Eds), Long-term Environmental Change in Arctic and Antarctic Lakes. Development in Paleoenvironmental Research 8, Springer, Dordrecht, pp. 53–87.

Lanés, S., Palma, R.M., 1998. Environmental implications of oncoids and associated sediments from the Remoredo Formation (Lower Jurassic), Mendoza, Argentina. Palaeogeography, Palaeoclimatology, Palaeoecology 140, 357–366.

Larsen, C.P.S., Pienitz, R., Smol, J.P., Moser, K.A., Cumming, B.F., Blais, J.M., Macdonald, G.M., Hall, R.I., 1998. Relations between lake morphometry and the presence of laminated lake sediments: a re-examination of Larsen and Macdonald (1993). Quaternary Science Reviews 17, 711–717.

Larsen, D., Crossey, L.J., 1996. Depositional environments and paleolimnology of an ancient caldera lake: Oligocene Creede Formation, Colorado. Geological Society of America Bulletin 108, 526–544.

Larsen, D., Crossey, L.J., 2000. Sedimentary petrology and authigenic mineral distributions in the Oligocene Creede Formation, Colorado, United States. In: Bethke, P.M., Hay, R.L. (Eds), Ancient Lake Creede: Its Volcano-Tectonic Setting, History of Sedimentation, and Relation to Mineralization in the Creede Mining District. Geological Society of America Special Paper 346, pp. 179–208.

Last, W.M., 1990a. Lacustrine dolomite – an overview of modern, Holocene, and Pleistocene occurrences. Earth-Science Reviews 27, 221–263.

Last, W.M., 1990b. Paleochemistry and paleohydrology of Ceylon Lake, a salt-dominated playa basin in the northern Great Plains, Canada. Journal of Paleolimnology 4, 219–238.

Last, W.M., 1992. Petrology of modern carbonate hardgrounds from East Basin Lake, a saline maar lake, southern Australia. Sedimentary Geology 81, 215–229.

Last, W.M., 1993. Geolimnology of Freefight Lake: an unusual hypersaline lake in the northern Great Plains of western Canada. Sedimentology 40, 431–448.
Last, W.M., 2001. Textural analysis of lake sediments. In: Last, W.M., Smol, P. (Eds), Tracking Environmental Change using Lake Sediments. Volume 2: Physical and Geochemical Methods. Kluwer Academic Publishers, Dordrecht, pp. 41–81.
Last, W.M., DeDeckker, P., 1990. Modern and Holocene carbonate sedimentology of two saline volcanic maar lakes, southern Australia. Sedimentology 37, 967–981.
Last, W.M., Deleqiat, J., Greengrass, K., Sukhan, S., 2002. Re-examination of the recent history of meromictic Waldsea Lake, Saskatchewan, Canada. Sedimentary Geology 148, 147–160.
Last, W.M., Ginn, F.M., 2005. Saline systems of the Great Plains of western Canada: an overview of the limnogeology and paleolimnology. Saline Systems 1, 10 (doi: 10.1186/1746-1448-1-10).
Last, W.M., Schweyen, T.H., 1985. Late Holocene history of Waldsea Lake, Saskatchewan, Canada. Quaternary Research 24, 219–234.
Last, W.M., Vance, R.E., 1997. Bedding characteristics of Holocene sediments from salt lakes of the northern Great Plains, western Canada. Journal of Paleolimnology 17, 297–318.
Last, W.M., Vance, R.E., 2002. The Holocene history of Oro Lake, one of the western Canada's longest continuous lacustrine records. Sedimentary Geology 148, 161–184.
Last, W.M., Vance, R.E., Wilson, S., Smol, J.P., 1998. A multi-proxy limnologic record of rapid early Holocene hydrologic change on the northern Great Plains, southwestern Saskatchewan, Canada. The Holocene 8, 503–520.
Leggitt, V.L., Biaggi, R.E., Buchheim, H.P., 2007. Palaeoenvironments associated with caddisfly-dominated microbial-carbonate mounds from the Tipton Shale Member of the Green River Formation: Eocene Lake Gosiute. Sedimentology 54, 661–699.
Leggitt, V.L., Cushman, R.A., Jr., 2001. Complex caddisfly-dominated bioherms from the Eocene Green River Formation. Sedimentary Geology 145, 377–396.
Leggitt, V.L., Loewen, M.A., 2002. Eocene Green River Formation "*Oocardium* tufa" reinterpreted as complex arrays of calcified caddisfly (Insecta: Trichoptera) larval cases. Sedimentary Geology 148, 139–146.
Lehman, J.T., 1975. Reconstructing the rate of accumulation of lake sediment: the effect of sediment focusing. Quaternary Research 5, 541–550.
Leng, M.J., Barker, P.A., 2006. A review of the oxygen isotope composition of lacustrine diatom silica for palaeoclimate reconstruction. Earth-Science Reviews 75, 5–27.
Leng, M.J., Lamb, A.L., Lamb, H.F., Telford, R.J., 1999b. Paleoclimatic implications of isotopic data from modern and early Holocene shells of the freshwater snail *Melanoides tuberculata*, from lakes in the Ethiopian rift valley. Journal of Paleolimnology 21, 97–106.
Leng, M.J., Marshall, J.D., 2004. Paleoclimate interpretation of stable isotope data from lake sediment archives. Quaternary Science Reviews 23, 811–831.
Leng, M.J., Roberts, N., Reed, J.M., Sloane, H.J., 1999a. Late Quaternary palaeohydrology of the Konya Basin, Turkey, based on isotope studies of modern hydrology and lacustrine carbonates. Journal of Paleolimnology 22, 187–204.
Lesack, L.F.W., Melack, J.M., 1991. The deposition, composition, and potential sources of major ionic solutes in rain for the central Amazon basin. Water Resources Research 27, 2953–2977.
Li, H.C., Ku, T.L., 1997. $\delta^{13}C-\delta^{18}O$ covariance as a paleohydrological indicator for closed-basin lakes. Palaeogeography, Palaeoclimatology, Palaeoecology 133, 69–80.
Lindberg, S.E., Lovett, G.M., Richter, D.D., Johnson, D.W., 1986. Atmospheric deposition and canopy interactions of major ions in a forest. Science 231, 141–145.
Lindqvist, J.K., 1994. Lacustrine stromatolites and oncoids: Manuherikia Group (Miocene), New Zealand. In: Bertrand-Sarfati, J., Monty, C. (Eds), Phanerozoic Stromatolites II. Kluwer Academic Publishers, Dordrecht, pp. 227–254.

Moscariello, A., Schneider, A.M., Filippi, M.L., 1998. Late glacial and early Holocene palaeoenvironmental changes in Geneva Bay (Lake Geneva, Switzerland). Palaeogeography, Palaeoclimatology, Palaeoecology 140, 51–73.

Muchane, M.W., 1996. Comparison of the isotope record in micrite, Lake Turkana, with the historical weather record over the last century. In: Johnson, T.C., Odada, E.O. (Eds), The Limnology, Climatology, and Paleoclimatology of East African Lakes. Gordon and Breach Publishers, Amsterdam, pp. 431–441.

Müller, G., 1969. Sedimentbildung im Platensee/Ungarn. Naturwissenschaften 56, 606–615.

Müller, G., 1971. Sediments of Lake Constance. In: Müller, G. (Ed.), Sedimentology of Parts of Central Europe. International Association of Sedimentologists Guidebook to Excursions held during the VIII International Sedimentological Congress 1971 (Heidelberg). Verlag Waldemar Kramer, Frankfurt am Main, pp. 237–252.

Müller, G., Irion, G., Förstner, U., 1972. Formation and diagenesis of inorganic Ca–Mg carbonates in the lacustrine environment. Naturwissenschften 59, 158–164.

Müller, G., Wagner, F., 1978. Holocene carbonate evolution in Lake Balaton (Hungary): a response to climate and impact of man. In: Matter, A., Tucker, M.E. (Ed.), Modern and Ancient Lake Sediments. International Association of Sedimentologists Special Publication 2, pp. 157–81.

Mullins, H.T., 1998. Environmental change controls of lacustrine carbonate, Cayuga Lake, New York. Geology 26, 443–446.

Murphy, D.H., Wilkinson, B.H., 1980. Carbonate deposition and facies distribution in a central Michigan marl lake. Sedimentology 27, 123–135.

Négrel, P., Casanova, J., Nicoud, G., 1997. Caractérization isotopique (O, C, Sr) des flux du basin versant du Lac d'Annecy (E France). Bulletin de la Société Géologique de France 168, 243–253.

Newton, M.S., Grossman, E.L., 1988. Late Quaternary chronology of tufa deposits, Walker Lake, Nevada. Journal of Geology 96, 417–433.

Nickel, E., 1985. Carbonates in alluvial fan systems. An approach to physiography, sedimentology, and diagenesis. Sedimentary Geology 42, 83–104.

Nicoud, G., Manalt, F., 2001. The lacustrine depression at Annecy (France), geological setting and Quaternary evolution. Journal of Paleolimnology 25, 137–147.

Niemi, T.M., 1997. Fluctuations of Pleistocene Lake Lisan in the Dead Sea rift. In: Niemi, T.M., Ben-Avraham, Z., and Gat, J.R. (Eds), The Dead Sea: The Lake and its Setting. Oxford University Press, New York, pp. 226–236.

Noffke, N., Gerdes, G., Klenke, T., Krumbein, W.E., 2001. Microbially induced sedimentary structures – a new category within the classification of primary sedimentary structures. Journal of Sedimentary Research 71, 649–656.

Nuhfer, E.B., Anderson, R.Y., Bradbury, J.P., Dean, W.E., 1993. Modern sedimentation in Elk Lake, Clearwater County, Minnesota. In: Bradbury, J.P., Dean, W.E. (Eds), Elk Lake, Minnesota: Evidence for Rapid Climate Change in the North-Central United States. Geological Society of America Special Paper 276, pp. 75–96.

Nuñez, R., Spiro, B., Pentecost, A., Kim, A., Coletta, P., 2002. Organo-chemical and stable isotope indicators of environmental change in a marl lake, Malham Tarn, North Yorkshire, UK. Journal of Paleolimnology 28, 403–417.

Oberhänsli, H., Allen, P.A., 1987. Stable isotopic signatures of Tertiary lake carbonates, eastern Ebro Basin, Spain. Palaeogeography, Palaeoclimatology, Palaeoecology 60, 59–75.

Ogorelec, B., Bole, B., Leonidakis, J., Cermelj, B., Misic, M., Faganeli, J., 2006. Recent sediment of Lake Bled (NW Slovenia): sedimentological and geochemical properties. Water, Air, and Soil Pollution Focus 6, 141–149.

Oremland, R.S., Cloern, J.E., Sofer, Z., Smith, R.L., Culbertson, C.W., Zehr, J., Miller, L., Cole, B., Harvey, R., Iversen, N., Klug, M., Des Marais, D.J., Rau, G., 1988. Microbial and biogeochemical processes in Big Soda Lake, Nevada. In: Fleet, A.J., Kelts, K., Talbot, M.R. (Eds), Lacustrine Petroleum Source Rocks, Geological Society (London) Special Publication 40, pp. 59–75.

Oren, A., 1997. Microbiological studies in the Dead Sea: 1892–1992. In: Niemi, T.M., Ben-Avraham, Z., and Gat, J.R. (Eds), The Dead Sea: The Lake and its Setting. Oxford University Press, New York, pp. 205–213.

Osborne, R.H., Licari, G.R., Link, M.H., 1982. Modern stromatolites, Walker Lake, Nevada. Sedimentary Geology 32, 39–61.

O'Sullivan, P.E., Reynolds, C.S. (Eds), 2004. The Lakes Handbook: Limnology and Limnetic Ecology, Volume 1. Blackwell Science, Malden, 699 pp.

Oviatt, C.G., 1997. Lake Bonneville fluctuations and global climate change. Geology 25, 155–158.

Oviatt, C.G., Habiger, G.D., Hay, J.E., 1994b. Variation in the composition of Lake Bonneville marl: a potential key to lake-level fluctuations and paleoclimate. Journal of Paleolimnology 11, 19–30.

Oviatt, C.G., McCoy, W.D., Nash, W.P., 1994a. Sequence stratigraphy of lacustrine deposits: a Quaternary example from the Bonneville basin, Utah. Geological Society of America Bulletin 106, 133–144.

Oviatt, C.G., Sack, D., Currey, D.R., 1994c. The Bonneville Basin, Quaternary, western United States. In: Gierlowski-Kordesch, E., Kelts, K. (Eds), Global Geological Record of Lake Basins. Cambridge University Press, Cambridge, pp. 371–375.

Oviatt, C.G., Thompson, R.S., Kaufman, D.S., Bright, J., Forester, R.M., 1999. Reinterpretation of the Burmester Core, Bonneville Basin, Utah. Quaternary Research 52, 180–184.

Owen, R.B., 2000. Late Cretaceous–Early Tertiary continental and lacustrine basins of Hong Kong and southeast China. In: Gierlowski-Kordesch, E.H., Kelts, K.R. (Eds), Lake Basins through Space and Time, AAPG Studies in Geology 46, AAPG, Tulsa, OK, pp. 329–334.

Pache, M., Reitner, J., Arp, G., 2001. Geochemical evidence for the formation of a large Miocene "travertine" mound at a sublacustrine spring in a soda lake (Wallerstein Castle Rock, Nördlinger Ries, Germany). Facies 45, 211–230.

Paik, I.S., 2005. The oldest record of microbial-caddisfly bioherms from the Early Cretaceous Jinju Formation, Korea: occurrence and palaeoenvironmental implications. Palaeogeography, Palaeoclimatology, Palaeoecology 218, 301–315.

Palacios-Fest, M.R., Cohen, A.S., Ruiz, J., Blank, B., 1993. Comparative paleoclimatic interpretations from nonmarine ostracodes using faunal assemblages, trace elements, shell chemistry, and stable isotope data. In: Swart, P.K., Lohmann, K.C., McKenzie, J., and Savin, S. (Eds), Climate Change in Continental Isotopic Records. American Geophysical Union Geophysical Monograph 78, pp. 179–190.

Palma, R.M., 2000. Lacustrine facies in the Upper Cretaceous Balbuena Subgroup (Salta Group): Andina Basin, Argentina. In: Gierlowski-Kordesch, E.H., Kelts, K.R. (Eds), Lake Basins through Space and Time, AAPG Studies in Geology 46, AAPG, Tulsa, OK, pp. 245–256.

Park, L.E., Gierlowski-Kordesch, E.H., 2007. Paleozoic lake faunas: establishing aquatic life on land. Palaeogeography, Palaeoclimatology, Palaeoecology 249, 160–179.

Park, L.E., Smith, A.J. (Eds) (2003). Bridging the Gap. Trends in the Ostracode Biological and Geological Sciences. Paleontological Society Papers 9, 290 pp.

Paul, J., Peryt, T.M., 2000. Kalkowsky's stromatolites revisited (Lower Triassic Buntsandstein, Harz Mountains, Germany). Palaeogeography, Palaeoclimatology, Palaeoecology 161, 435–458.

Pedone, V.A., Folk, R.L., 1996. Formation of aragonite cement by nanobacteria in the Great Salt Lake, Utah. Geology 24, 763–765.

Rogers, D.A., Astin, T. R., 1991. Ephemeral lakes, mud pellet dunes and wind-blown sand and silt: reinterpretation of Devonian lacustrine cycles in north Scotland. In: Anadón, P., Cabrera, Ll., Kelts, K. (Eds), Lacustrine Facies Analysis. International Association of Sedimentologists Special Publication 13, pp. 199–221.
Romero-Viana, L., Julià, R., Camacho, A., Vicente, E., Miracle, M.R., 2008. Climate signal in varve thickness: Lake La Cruz (Spain), a case study. Journal of Paleolimnology 40, 703–714.
Rosen, M.R., 1994. The importance of groundwater in playas: a review of playa classifications and the sedimentology and hydrology of playas. In: Rosen, M.R. (Ed.), Paleoclimate and Basin Evolution of Playa Systems. Geological Society of America Special Paper 289, pp. 1–18.
Rosen, M.R., Arehart, G.B., Lico, M.S., 2004. Exceptionally fast growth rate of <100-year-old microbially-mediated tufa, Big Soda Lake, Nevada. Geology 32, 409–412.
Rosenbaum, J.G., Kaufman, D.S. (Eds), 2009. Paleoenvironments of Bear Lake, Utah and Idaho, and its Catchment. Geological Society of America Special Paper 450, 351 pp.
Rouchy, J.M., Servant, M., Fournier, M., Causse, C., 1996. Extensive carbonate algal bioherms in upper Pleistocene saline lakes of the central Altiplano of Bolivia. Sedimentology 43, 973–993.
Rowlands, D.L.G., Webster, R.K., 1971. Precipitation of vaterite in lake water. Nature Physical Science 229, 158.
Ruble, T.E., Philp, R.P., 1998. Stratigraphy, depositional environments, organic geochemistry of source rocks in the Green River petroleum system, Uinta Basin, Utah. In: Pitman, J.K., Carroll, A.R. (Eds), Modern and Ancient Lake Systems. Utah Geological Association Guidebook 26, C&M Press, Denver, CO, pp. 289–321.
Rudra, D.K., Maulik, P.K., 1994. Lower Jurassic Kota Limestone of India. In: Gierlowski-Kordesch, E., Kelts, K. (Eds), Global Geological Record of Lake Basins. Cambridge University Press, Cambridge, pp. 185–191.
Ryder, R.T., Fouch, T.D., Elson, J.H., 1976. Early Tertiary sedimentation in the western Uinta Basin, Utah. Geological Society of America Bulletin 87, 496–512.
Rye, R.O., Bethke, P.M., Finkelstein, D.B., 2000. Stable isotope evolution and paleolimnology of ancient Lake Creede. In: Bethke, P.M., Hay, R.L. (Eds), Ancient Lake Creede: Its Volcano-Tectonic Setting, History of Sedimentation, and Relation to Mineralization in the Creede Mining District. Geological Society of America Special Paper 346, pp. 233–265.
Sabato, L., Bertini, A., Masini, F., Alvianelli, A., Napoleone, G., Pieri, P., 2005. The lower and middle Pleistocene geological record of the San Lorenzo lacustrine succession in the Sant'Arcangelo Basin (southern Apennines, Italy). Quaternary International 131, 59–69.
Sáez, A., Cabrera, Ll., 2002. Sedimentological and palaeohydrological responses to tectonics and climate in a small, closed, lacustrine system: Oligocene As Pontes basin (Spain). Sedimentology 49, 1073–1094.
Sáez, A., Cabrera, Ll., Jensen, A., Chong, G., 1999. Late Neogene lacustrine record and palaeogeography in the Quillagua-Llamara basin, Central Andean fore-arc (northern Chile). Palaeogeography, Palaeoclimatology, Palaeoecology 151, 5–37.
Sáez, A., Valero-Garcés, B.L., Moreno, A., Bao, R., Pueyo, J.J., González-Sampériz, P., Giralt, S., Taberner, C., Herrera, C., Gibert, R.O., 2007. Lacustrine sedimentation in active volcanic settings: the Late Quaternary depositional evolution of Lake Chungará (northern Chile). Sedimentology 54, 1191–1222.
Sánchez-Román, M., Vasconcelos, C., Schmid, T., Dittrich, M., McKenzie, J.A., Zenobi, R., Rivadeneyra, M.A., 2008. Aerobic microbial dolomite at the nanometer scale: implications for the geologic record. Geology 36, 879–882.

Sanz-Montero, M.E., Rodríguez-Aranda, J.P., Calvo, J.P., 2006. Mediation of endoevaporitic microbial communities in early replacement of gypsum by dolomite: a case study from Miocene lake deposits from the Madrid Basin, Spain. Journal of Sedimentary Research 76, 1257–1266.

Sanz-Montero, M.E., Rodríguez-Aranda, J.P., García del Cura, M.A., 2008. Dolomite-silica stromatolites in Miocene lacustrine deposits from the Duero Basin, Spain: the role of organotemplates in the precipitation of dolomite. Sedimentology 55, 729–750.

Sanz-Rubio, E., Sánchez-Moral, S., Cañaveras, J.C., Calvo, J.P., Rouchy, J.M., 2001. Calcitization of Mg–Ca carbonate and Ca sulphate deposits in a continental Tertiary basin (Calatayud Basin, NE Spain). Sedimentary Geology 140, 123–142.

Sapozhnikov, D.G., Tsvetkov, A.I., 1959. Precipitation of hydrous calcium carbonate on the bottom of Lake Issyk-kul. Doklady of the Academy of Sciences of the USSR (Doklady Akademia Nauk SSR) 124, 402–405.

Schäfer, A., Stapf, K.R.G., 1978. Permian Saar-Nahe Basin and Recent Lake Constance (Germany): two environments of lacustrine algal carbonates. In: Matter, A., Tucker, M.E. (Eds), Modern and Ancient Lake Sediments. International Association of Sedimentologists Special Publication 2, pp. 83–107.

Schelske, C.L., Hodell, D.A., 1991. Recent changes in productivity and climate of Lake Ontario detected by isotopic analysis of sediments. Limnology and Oceanography 36, 961–975.

Schlische, R.W., Olsen, P.E., 1990. Quantitative filling model for continental extensional basins with applications to Early Mesozoic rifts of eastern North America. Journal of Geology 98, 135–155.

Schmeider, K., Schünemann, B., Schröder, H.G., 2004. Spatial patterns of surface sediment variables in the littoral zone of Lake Constance (Germany). Archiv für Hydrobiologie 161, 455–468.

Schneider, J., 1977. Carbonate construction and decomposition by epilithic and endolithic micro-organisms in salt- and freshwater. In: Flügel, E. (Ed.), Fossil Algae: Recent Results and Developments. Springer-Verlag, Berlin, pp. 248–260.

Schneider, J., Schröder, H.G., Le Campion-Alsumard, Th., 1983. Algal micro-reefs – coated grains from freshwater environments. In: Peryt, T.M. (Ed.), Coated Grains. Springer-Verlag, Berlin, pp. 285–298.

Schneider, J., Siegesmund, S., Gebhardt, U., 1984. Paläontologie und Genese limnischer Schill- und Algenkarbonaten in der Randfazies der kohleführenden Wettiner Schichten (Oberkarbon, Stefan C) des NE-Saaletroges. Hallesches Jahrbuch für Geowissenschaften 9, 35–51.

Schnellman, M., Anselmetti, F.S., Giardini, D., McKenzie, J.A., 2006. 15,000 years of mass movement history in Lake Lucerne: implications for seismic and tsunami hazards. Eclogae Geologicae Helvetiae 99, 409–428.

Scholl, D.W., 1960. Pleistocene algal pinnacles at Searles Lake, California. Journal of Sedimentary Petrology 30, 414–431.

Scholl, D.W., Taft, W.H., 1964. Algae, contributors to the formation of calcareous tufa, Mono Lake, California. Journal of Sedimentary Petrology 34, 309–319.

Schudack, M.E., 1993. Die Charophyten in Oberjura und Unterkreide Westeuropas mit einer phylogenetischen Analyse der Gesamtgruppe. Berliner Geowissenschaftliche Abhandlungen 8E, 1–209.

Schwalb, A., Dean, W.E., 1998. Stable isotopes and sediments from Pickerel Lake, South Dakota, USA: a 12ky record of environmental changes. Journal of Paleolimnology 20, 15–30.

Schwalb, A., Dean, W.E., 2002. Reconstruction of hydrological changes and response to effective moisture variations from north-central USA lake sediments. Quaternary Science Reviews 21, 1541–1554.

Schwalb, A., Lister, G., Kelts, K., 1994. Ostracode carbonate $\delta^{18}O$- and $\delta^{13}C$-signatures of hydrological and climatic changes affecting Lake Neuchâtel, Switzerland, since the latest Pleistocene. Journal of Paleolimnology 11, 3–17.

Schwalb, A., Locke, S.M., Dean, W.E., 1995. Ostracod $\delta^{18}O$ and $\delta^{13}C$ evidence of Holocene environmental changes in the sediments of two Minnesota lakes. Journal of Paleolimnology 14, 281–296.

Shanahan, T.M., Overpeck, J.T., Beck, J.W., Wheeler, C.W., Peck, J.A., King, J.W., Scholz, C.A., 2008. The formation of biogeochemical laminations in Lake Bosumtwi, Ghana, and their usefulness as indicators of past climatic change. Journal of Paleolimnology 40, 339–355.

Shteinman, B., Eckert, W., Kaganowsky, S., Zohary, T., 1997. Seiche-induced resuspension in Lake Kinneret: a fluorescent tracer experiment. Water, Air, and Soil Pollution 99, 123–131.

Simmons, G.M., Jr., Vestal, J.R., Wharton, R.A., Jr., 1993. Environmental regulators of microbial activity in continental Antarctic lakes. In: Green, W.J., Friedmann, E.I. (Eds), Physical and Biogeochemical Processes in Antarctic Lakes, Vol. 59. American Geophysical Union Antarctic Research Series, Washington, pp. 165–195.

Sly, P.G., 1978. Sedimentary processes in lakes. In: Lerman, A. (Ed.), Lakes: Chemistry, Geology, Physics. Springer, Berlin, pp. 65–89.

Smith, M.A., Hollander, D.J., 1999. Historical linkage between atmospheric circulation patterns and the oxygen isotopic record of sedimentary carbonates from Lake Mendota, Wisconsin, U.S.A.. Geology 27, 589–592.

Smith, M.E., Carroll, A.R., Singer, B.S., 2008. Synoptic reconstruction of a major ancient lake system: Eocene Green River Formation, western United States. Geological Society of America Bulletin 120, 54–84.

Smoot, J.P., 1983. Depositional subenvironments in an arid closed basin; the Wilkins Peak Member of the Green River Formation (Eocene), Wyoming, U.S.A. Sedimentology 30, 801–827.

Smoot, J.P., Benson, L.V., 1998. Sedimentary structures as indicators of paleoclimatic fluctuations: Pyramid Lake, Nevada. In: Pitman, J.K., Carroll, A.R. (Eds), Modern and Ancient Lake Systems. Utah Geological Association Guidebook, Vol. 26. C&M Press, Denver, CO, pp. 131–161.

Smoot, J.P., Lowenstein, T.K., 1991. Depositional environments of non-marine evaporites. In: Melvin, J.L. (Ed.), Evaporites, Petroleum and Mineral Resources. Elsevier, Amsterdam, pp. 189–347.

Soreghan, M.J., 1998. Facies distributions within an ancient asymmetric lake basin: the Apache Canyon Formation, Bisbee Basin, Arizona. In: Pitman, J.K., Carroll, A.R. (Eds), Modern and Ancient Lake Systems. Utah Geological Association Guidebook 26, C&M Press, Denver, CO, pp. 163–190.

Spencer, R.J., Baedecker, M.J., Eugster, H.P., Forester, R.M., Goldhaber, M.B., Jones, B.F., Kelts, K., McKenzie, J., Madsen, D.B., Rettig, S.L., Rubin, M., Bowser, C.J., 1984. Great Salt Lake, and precursors, Utah: the last 30,000 years. Contributions to Mineralogy and Petrology 86, 321–334.

Stabel, H.-H., 1985. Mechanisms controlling the sedimentation sequence of various elements in prealpine lakes. In: Stumm, W. (Ed.), Chemical Processes in Lakes. Wiley, New York, pp. 143–167.

Stabel, H.-H., 1986. Calcite precipitation in Lake Constance – chemical equilibrium, sedimentation, and nucleation by algae. Limnology and Oceanography 31, 1081–1093.

Stankovic, S., 1960. The Balkan Lake Ohrid and its Living World. Monographiae Biologicae 9. Dr. W. Junk Publisher, The Hague, 357 pp.

Starratt, S.W. (Ed.), 2007a. Pond Scum to Carbon Sink: Geological and Environmental Applications of the Diatoms. Paleontological Society Short Course, Paleontological Society Papers 13, 193 pp.

Starratt, S.W., 2007b. Long-term records of continental diatoms: paleolimnology and biochronology. In: Starratt, S.W. (Ed.), Pond Scum to Carbon Sink: Geological and Environmental Applications of the Diatoms. Paleontological Society Short Course, Paleontological Society Papers 13, pp. 111–120.
Stein, M., 2001. The sedimentary and geochemical record of Neogene–Quaternary water bodies in the Dead Sea Basin – inferences for the regional paleoclimatic history. Journal of Paleolimnology 26, 271–282.
Stein, M., Starinsky, A., Goldstein, S.L., Katz, A., Machlus, M., Schramm, A., 1997. Strontium isotopic, chemical, and sedimentological evidence for the evolution of Lake Lisan and the Dead Sea. Geochimica et Cosmochimica Acta 61, 3975–3992.
Stiller, M., 1977. Origin of sedimentation components in Lake Kinneret traced by their isotopic composition. In: Golterman, H.L. (Ed.), 1st International Proceedings on the Interactions between Sediments and Freshwater, Amsterdam (6–10 Sept.). Dr. W. Junk B.V. Publishers, Wageningen, pp. 57–64.
Stiller, M., Kaufman, A., 1985. Paleoclimatic trends revealed by the isotopic composition of carbonates in Lake Kinneret. Zeitschrift für Gletscherkunde und Glaziologie 21, 79–87.
Stoermer, E.F., Smol, J.P. (Eds), 1999. The Diatoms: Applications for the Environmental and Earth Sciences. Cambridge University Press, Cambridge, 469 pp.
Stoffers, P., Hecky, R.E., 1978. Late Pleistocene-Holocene evolution of the Kivu-Tanganyika basin. In: Matter, A., Tucker, M.E. (Eds), Modern and Ancient Lake Sediments. International Association of Sedimentologists Special Publication 2, pp. 43–55.
Strong, A.E., Eadie, B.J., 1978. Satellite observations of calcium carbonate precipitations in the Great Lakes. Limnology and Oceanography 23, 877–887.
Stuiver, M., 1970. Oxygen and carbon isotope ratios of freshwater carbonates as climatic indicators. Journal of Geophysical Research 75, 5247–5257.
Stumm, W., 2004. Chemical processes regulating the composition of lake waters. In: O'Sullivan, P.E., Reynolds, C.S. (Eds), The Lakes Handbook, Volume 1 – Limnology and Limnetic Ecology. Blackwell Publishing, Malden, pp. 79–106.
Sturm, M., Matter, A., 1978. Turbidites and varves in Lake Brienz (Switzerland): deposition of clastic detritus by density currents. In: Matter, A., Tucker, M.E. (Ed.), Modern and Ancient Lake Sediments. International Association of Sedimentologists Special Publication 2, pp. 147–168.
Surdam, R.C., Stanley, K.O., 1979. Lacustrine sedimentation during the culminating phase of Eocene Lake Gosiute, Wyoming (Green River Formation). Geological Society of America Bulletin 90, 93–110.
Surdam, R.C., Wray, J.L., 1976. Lacustrine stromatolites, Eocene Green River Formation, Wyoming. In: Walter, M.R. (Ed.), Stromatolites. Elsevier, Amsterdam, pp. 535–541.
Swirydczuk, K., Wilkinson, B.H., Smith, G.R., 1979. The Pliocene Glenns Ferry oolite: lake margin carbonate deposition in the southwestern Snake River Plain. Journal of Sedimentary Petrology 49, 995–1004.
Swirydczuk, K., Wilkinson, B.H., Smith, G.R., 1980. The Pliocene Glenns Ferry oolite – II. Sedimentology of oolitic lacustrine terrace deposits. Journal of Sedimentary Petrology 50, 1237–1248.
Talbot, M.R., 1990. A review of palaeohydrological interpretation of carbon and oxygen isotopic ratios in primary lacustrine carbonates. Chemical Geology 80, 261–279.
Talbot, M.R., 1994. Paleohydrology of the late Miocene Ridge Basin lake, California. Geological Society of America Bulletin 106, 1121–1129.
Talbot, M.R., Holm, K., Williams, M.A.J., 1994. Sedimentation in low gradient desert margin systems: a comparison of the late Triassic of north-west Somerset (England) and the late Quaternary of east-central Australia. In: Rosen, M.R. (Ed.), Paleoclimate and Basin Evolution of Playa Systems. Geological Society of America Special Paper 289, pp. 97–117.

Talbot, M.R., Kelts, K., 1986. Primary and diagenetic carbonates in the anoxic sediments of Lake Bosumtwi, Ghana. Geology 14, 912–916.

Talbot, M.R., Kelts, K., 1990. Paleolimnological signatures from carbon and oxygen isotopic ratios in carbonates from organic carbon-rich lacustrine sediments. In: Katz, B.J. (Ed.), Lacustrine Basin Exploration: Case Studies and Modern Analogs, AAPG Memoir 50, pp. 88–112.

Talbot, M.R., Livingstone, D.A., Palmer, P.G., Maley, J., Melack, J.M., Delibrias, G., Gullik-Sen, S., 1984. Preliminary results from sediment cores from Lake Bosumtwi, Ghana. Palaeoecology of Africa 16, 173–192.

Teller, J.T., Last, W.M., 1990. Paleohydrological indicators in playas and salt lakes, with examples from Canada, Australia, and Africa. Palaeogeography, Palaeoclimatology, Palaeoecology 76, 215–240.

Teranes, J.N., McKenzie, J.A., Bernasconi, S.M., Lotter, A.F., Sturm, M., 1999. A study of oxygen isotopic fractionation during bio-induced calcite precipitation in eutrophic Baldeggersee, Switzerland. Geochimica et Cosmochimica Acta 63, 1981–1999.

Tevesz, M.J.S., Smith, J.E., Coakley, J.P., Risk, M.J., 1997. Stable carbon and oxygen isotope records from Lake Erie sediment cores: mollusc aragonite 4600BP–200BP. Journal of Great Lakes Research 23, 307–316.

Thiel, V., Merz-Preiss, M., Reitner, J., Michaelis, W., 1997. Biomarker studies on microbial carbonates: extractable lipids of a calcifying cyanobacterial mat (Everglades, USA). Facies 36, 163–172.

Thompson, J.B., Ferris, F.G., 1990. Cyanobacterial precipitation of gypsum, calcite, and magnesite from natural alkaline lake water. Geology 18, 995–998.

Thompson, J.B., Ferris, F.G., Smith, D.A., 1990. Geomicrobiology and sedimentology of the mixolimnion and chemocline in Fayetteville Green Lake, New York. Palaios 5, 52–75.

Thompson, J.B., Schultze-Lam, S., Berveridge, T.J., Des Marais, D.J., 1997. Whiting events: biogenic origin due to the photosynthetic activity of cyanobacterial picoplankton. Limnology and Oceanography 42, 133–141.

Thompson, R., Kelts, K., 1974. Holocene sediments and magnetic stratigraphy from Lakes Zug and Zürich, Switzerland. Sedimentology 21, 577–596.

Thunmark, S., 1937. Ueber die regionale Limnologie von Südschweden. Sveriges Geologiska Undersökning Årsbok 31, 1–160.

Tiercelin, J.-J., Cohen, A.S., Soreghan, M.J., Lezzar, K.-E., 1994. Pleistocene-Modern deposits of the Lake Tanganyika rift basin, East Africa: a modern analog for lacustrine source rocks and reservoirs. In: Lomando, A.J., Schreiber, B.C., Harris, P.M. (Eds), Lacustrine Reservoirs and Depositional Systems. SEPM Core Workshop 19, Denver, CO, pp. 37–59.

Tiercelin, J.-J., Pflumio, C., Castrec, M., Boulègue, J., Gente, P., Rolet, J., Coussement, C., Stetter, K.O., Huber, R., Buku, S., Mifunu, W., 1993. Hydrothermal vents in Lake Tanganyika, East African rift system. Geology 21, 499–502.

Treese, K.L., Wilkinson, B.H., 1982. Peat-marl deposition in a Holocene paludal–lacustrine basin – Sucker Lake, Michigan. Sedimentology 29, 375–390.

Trewin, N.H., Davidson, R.G., 1999. Lake-level changes, sedimentation, and faunas in a Middle Devonian basin-margin fish bed. Journal of the Geological Society London 156, 535–548.

Tucker, M.E., 1978. Triassic lacustrine sediments from South Wales: shore-zone clastics, evaporites, and carbonates. In: Matter, A., Tucker, M.E. (Eds), Modern and Ancient Lake Sediments. International Association of Sedimentologists Special Publication 2, pp. 205–224.

Tucker, M.E., Wright, V.P., 1990. Carbonate Sedimentology. Blackwell Scientific Publications, Oxford, 482 pp.

Turner, B.F., Gardner, L.R., Sharp, W.E., Blood, E.R., 1996. The geochemistry of Lake Bosumtwi, a hydrologically closed basin in the humid zone of tropical Ghana. Limnology and Oceanography 41, 1415–1424.

Tütken, T., Vennemann, T.W., Janz, H., Heizmann, E.P.J., 2006. Palaeoenvironment and palaeoclimate of the Middle Miocene lake in the Steinheim basin, SW Germany: a reconstruction from C, O, and Sr isotopes of fossil remains. Palaeogeography, Palaeoclimatology, Palaeoecology 241, 457–491.

Udomkan, B., Ratanasthien, B., Takayasu, K., Fyfe, W.S., Sato, S., Kandharosa, W., Wongpornchai, P., Kusakabe, M., 2003. Fluctuation of depositional environment in the Ban Mark Coal deposit, Krabi Mine, southern Thailand: stable isotope implication. ScienceAsia 29, 307–317.

Utrilla, R., Vázquez, A., Anadón, P., 1998. Paleohydrology of the Upper Miocene Bicorb lake (eastern Spain) as inferred from stable isotopic data from inorganic carbonates. Sedimentary Geology 121, 191–206.

Valero Garcés, B.L., 1993. Lacustrine deposition and related volcanism in a transtensional tectonic setting: Upper Stephanian–Lower Autunian in the Aragón–Béarn Basin, western Pyrenees (Spain–France). Sedimentary Geology 83, 133–160.

Valero Garcés, B.L., 1994. Carbonate lacustrine phases in the continental Permian Aragón–Béarn Basin (western Pyrenees). In: Gierlowski-Kordesch, E., Kelts, K. (Eds), Global Geological Record of Lake Basins. Cambridge University Press, Cambridge, pp. 107–119.

Valero Garcés, B.L., Gierlowski-Kordesch, E.H., Bragonier, W.A., 1994. Lacustrine facies model for nonmarine limestone within cyclothems in the Pennsylvanian (Upper Freeport Formation, Appalachian Basin) and its implications. In: Lomando, A.J., Schreiber, B.C., Harris, P.M. (Eds), Lacustrine Reservoirs and Depositional Systems. SEPM Core Workshop 19, Denver, CO, pp. 321–381.

Valero Garcés, B.L., Gierlowski-Kordesch, E.H., Bragonier, W.A., 1997. Pennsylvanian continental cyclothem development: no evidence of direct climatic control in the Upper Freeport Formation (Allegheny Group) of Pennsylvania (northern Appalachian basin). Sedimentary Geology 109, 305–320.

Valero Garcés, B.L., Grosjean, M., Kelts, K., Schreier, H., Messerli, B., 1999. Holocene lacustrine deposition in the Atacama Altiplano: facies models, climate and tectonic forcing. Palaeogeography, Palaeoclimatology, Palaeoecology 151, 101–125.

Valero Garcés, B.L., Grosjean, M., Messerli, B., Schwalb, A., Kelts, K., 2000. Late Quaternary lacustrine deposition in the Chilean Altiplano (18°–28°S). In: Gierlowski-Kordesch, E.H., Kelts, K.R. (Eds), Lake Basins through Space and Time, AAPG Studies in Geology 46, AAPG, Tulsa, OK, pp. 625–636.

Valero Garcés, B.L., Kelts, K., Ito, E., 1995. Oxygen and carbon isotope trends and sedimentological evolution of a meromictic and saline lacustrine system: the Holocene Medicine Lake basin, North American Great Plains, U.S.A.. Palaeogeography, Palaeoclimatology, Palaeoecology 117, 253–278.

Verrecchia, E.P., 2007. Lacustrine and palustrine geochemical sediments. In: Nash, D.J., McLaren, S.J. (Eds), Geochemical Sediments and Landscapes. Blackwell Publishing, Malden, USA, pp. 298–329.

Veski, S., Seppä, H., Ojala, A.E.K., 2004. Cold event at 8200 yr. B.P. recorded in annually laminated lake sediments in eastern Europe. Geology 32, 681–684.

Vincens, A., Casanova, J., Tiercelin, J.-J., 1986. Palaeolimnology of Lake Bogoria (Kenya) during the 4500 BP high lacustrine phase. In: Frostick, L.E., Renaut, R.W., Reid, I., Tiercelin, J.-J. (Eds), Sedimentation in the African Rifts. Geological Society (London) Special Publication 25, pp. 323–330.

CHAPTER 2

Palustrine Carbonates

A.M. Alonso-Zarza[1,*] *and* V.P. Wright[2]

Contents

1. Introduction 103
2. The Sedimentary (Ancient) Record of Palustrine Carbonates 105
   2.1. The primary lacustrine mud 105
   2.2. Carbonate palustrine facies in the sedimentary record 107
   2.3. Genetic aspects of the palustrine carbonate facies 114
3. Wetlands: The Recent Analogues for Ancient Palustrine Environments? 115
4. Environmental Controls on Palustrine Deposition 119
5. Conclusions 125
Acknowledgements 126
References 126

1. Introduction

Palustrine carbonates can be regarded as part of a complex continuum of chemical sediments extending from groundwater calcretes and dolocretes, through tufas and travertines, as well as carbonates formed in discharge wetlands, riparian wetlands, ponds and lakes. These carbonates form by the precipitation of calcium carbonate from groundwaters, surface or groundwater-fed rivers and lakes with a high content of calcium bicarbonate. They are indicators of special hydrological conditions, and the remarkable

*Corresponding author.
E-mail address: alonsoza@geo.ucm.es

[1] Dpto Petrología y Geoquímica, Fac. CC. Geológicas, IGE-CSIC, Universidad Complutense de Madrid, 28040 Madrid, Spain
[2] BG Group, 100 Thames Valley Park Drive, Reading RG6 1PT; School of Earth & Ocean Sciences, Cardiff University, Cardiff CF10 3YE, UK

Developments in Sedimentology, Volume 61
ISSN 0070-4571, DOI 10.1016/S0070-4571(09)06102-0

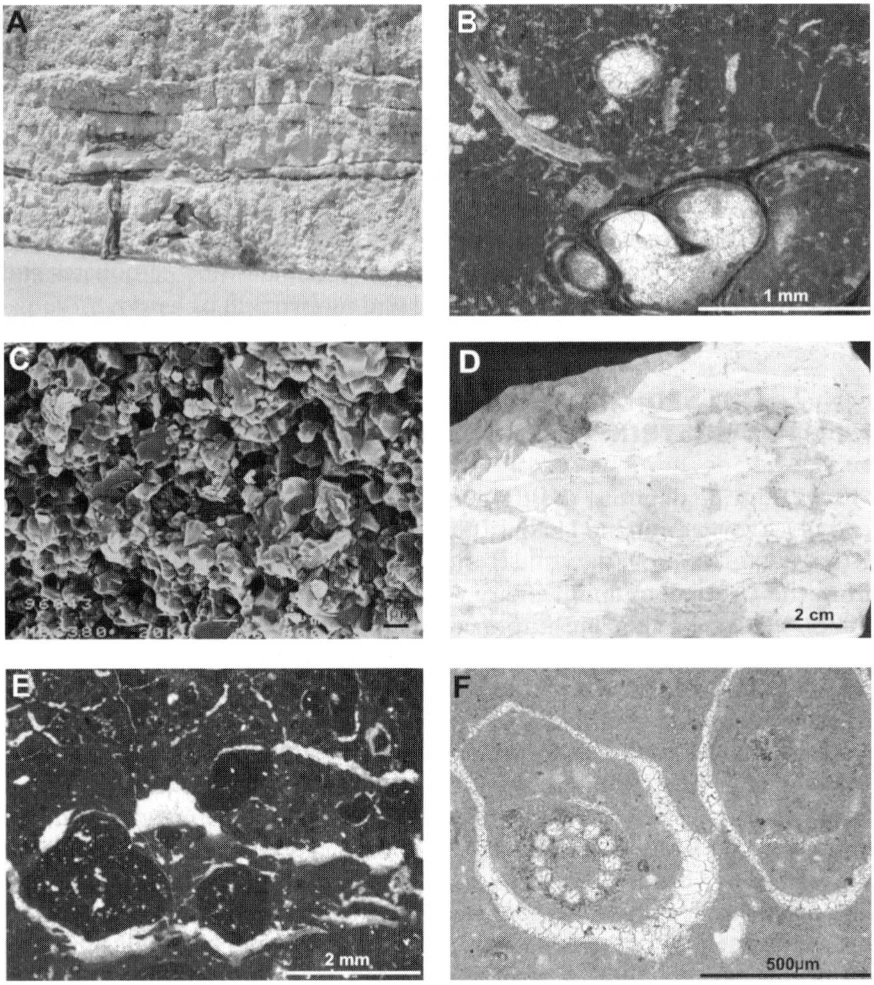

Figure 1 Ancient lacustrine and palustrine facies. (A) Nodular palustrine carbonates showing irregular tops and some interbedded marls. The figure is 1.60 m high. (B) Lacustrine wackestone with gastropods and charophytes. This is the common aspect of the initial lacustrine muds under microscope. (C) SEM image of the lacustrine mud. The crystals are fine (about 1 µm) and show different morphologies, commonly sub-euhedral. (D) Polished hand sample of a mottled and '*in situ*' brecciated limestone with irregular darker haloes. Desiccation cracks are horizontal, following the bedding planes; they are filled by micrite and vadose silt. (E) Photomicrograph of a mottled limestone showing zones with preferential Fe-staining and planar irregular desiccation cracks, filled by vadose silt and later coarse spar. (F) Photomicrograph of a nodular limestone with circumgranular cracks, which in this case enclose a charophyte stem.

pedogenesis and diagenesis. In ancient low-gradient and low-energy lakes, the primary carbonate deposit is composed of mud-size sediment, bioclasts (Figure 1B), and other grains such as peloids, coated grains (ooids) or intraclasts. Besides the direct contribution of skeletal (shell) material, much of this carbonate is/was induced by the metabolic activities of microbial biofilms and algae, including charophytes. In recent fresh-water lakes, the mineralogical composition of the primary carbonates is mostly low-magnesian calcite (<4 mol% Mg) (LMC) or high-magnesian calcite (>4 mol% Mg) (HMC), more rarely aragonite or protodolomite (Tullner and Cserny, 2003). Following diagenesis and mineralogical stabilisation, the mineralogy of the ancient fresh-water carbonates is low-magnesium calcite or dolomite (Wright et al., 1997; Anadón et al., 2000); this affects not only the mud-size carbonate, but also the aragonitic bioclasts, particularly molluscs. Detailed microfabric comparison, based on SEM studies, between ancient and recent carbonate muds shows their striking similarities, both being composed of sub- to euhedral calcite crystals about 1–2 μm across (Figure 1C). More rounded crystal morphologies recognised in recent deposits usually show higher amounts of Mg and are interpreted as high-magnesium calcite or protodolomite (Alonso-Zarza et al., 2006). This agrees with the findings in Lake Balaton. The origin of the carbonate precipitated in the lake is also discussed by Gierlowski-Kordesch (2010– this volume).

2.2. Carbonate palustrine facies in the sedimentary record

Palustrine deposits can be recognised by identifying the features described by Freytet and Verrecchia (2002), including marmorisation, nodulisation and pedogenic recrystallisation, clay authigenesis, desiccation cracks, traces of burrows and roots, and the so-called pseudomicrokarst. Additionally, root moulds, fenestral and alveolar structures, indicating the influence of the vegetation cover, are common in many palustrine limestones. All these features impart on lacustrine deposits a distinctive set of facies described by Alonso-Zarza (2003) and detailed below. Facies 1–6 are probably the more characteristic palustrine facies, but some other facies are closely associated and may give good indications of the general environmental setting within the overall deposystem (facies 7–12).

2.2.1. Facies 1: mottled limestones

Facies 1 consists of mottled limestones that are micrites with minor amounts of detrital grains. They show a strong yellow-orange-red mottling (Figure 1D) which, under the microscope, is seen as very diffuse, relatively dark haloes. Charophytes, gastropods, ostracods or any other fossil remains are relatively rare in this facies. The mottled areas may be outlined by desiccation cracks, which show different morphologies such as circumgranular, planar or irregular. These cracks are filled by both microsparitic silt and blocky sparry calcite (Figure 1E). It is not uncommon to see evidence of several stages of cementation.

Mottled or marmorised limestones indicate the remobilisation of iron due to changes in the Eh of the groundwater when the water table oscillates (Freytet, 1973). This occurs when the iron content of the sediment is less than 2%. Manganese and calcium can also migrate with iron, enabling the formation of ferruginous nodules, tubular voids, concretions and iron crusts (Freytet and Verrecchia, 2002). Freytet (1973) and Freytet and Plaziat (1982) provided clear descriptions and interpretations of all these features. The mottling is similar to that recognised in poorly drained (gley) palaeosols, as described by PiPujol and Buurman (1997) from the Eocene of the Ebro Basin.

2.2.2. Facies 2: nodular and brecciated limestones

The nodular and brecciated limestones consist of centimetre-scale irregular micrite nodules embedded in a softer chalky matrix, or separated by different types of cracks that may remain empty or become filled with microspar and/or sparry calcite. The morphology of the nodules varies from more or less round to angular, forming a breccia. The micrite may contain detrital grains and the debris of charophytes (Figure 1F), ostracods or molluscs. Nodular limestones may be light in colour, but mottling (pseudogley) is also common. The nodulisation process has been explained clearly by Freytet (1973), and is mainly due to desiccation and the subsequent formation of planar to curved fissures that are later filled under vadose and phreatic oxidising conditions.

These limestones can form single, metre-thick beds, indicating that the whole lacustrine succession was subaerially exposed, during a single or multiple events. In other cases only the top or base of lacustrine deposit is nodulised or brecciated, indicating less exposure and the evolution of the lake level. Irregular fenestrae are also present, although not restricted to this facies.

Similar features are most commonly found in intertidal successions where drying and wetting cycles are frequent. Their presence in some palustrine successions is evidence of short periods of exposure and rewetting along shorelines or in wetlands.

2.2.3. Facies 3: limestones with vertical root cavities

These are matrix-rich limestones with gastropod shells, charophytes, ostracods, desiccation cracks, and fenestral and alveolar structures (Figure 2A). Root cavities are large, irregular, vertical cavities several centimetres wide and with lengths up to decimetre scale. They are commonly wider at the top of the beds and taper downwards. These cavities may be empty or partially filled with a looser micrite matrix, microspar peloids, intraclasts, bioclasts and spar cement. Limestones with vertical root cavities either occur as tabular beds or display a wavy, convex-up upper surface (Figure 1A). Another characteristic of this facies is that it is commonly more indurated than any other palustrine or lacustrine deposits, and is therefore very prominent at outcrop scale.

Limestones with vertical root cavities commonly display a columnar structure. The vertically elongate columns can be up to 10 cm in diameter and may reach 1 m in length; horizontal cracks are a common feature. The columns are thought to form by calcification around vertically penetrating roots. Similar to the prismatic structures recognised in calcretes (Esteban and Klappa, 1983), these columns may be found in either clayey or softer carbonate sediments. Roots and associated micro-organisms can also brecciate and nodulise the host carbonate mud in areas of rhizosphere influence. This effect is especially visible in marly sediments.

2.2.4. Facies 4: pseudomicrokarst

Facies 4 consists of limestones with pseudomicrokarst. Plaziat and Freytet (1978) introduced this term to describe limestones with irregular and complex cavities that resemble a karst system in which the cavities are small, mostly cylindrical, and vertically elongated. Individual cavities are only a few centimetres long, but are associated with larger networks on the decimetre scale. The prefix 'pseudo' is used because the enlargement of the cavities is mostly mechanical (root activity and desiccation); dissolution is only a minor process. The cavities show sharp boundaries and are commonly very irregular in form (Figure 2B). However, in some cases, the margins are rounded, indicating that some dissolution has also occurred (Platt, 1989).

The pseudomicrokarst networks appear to have developed from vertical root cavities which have become connected to each other horizontally; this is especially so at the top of the beds. Desiccation cracks are common and arranged in an orthogonal network; they may cut root cavities. Cavity fills are complex and include peloids, coated grains and intraclasts (see below) as well as different types of cements from vadose to coarse blocky spar (Figure 2C). The succession of these cements is a criterion for deciphering the movements of the water table (Freytet and Plaziat, 1982).

2.2.5. Facies 5: peloidal, coated grain and/or intraclastic limestones

Peloidal, coated grain and/or intraclastic limestones are important components of palustrine limestones. Peloids are more or less rounded grains formed by micrite including some clay, pseudospar or compound micritic grains, and may be up to several millimetres in diameter (Figure 2D). They are commonly coated with irregular laminae formed of dark micrite alternating with lighter micrite layers that may include smaller peloids. SEM

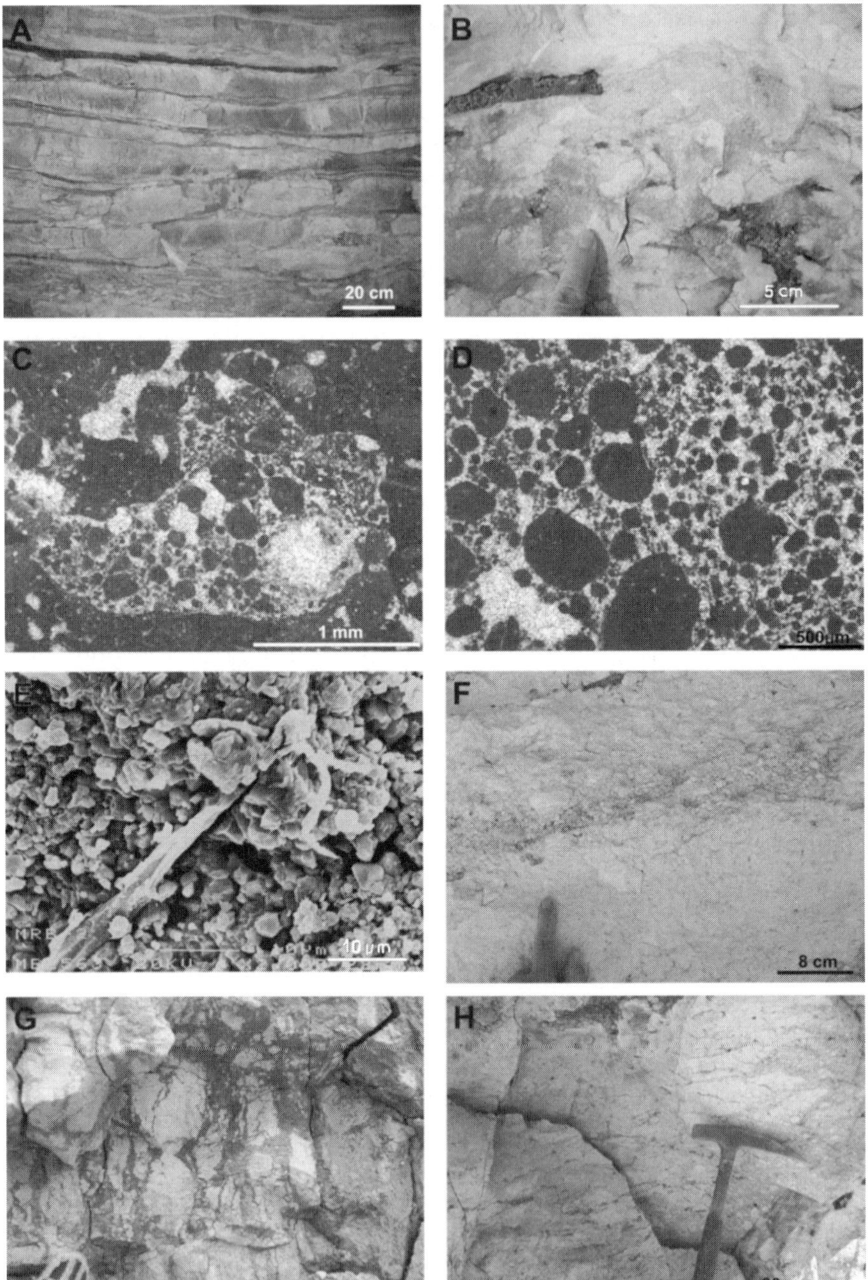
A
20 cm
B
5 cm
C
1 mm
D
500µm
E
10 µm
F
8 cm
G
3 cm
H
20 cm

observation of the latter (Alonso-Zarza et al., 1992) reveals that a network of fungal filaments arranged within micrite crystals forms the coatings.

Intraclasts are more varied in size, ranging from less than a millimetre to several centimetres long; they are rounded to angular and are commonly poorly sorted. The intraclasts consist of micrite with scarce fossil debris. Root moulds, alveolar septal structures (Wright, 1986), organic filaments (Figure 2E) and different types of desiccation cracks are easily recognised within the intraclasts as well as in the intragranular pores. The peloidal, coated grains and intraclastic sediments are commonly cemented by calcite mosaics, the sizes of which vary between that of microspar and coarsely crystalline. These sediments occur in different types of cavities and form a complex network. This facies has also been named granular limestones and may form through the process of grainification (Mazzullo and Birdwell, 1989; Wright, 1990), which also occurs in peritidal settings. These appear as beds formed entirely of this facies at the top or base of lacustrine and peritidal deposits, or filling different types of cavities. A characteristic is that both the grains themselves and the cavities and surrounding matrices show clear evidence of multiple phases of fragmentation (grainification), coating, internal sediment, deposition of microsparitic silt and blocky calcite cementation. A common feature of these cavity fills is inverse grading.

2.2.6. Facies 6: Laminar calcretes

Laminar calcretes are a common feature and consist of centimetre-thick sheet-like or stomatolitic, very irregularly laminated micrites. Some forms are densely laminated but others are fenestral. The former are similar to crusts associated with calcrete profiles (see Alonso-Zarza and Wright, 2010 – this volume), but the latter are calcified root mats (Wright et al., 1988). In all cases the micromorphology shows a wide range of biogenic structures such as calcified organic filaments, alveolar septal structures,

Figure 2 Ancient palustrine facies. (A) Decimetre-thick beds of palustrine carbonates, the tops of which contain vertical roots. (B) Top-most part of a palustrine bed with highly irregular pseudomicrokarst filled by micritic clasts. The cavities are very irregular. (C) Photomicrograph of a pseudomicrokarst infill consisting of micritic grains embedded in calcite cement. (D) Peloidal limestone. The peloids are more or less rounded and show poor sorting. In this case, they do not show micritic coatings. (E) SEM view of the peloidal limestones. The micrite crystals are very fine. Organic filaments, contributing to the breaking of the micrite, are visible. (F) Conglomerates consisting of limestone clasts derived from the lower palustrine limestones with desiccation and root traces (above finger). (G) Green marlstone beds with various types of desiccation and root cracks resulting from *in situ* brecchiation of the bed; intraclasts are visible. (H) Palustrine sepiolite bed, showing the same features as the palustrine carbonates, such as desiccation, formation of intraclasts and some reworking. In this case, the bed is silicified.

2.3. Genetic aspects of the palustrine carbonate facies

The facies described above clearly show the different intensities with which fine-grained subaqueously formed continental carbonates are affected by pedogenesis and reworking. The variability of the processes involved in each palustrine microenvironment (Figure 3) can be related to the proportion of time spent subaerially exposed (referred to as the 'exposure index') (Platt and Wright, 1992). A continuum from less pedogenically modified limestones to those that are totally modified can be identified, taking into account mostly the palustrine facies (1 to 5) described above (facies 4–8 of Figure 3). This continuum reflects not only the degree of pedogenic modification but also the relative influence of physicochemical versus biogenic processes.

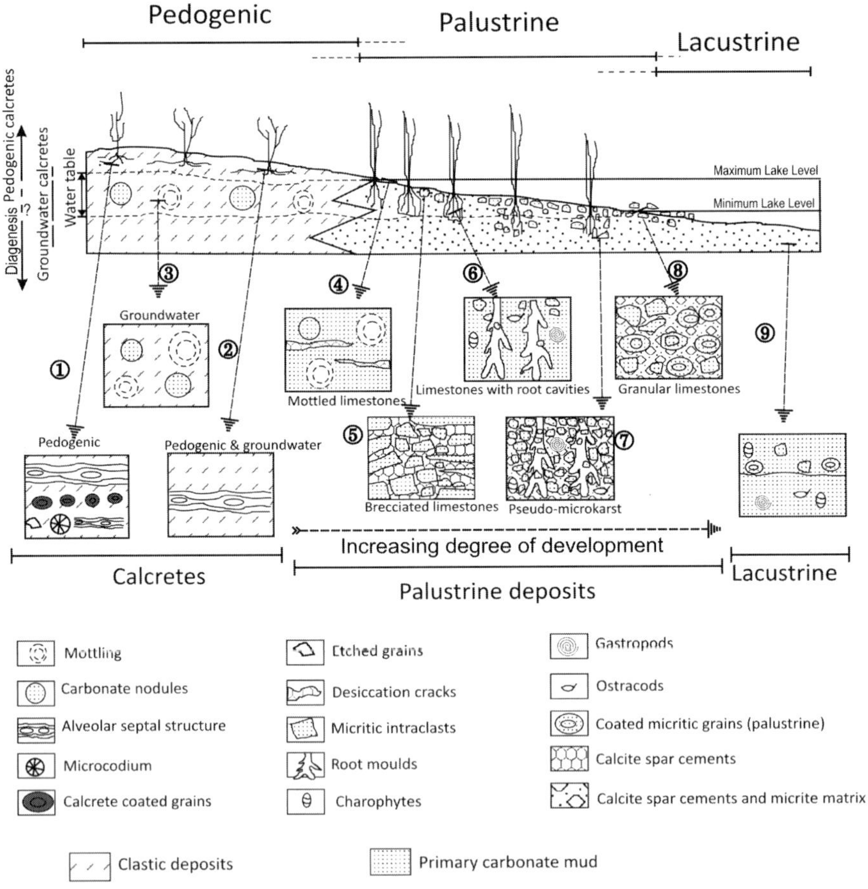

Figure 3 The various palustrine-calcrete facies associations. Common petrographic aspects of these carbonates are also shown. From Alonso-Zarza (2003). 1–3: Calcretes; 4–8 (1–5 in the text): different palustrine deposits from the less (4) to the more developed ones (8); 9: non-pedogenically modified lacustrine deposits.

Mottled and nodular limestones mostly reveal the influence of physicochemical processes such as desiccation and iron mobilisation; both limestone types may be considered as less developed palustrine limestones. Limestones with root cavities, as well as those with a prismatic structure, indicate the presence of a well-established vegetation cover. Roots act in two different ways: (1) by penetrating the carbonate muds and/or (2) inducing biochemical precipitation of carbonate around the rhizosphere. In either case, the carbonates formed under these conditions indicate more intense soil modification and longer intervals of exposure.

The more developed palustrine limestones are the granular limestones (peloidal or intraclastic) and those carbonates with pseudomicrokarst. In both, the result is an important loss of the primary muddy texture of the deposits. These processes of formation of syngenetic grainstones, or to use a wider term, granular limestones, have been extensively described not only from palustrine environments (Freytet and Plaziat, 1982; Alonso-Zarza et al., 1992; Armenteros et al., 1997), but also from peritidal settings (Mazzullo and Birdwell, 1989). The formation of these textures is driven by the repeated wetting and drying of the carbonate mud and by root systems that penetrated the recently deposited mud to reach the water table. This, together with the desiccation of the surface of the sediment, contributes to intense fragmentation of the mud. The intensity and duration of these processes, as well as any later rise in the water table during wetter periods, can cause the reworking, concentration and coating of the mud fragments. Movement of grains over the sediment surface gives rise to the formation of beds of mostly (but not necessarily) coated micrite grains that accumulate at some distance from the place where the fragmentation originally occurred.

In contrast, pseudomicrokarst and brecciated limestones are formed '*in situ*'. In both cases, the activity of micro-organisms such as fungi and bacteria plays a significant role, especially in the coating of some grains (Alonso-Zarza et al., 1992).

3. Wetlands: The Recent Analogues for Ancient Palustrine Environments?

The search for a recent sedimentary environment where the deposits show similar features and composition as do ancient palustrine carbonates has been the aim of many studies during the last 15 years. Narrowing palustrine deposits to a specific setting is difficult because they occur in a broad spectrum of depositional environments including: (1) coastal plains, such as the Florida Everglades (Platt and Wright, 1992), and the Lower Cretaceous of Croatia (Dini et al., 1998), (2) alluvial settings, with examples from the Upper Jurassic Western Interior, USA (Dunagan and Turner, 2004), the Tertiary of Wyoming (Bowen and Bloch, 2002) and the

Miocene of Spain (Armenteros and Huerta, 2006) and (3) marginal areas of larger ancient (Freytet and Plaziat, 1982) and recent lacustrine systems, as the previously mentioned Lake Balaton in Hungary (Tullner and Cserny, 2003).

It seems to be in common agreement that wetlands constitute the closest recent analogue to the ancient environment in which palustrine deposits accumulated; in fact, Wright and Platt (1995) noted that the abundance of ancient palustrine carbonates is a testimony to the widespread development of seasonal wetlands in the past, which is consistent with the widespread occurrence today of wetlands, covering some 6% of the land surface (Marty and Meyer, 2006). They have been recognised in all climatic zones and on all continents except Antarctica (Mitsch and Gosselink, 1993). The variability of modern wetlands makes it difficult to establish an all-encompassing sedimentary definition because there are a number of ecological, hydrological and sedimentary parameters that should be considered.

Following the definition by Cowardin et al. (1979), palustrine wetlands must fulfil one or more of the following requirements: the land supports predominantly hydrophytes, at least periodically; the substrate is predominantly undrained hydric soil, and the substrate is non-soil and is saturated with a water cover or covered by shallow water at some time during the growing season. Moreover, there are some specific requirements for water depth and composition, that is, less than 2 m deep at low water, and salinity from ocean-derived salts of less than 0.5 ppt. Additionally, there is a lack of wave-formed features on the shorelines (Cowardin et al., 1979). More recently, wetlands have been defined as landscape units with a spatially and temporally positive hydrologic anomaly in relation to the adjacent drier land without being rivers, lakes or marine environments (DGOH, 1991).

The study of the sediments recently deposited in wetlands, especially if they are carbonates, is crucial to determine if modern wetlands constitute the most likely analogue for palustrine carbonate successions. In recent years, the use of different technologies to core lake systems has contributed to a better understanding of these sediments; most of the work, however, is focused on siliciclastics, such as the Loobi Swamp in Kenya (Ashley et al., 2004). Some detailed studies of carbonate deposits in marginal areas of lakes are being carried on in Lake Balaton (Tullner and Cserny, 2003). Apart from the lack of studies on recent carbonate wetlands (or even lakes), one additional problem is diagenesis, which may transform the mineralogy and texture of the initial sediments to make them more stable and lithified.

The Las Tablas wetlands are located in the southern Meseta of Spain and were formed initially by the flooding of the Guadiana and Gigüela rivers, together with the upwelling of groundwater through many springs. The relative contribution of surface waters versus groundwater varies in space and time. The total wetland area was about 60 km^2 30 years ago, but today it is only of 16.75 km^2, because of increased water usage by Man. The Las Tablas wetlands are shallow, usually less than 1 m deep (Figure 4A and B). The aquatic environment is highly turbid, with high sedimentation rates. The dominant anions are carbonate and sulphate. The climate is classified as cold, temperate continental with a dry season. Isotherms range from 12 to 14°C, rainfall is from 400 to 500 mm year^{-1}, and potential evaporation averages 778 mm year^{-1}. All these data are extensively discussed in Álvarez-Cobelas and Cirujano (1996) and Dorado-Valiño et al. (2002). The Las Tablas wetlands are now a National Park under a strong preservation policy.

The core of the Las Tablas de Daimiel wetlands shows a 38.5-m thick succession of sediments that range from the Middle Pleistocene (even Early?) to the Holocene. The study of the core (Alonso-Zarza et al., 2006) has shown that the more indurated sediments exhibit clear palustrine features all through the core, but specially in the upper Holocene part, where the sediments consist mostly of low-Mg calcite (45–100%), with varying amounts of quartz (0–35%), clays (0–20%) and aragonite from shells (0–5%). In this part, no traces can be found of dolomite, opal, high-Mg calcite or organic matter, which occur in the Middle Pleistocene part. Three main carbonate facies have been recognised: (1) massive chalky carbonates; (2) indurated wackestones to packstones with gastropods, oncoids, ostracods and charophytes (Figure 4C), at some intervals showing desiccation cracks, alveolar septal structures, pseudomicrokarst and mottling; and (3) packstones of intraclasts containing some quartz grains (Figure 4D). The massive chalky carbonates represent deposits that have not been modified by pedogenesis. The indurated limestones formed in the shallower areas, as indicated not only by the sedimentary features but also by the low aquatic pollen diversity and the emergent pollen taxa recorded in the sediments (Valdeolmillos Rodríguez, 2005). In this environment, the sediments were intensively modified, allowing the micrite crystals to change their size (larger crystals) and shape (Figure 4E). In addition, phreatic and vadose cementation also contributed to the lithification of the sediments. Packstones of intraclasts in Las Tablas have been interpreted as the result of desiccation and reworking events of the slightly lithified wackestones, and reflect more humid periods that caused the rise of the lake water. Some of the carbonate deposits show calcified microbial tubes similar to those recognised in calcretes (Figure 4F). Apart from the carbonate deposits, peat beds are also common. The stable-isotope data of the carbonates are also very similar to those recorded in other palustrine deposits (Arenas et al., 1997; Alonso-Zarza and Calvo, 2000; Tandon and Andrews, 2001; Dunagan and Turner, 2004), with a lack of covariance of C and O stable isotopes indicating an open system. With all these data in mind, it is likely that this wetland area can be considered as the best modern analogue for ancient palustrine carbonates.

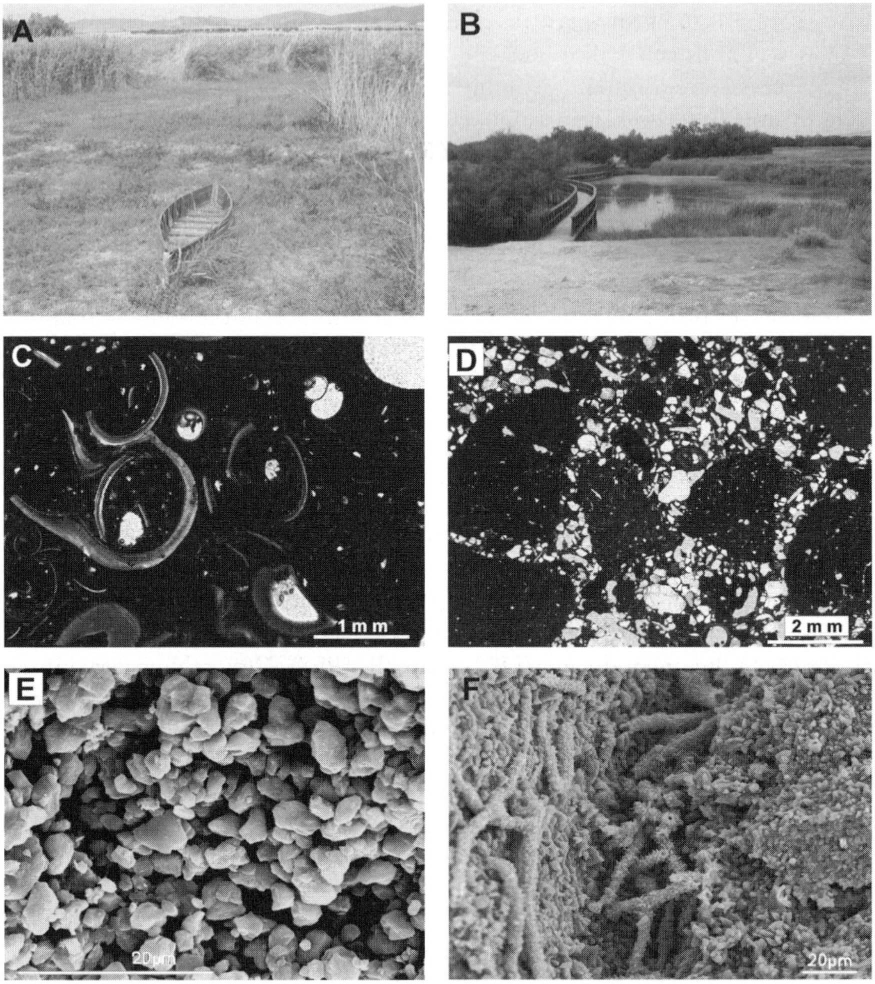

Figure 4 Palustrine facies and environments from the Las Tablas de Daimiel wetland. (A) View of a dry area of Las Tablas. The area around the boat was wet few years ago. Reeds are the most characteristic vegetation. (B) View of an active, flooded area of Las Tablas. The depth of water is less than 1 m. (C) Indurated wackestones with bioclasts. (D) Packstone of intraclasts, formed by desiccation of the previous limestones and their later reworking. Quartz grains are also common. This facies probably indicates prolonged subaerial exposure. (E) SEM image of the palustrine mudstone facies, which are very porous and consist of very fine sub-euhedral crystals. (F) SEM image of microbial calcified filaments in a palustrine limestone, indicating the role of vegetation in the formation of these facies.

However, more studies are needed in order to increase the insight into wetland deposition.

4. Environmental Controls on Palustrine Deposition

Tectonism, climate and eustasy are the main factors operating in the marine realm. In terrestrial environments, eustasy drives the base level which is critical to fluvial systems, erosion and sedimentation rates. Eustasy is important to basins even far away from, but connected with, the ocean, and some examples will be discussed later. However, eustasy does not directly control sedimentary deposition in a number of examples, including sedimentation of palustrine deposits. In basins lacking a clear connection to the oceans, two additional factors, viz. hydrology and supply/sources of carbonates, have to be considered. Because these factors are integrated, it is difficult in many cases to distinguish the specific effects of climate, hydrology and tectonism.

Palustrine carbonates are obviously sensitive to climate. There must be sufficient rainfall to carry the carbonate-rich solutions either in the surface waters or in the groundwater (De Wet et al., 1998). Very arid or very humid climates do not favour carbonate deposition in lakes (Cecil, 1990), whereas semi-arid to sub-humid climates with marked seasonality are more appropriate (Platt and Wright, 1991; Sanz et al., 1995; Gierlowski-Kordesch, 1998). Palustrine carbonates are sensitive to fluctuations in humidity. Therefore, palustrine facies and sequences may vary according to the climate regime (Platt and Wright, 1992). In sub-humid climates, palustrine deposits contain more organic matter – such as in the Miocene of the Teruel Graben (Alonso-Zarza and Calvo, 2000) – which may develop into coals, such as in the Oligocene of the Ebro Basin (Cabrera and Sáez, 1987). In contrast, semi-arid climates are characterised by wide pseudomicrokarst development on top of the successions, and organic matter is rarely preserved, as in the Upper Jurassic of the Morrison Formation (Dunagan and Turner, 2004). More arid climates favour the presence of evaporite nodules within the palustrine carbonates, or palustrine carbonates dominated by dolomite (Sanz et al., 1999).

Climate also affects the carbonate precipitation rate, both of inorganic precipitation (because of temperature effects in the water body and evaporation) and of biogenically produced carbonate (Kelts and Hsü, 1978; Platt and Wright, 1991; Verrecchia, 2007). In palustrine environments, the climate also plays a role in the run-off in the catchment area and hence in sediment supply, which is typically siliciclastic in composition. Thus climate change will impact upon the short- to medium-term sedimentary

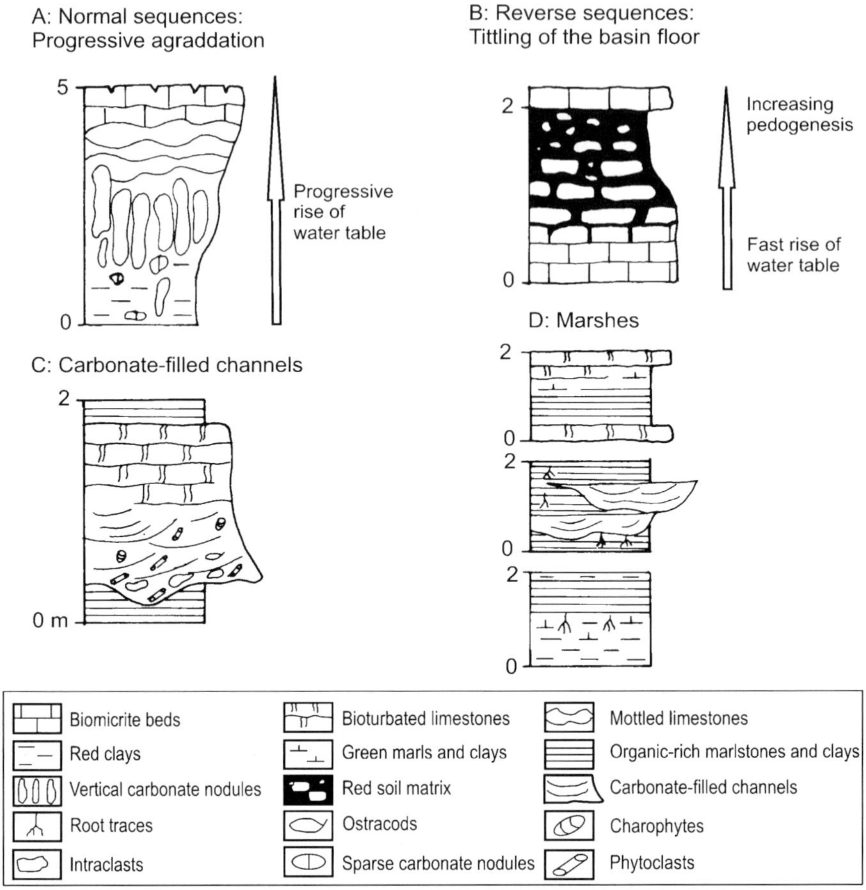

Figure 5 Palustrine deposits from the Teruel Basin. (A) and (B) show different arrangements of facies, depending on how fast the water table reaches the surface. Some carbonate-filled channels (C) and marshes (D) are commonly associated with these palustrine-wetland environments.

to wetter conditions during the Late Holocene favoured the development and growth of the wetland. However, the lowering of the surface near or at the water table may be controlled by tectonic subsidence pulses, as seen at a small scale (metres) in the Neogene deposits of the Teruel Graben (Alonso-Zarza and Calvo, 2000), or at a large scale (a hundred metres) in the Cretaceous of the Aguilar de Campo Formation (Hernández et al., 1998).

The possibility of identifying climatic signals in short-term lacustrine sequences has been the objective of several recent studies (e.g., Abdul-Aziz et al., 2000), which conclude that some sequences may correspond to astronomical precession cycles. In a study of the cyclicity of Tertiary deposits of the Ebro Basin, Luzón et al. (2002) showed that marginal lacustrine deposits are not suitable for identifying the effects of climatic cycles because non-periodic processes can play a more important role in the sedimentation, modifying the original sedimentological features. In open lake areas of the same lacustrine system, spectral analysis clearly indicates precession-related climatic cyclicity. Although there is common agreement on the role of climate in creating short- to medium-term (up to 10–20 m thick) sedimentary lacustrine sequences (Arenas and Pardo, 1999; Armenteros and Huerta, 2006, amongst many others), care must be taken when interpreting the causes of the cyclicity, because short-term subsidence episodes may also give place to the formation of metre-scale sequences (Huerta and Armenteros, 2005).

Cyclicity can also reflect groundwater activity. One interesting example comes from the Lower Pliocene Hagul Formation of the East Cairo District, Egypt, described by Khalaf and Gaber (2008). Here, more than 10 depositional cycles are recognised which represent alluvial sedimentation, followed by calcretisation, and formation of palustrine carbonates. The distal alluvial plain setting received alluvial sediment during more humid intervals, with pedogenic calcretes forming during semi-arid climates with episodic precipitation. During sub-humid phases, the groundwater levels gradually rose and groundwater calcrete accumulated, followed by shallow palustrine carbonate wetlands.

Climate and tectonism also determine the topography, the hydrology of the lake (open/closed), the mechanisms of water supply and the sources of the carbonate and calcium ions within the lake.

In the Cenozoic deposits of the Ebro Basin, the asymmetry in the distribution of shallow-lake deposits, which are dominated by evaporites in the southern Iberian margin and palustrine carbonates in the North (Pyrenean margins), reflects the strong hydrological contrast and different topographic gradients. The Iberian margin has a low gradient and smaller water supply, whereas the Pyrenean margins have steep gradients and a permanent water supply (Arenas and Pardo, 1999). The presence within lake basins of topographic highs can divide the basins and favour the development of palustrine deposits in the shallower parts; in the Cañadón Asfalto Formation in Argentina, for example, a bioherm ridge acted as a physiographic barrier (Cabaleri and Armella, 2005).

The mechanism of water supply to the lake plays an important role in determining the lake chemistry and therefore the mineralogy of the sediments, as well as later transformations during early diagenesis. The relative contribution of groundwater and surface water to the inundated area is considered important for distinguishing between lakes and wetlands (Currey, 1990). According to Dunagan and Turner (2004), many wetlands seem to be fed primarily by groundwater, but may receive surface water by sheet flows; in contrast, lakes are fed mainly by surface water from run-off, rivers and streams that enter the basin, with a secondary groundwater contribution. However, even in modern environments this differentiation is difficult to make because the nature of the water supply may vary with time as a consequence of climate and even tectonism. Some recent wetlands such as the Loboi Swamp are fed by springs (Ashley et al., 2004); in addition, the Las Tablas wetlands are fed by a combination of springs, rivers and run-off (Álvarez-Cobelas and Cirujano, 1996).

Commonly, as in the Eocene of South Dakota, palustrine limestones are found near fault zones spatially associated with palaeogroundwater or spring deposits. Some of these carbonates formed behind tufa barrages (Evans, 1999), and their occurrence supports the importance of groundwater supply in shallow lacustrine–palustrine environments. During the Upper Triassic, the discharge of water through fault-bounded spring systems favoured the development of an extensive lake system in Poland; travertines formed near the springs, whereas palustrine carbonates were deposited in more distal areas (Szulc et al., 2006). Hernández et al. (1998) have recognised hot-water spring travertines fed through fault systems in the NE of Spain. This is one of the main feeding mechanisms to the palustrine–lacustrine system. Unfortunately, the preservation potential of the spring conduits and even of their deposits is limited, and proving a spring-fed source for the palustrine carbonates is therefore difficult in many cases.

A purely meteoric, run-off supply gives rise to lake waters with a low solute content. If there is any carbonate precipitation in such lakes, it is usually mainly calcite. However, chemical precipitation may be inhibited or diluted by clastic input that may constitute the main infill of the lake. If the lake is mainly fed by groundwater, directly or via springs, a more complex lake geochemistry can be expected, and the mineralogy of the palustrine deposits will depend on the composition of the groundwaters. Factors such as the distance that the groundwater has flowed, the composition of the catchment areas and the ratio of rock/water interaction and evaporation rates all control groundwater composition and therefore the mineralogy of the primary precipitates and their possible transformation during early diagenesis (Arakel and McConchie, 1982).

Low-Mg calcite is a common precipitate if the groundwater flow path is relatively short, or if the catchment area is dominated by low-Mg calcite, and little evaporation occurs. Dolomite and gypsum may form from more evolved groundwater due to removal of less-soluble phases during flow towards the lake or within the lake itself (Wright and Sandler, 1994; Calvo et al., 1995). However, the composition of groundwaters and the formation of other precipitates such as dolomite, gypsum or Mg clays depend not only on the hydrology but also on the climate and/or the presence of source rocks containing evaporites.

Many of the documented examples of palustrine carbonates developed in basins surrounded by hinterlands dominated by carbonate rocks. There are a number of examples on the Iberian Peninsula (Platt, 1989; Alonso-Zarza et al., 1992, amongst many others) and in the Triassic of Pennsylvania (De Wet et al., 1998). However, this is not always a prerequisite for lacustrine–palustrine carbonate sedimentation. For example, in the Upper Palaeozoic of the Pyrenees, lacustrine carbonates are interbedded with pyroclastic layers (Valero, 1993). The weathering of calc-alkaline volcanic rocks in the catchment area favoured low to moderate calcite production in adjacent lakes. Nevertheless, the presence of carbonates underlying the lakes or in the basin margins notably contributes to carbonate precipitation within the lakes for two main reasons: (1) it favours high carbonate concentrations in the surface and groundwater (Gierlowski-Kordesch, 1998) and (2) it favours a low siliciclastic input; both control the carbonate production within the lake (Cohen, 1989). Some well-known lacustrine and palustrine deposits and even palustrine-saline and carbonate lakes are found in association with karstic systems, either in depressions of the karst or in areas with supplies by the output of the karstic waters (Poyato et al., 1998; Pérez et al., 2002; Alonso-Zarza et al., 2006).

Palustrine carbonates have also been recorded in marginal-marine settings, specifically in coastal-plain systems, as in the Eocene of the Isle of Wight off England (Armenteros and Daley, 1998), and in the Cretaceous of Croatia (Dini et al., 1998) and Spain (Martín-Chivelet and Giménez, 1992); they may cover thousands of square kilometres (MacNeil and Jones, 2006). It has been argued that the Florida Everglades may represent a possible modern analogue for palustrine deposition (Platt and Wright, 1992). However, there are few documented examples of such carbonates and it is possible that other cases exist that have been attributed instead to calcretes or marine deposits. An additional complication might be that palustrine carbonates may overprint marine macrophytic deposits (such as mangrove facies) during shallowing-upwards intervals (Wright and Azeredo, 2006). In their study of palustrine carbonates from the Devonian of the Northwest Territories of Canada, MacNeil and Jones (2006) stressed the importance of these coastal palustrine carbonates because of their implications for sequence-stratigraphic analysis; that is, they indicate subaerial unconformities and sequence boundaries, similar to calcretes and karsts.

5. Conclusions

Palustrine deposits are a distinctive type of non-marine carbonates that are associated with hard-water lake-margin and wetland settings. They represent the modification of carbonate muds by wetting and drying and more prolonged subaerial exposure, and pedogenesis associated with macrophytes, and so are effectively a post-Silurian phenomenon. They are part of a continuum of carbonate facies related to hydrological systems, closely related to groundwater calcretes, tufas and travertines, wetlands and shallow-lake deposits. Palustrine carbonates are commonly, but not always, associated with the late stages of basin filling, reflecting limited tectonic relief development, limited accommodation-space creation, limited terrigenous sediment input and low-relief landscapes where the position of the water table is a critical control on carbonate accumulation. They are sensitive to subtle hydrological changes in basins and so respond to both climatic and tectonic effects, resulting in the need to critically separate such effects when interpreting climatic signals. Despite their relatively common occurrence in the stratigraphic record, modern analogues are very rare. Therefore, it is imperative that we track down active examples in order to release all the data locked in the ancient palustrine record.

ACKNOWLEDGEMENTS

This work was funded by project CGL-2008-05584-C02-02 from the Spanish *Ministerio de Ciencia e Innovación* and CCG07-UCM/AMB2299-910404 from CM-UCM. Dr. Dunagan is thanked for his critical revision of the manuscript. L. H. Tanner and T. van Loon have contributed to the careful editing of this contribution.

REFERENCES

Abdul-Aziz, H., Hilgen, F., Krijgsman, W., Sanz, E., Calvo, J.P., 2000. Astronomical forcing of sedimentary cycles in the middle to late Miocene continental Calatayud Basin (NE Spain). Earth and Planetary Science Letters 177, 9–22.

Alonso-Zarza, A.M., 2003. Palaeoenvironmental significance of palustrine carbonates and calcretes in the geological record. Earth-Science Reviews 60, 261–298.

Alonso-Zarza, A.M., Calvo, J.P., 2000. Palustrine sedimentation in an episodically subsiding basin: the Miocene of the northern Teruel Graben (Spain). Palaeogeography, Palaeoclimatology, Palaeoecology 160, 1–21.

Alonso-Zarza, A.M., Calvo, J.P., García del Cura, M.A., 1992. Palustrine sedimentation and associated features – grainification and pseudo-microkarst – in the Middle Miocene (Intermediate Unit) of the Madrid Basin, Spain. Sedimentary Geology 76, 43–61.

Alonso-Zarza, A.M., Dorado-Valiño, M., Valdeolmillos-Rodríguez, A., Ruiz-Zapata, M.B., 2006. A recent analogue for palustrine carbonate environments: the Quaternary deposits of Las Tablas de Daimiel wetlands, Ciudad Real, Spain. In: Alonso-Zarza, A.M., Tanner, L.H. (Eds), Paleoenvironmental Record and Applications of Calcretes and Palustrine Carbonates. Geological Society of America, Special Paper 416, pp. 153–168.

Alonso-Zarza, A.M., Tanner, L.H. (Eds), 2006. Paleoenvironmental Record and Applications of Calcretes and Palustrine Carbonates. Geological Society of America, Special Paper 416, 239 pp.

Alonso-Zarza, A.M., Wright, V.P., 2010. Calcretes. In: Alonso-Zarza, A.M., Tanner, L.H. (Eds), Continental Carbonates. Developments in Sedimentology, 61, pp. 225–268.

Álvarez-Cobelas, M., Cirujano, S. (Eds), 1996. Las Tablas de Daimiel. Ecología Acuática y Sociedad. Organismo Autónomo Parques Naturales, Madrid, 368 pp.

Anadón, P., Utrilla, R., Vázquez, A., 2000. Use of charophyte carbonates as proxy indicators of subtle hydrological and chemical changes in marl lakes: example from the Miocene Bicorb Basin, eastern Spain. Sedimentary Geology 133, 325–347.

Arakel, A.V., McConchie, D., 1982. Classification and genesis of calcrete and gypsite lithofacies in palaeodrainage basins of inland Australia and their relationship to carnotite mineralization. Journal of Sedimentary Petrology 52, 1147–1170.

Arenas, C., Cabrera, L., Ramos, L., 2007. Sedimentology and tufa facies and continental microbialites from the Palaeogene of Mallorca Island (Spain). Sedimentary Geology 197, 1–22.

Arenas, C., Casanova, J., Pardo, G., 1997. Stable-isotope characterization of the Miocene lacustrine systems of Los Monegros (Ebro Basin, Spain): palaeogeographic and palaeoclimatic implications. Palaeogeography, Palaeoclimatology, Palaeoecology 128, 133–155.

Arenas, C., Pardo, G., 1999. Latest Oligocene–Late Miocene lacustrine systems of the north-central part of the Ebro Basin (Spain): sedimentary facies model and palaeogeographic synthesis. Palaeogeography, Palaeoclimatology, Palaeoecology 151, 127–148.

Armenteros, I., Daley, B., 1998. Pedogenic modification and structure evolution in palustrine facies as exemplified by Bembridge Limestone (late Olocene) of Isle of Wight, southern England. Sedimentary Geology 119, 275–295.

Armenteros, I., Daley, B., García, E., 1997. Lacustrine and palustrine facies in the Bembridge Limestone (late Eocene, Hamshire Basin) of Isle of Wight, southern England. Palaeogeography, Palaeoclimatology, Palaeoecology 128, 111–132.

Armenteros, I., Huerta, P., 2006. The role of clastic sediment influx in the formation of calcrete and palustrine facies: a response to paleogeographic and climatic conditions in the southeastern Tertiary Duero basin (northern Spain). In: Alonso-Zarza, A.M., Tanner, L.H. (Eds), Paleoenvironmental Record and Applications of Calcretes and Palustrine Carbonates. Geological Society of America, Special Paper 416, pp. 119–132.

Ashley, G.M., Maitima Mworia, J., Muasya, A.M., Owens, R.B., Driese, S.G., Hover, V.C., Renaut, R.W., Goman, M.F., Mathai, S., Blatt, S.H., 2004. Sedimentation and recent history of a freshwater wetland in a semi-arid environment: Loboi Swamp, Kenya, East Africa. Sedimentology 51, 1–21.

Bohacs, K.M., Carroll, A.R, Neal, J.E., Mankiewicz, P.J., 2000. Lake-basin type, source potential, and hydrocarbon character: an integrated-sequence-stratigraphic-geochemical framework. In: Gierlowski-Kordesch, E.H., Kelts, K.R. (Eds), Lake Basins through Space and Time. AAPG Studies in Geology 46, pp. 3–34.

Bowen, G.J., Bloch, J.I., 2002. Petrography and geochemistry of floodplain limestones from the Clarks Fork Basin, Wyoming, USA: carbonate deposition and fossil accumulation on a Paleocene–Eocene floodplain. Journal of Sedimentary Research 72, 46–58.

Bustillo, M.A., Alonso-Zarza, A.M., 2007. Overlapping of pedogenesis and meteoric diagenesis in distal alluvial and shallow lacustrine deposits in the Madrid Miocene Basin, Spain. Sedimentary Geology 198, 255–271.

Cabaleri, N.G., Armella, C., 2005. Influence of a biohermal belt on the lacustrine sedimentation of the Cañadón Asfalto Formation (Upper Jurassic, Chubut province, Southern Argentina). Geologica Acta 3, 205–214.

Cabrera, L., Sáez, A., 1987. Coal deposition in carbonate-rich shallow lacustrine systems: the Calaf and Mequinenza sequences (Oligocene, eastern Ebro Basin, NE Spain). Journal of the Geological Society 144, 451–461.

Calvo, J.P., Jones, B.F., Bustillo, M., Fort, R., Alonso-Zarza, A.M., Kendall, C., 1995. Sedimentology and geochemistry of carbonates from lacustrine sequences in the Madrid Basin, Central Spain. Chemical Geology 123, 173–191.

Andean forearc, northern Chile. Palaeogeography, Palaeoclimatology, Palaeoecology 151, 55–77.
Mazzullo, S.J., Birdwell, B.A., 1989. Syngenetic formation of grainstones and pisolites from fenestral carbonates in peritidal settings. Journal of Sedimentary Petrology 59, 605–611.
Mitsch, W.J., Gosselink, J.G., 1993. Wetlands (2nd Edition). New York, Van Nostrand Reinhold, 722 pp.
Mount, J.F., Cohen, A.S., 1984. Petrology and geochemistry of rhizoliths from Plio-Pleistocene fluvial and marginal lacustrine deposits, east Lake Turkana, Kenya. Journal of Sedimentary Petrology 54, 263–275.
Nickel, E., 1982. Alluvial-fan-carbonate facies with evaporites, Eocene Guarga Formation, Southern Pyrenees, Spain. Sedimentology 29, 761–796.
Pérez, A., Luzón, A., Roc, A.C., Soria, A.R., Mayayo, M.J., Sánchez, J.A., 2002. Sedimentary facies distribution and genesis of a recent carbonate-rich saline lake: Gallocanta Lake, Iberian Chain, NE Spain. Sedimentary Geology 148, 185–202.
PiPujol, M.D., Buurman, P., 1997. Dynamics of iron and calcium carbonate redistribution and palaeohydrology in middle Eocene alluvial paleosols of the southeast Ebro Basin margin. Palaeogeography, Palaeoclimatology, Palaeoecology 134, 87–107.
Platt, N.H., 1989. Lacustrine carbonates and pedogenesis: sedimentology and origin of palustrine deposits from the Early Cretaceous Rupelo Formation, W Cameros Basin, N Spain. Sedimentology 36, 665–684.
Platt, N.H., Wright, V.P., 1991. Lacustrine carbonates: facies models, facies distribution and hydrocarbon aspects. In: Anadón, P., Cabrera, L., Kelts, K. (Eds), Lacustrine Facies Analysis. Special Publication International Association of Sedimentologists 13, pp. 57–74.
Platt, N.H., Wright, V.P., 1992. Palustrine carbonates at the Florida Everglades: towards an exposure index for the fresh-water environment. Journal of Sedimentary Petrology 62, 1058–1071.
Plaziat, J.C., Freytet, P., 1978. Le pseudo-microkarst pédologique: un aspect particulier des paléo-pédogenèses développées sur les dépôts calcaires lacustres dans le Tertiaire du Languedoc. Comptes Rendus de l'Academie des Sciences, Paris 286, 1661–1664.
Poyato, F.J., Talbot, M.R., Fregenal-Martínez, M.A., Meléndez, N., Wenz, S., 1998. First isotopic and multidisciplinary evidence for nonmarine coelacanths and pycnodontiform fishes: palaeoenvironmental implications. Palaeogeography, Palaeoclimatology, Palaeoecology 144, 65–84.
Rodríguez-Aranda, J.P., Calvo, J.P., 1998. Trace fossils and rhizoliths as a tool for sedimentological and palaeoenvironmental analysis of ancient continental evaporite succesions. Palaeogeography, Palaeoclimatology, Palaeoecology 140, 383–399.
Sáez, A., Cabrera, L., 2002. Sedimentological and palaeohydrological responses to tectonics and climate in a small, closed, lacustrine system: Oligocene As Pontes Basin (Spain). Sedimentology 49, 1073–1094.
Sanz, E., Hoyos, M., Calvo, J.P., Rouchy, J.M., 1999. Nodular anhydrite growth controlled by pedogenic structures in evaporite lake formations. Sedimentary Geology 125, 195–203.
Sanz, M.E., Alonso-Zarza, A.M., Calvo, J.P., 1995. Carbonate pond deposits related to semi-arid alluvial systems: examples from the Tertiary Madrid Basin, Spain. Sedimentology 42, 437–452.
Szulc, J., Gradzinski, M., Lewandowska, A., Heunisch, C., 2006. The Upper Triassic crenogenic limestones in Upper Silesia (southern Poland) and their paleoenvironmental context. In: Alonso-Zarza, A.M., Tanner, L.H. (Eds), Paleoenvironmental Record and Applications of Calcretes and Palustrine Carbonates. Geological Society of America, Special Paper 416, pp. 153–168.

Tandon, S.K., Andrews, J.E., 2001. Lithofacies associations and stable isotopes of palustrine and calcrete carbonates: examples from an Indian Maastrichtian regolith. Sedimentology 48, 339–355.

Tullner, T., Cserny, T., 2003. New aspects of lake-level changes: Lake Balaton, Hungary. Acta Geologica Hungarica 46, 215–238.

Valdeolmillos Rodríguez, A., 2005. Registro paleoclimático y paleoambiental de los últimos 350.000 años en el Parque Nacional de Las Tablas de Daimiel (Ciudad Real). Ph.D. Thesis, Alcalá de Henares Madrid, Universidad de Alcalá, 308 pp.

Valero, B.L., 1993. Lacustrine deposition and related volcanism in a transtensional tectonic setting: Upper Stephanian–Lower Autunian in the Aragón-Béarn Basin, western Pyrenees (Spain-France). Sedimentary Geology 83, 133–160.

Vázquez-Urbez, M., Arenas, C., Pardo, G., 2002. Facies fluvio-lacustres de la Unidad superior de la Muela de Borja (Cuenca del Ebro): modelo sedimentario. Revista de la Sociedad Geológica de España 15, 41–54.

Verrecchia, E.P., 2007. Lacustrine and palustrine geochemical sediments. In: Nash, D.J., McLaren, S.J. (Eds), Geochemical Sediments and Landscapes. Blackwell Publishing, Oxford, pp. 298–329.

Wright, V.P., 1986. The role of fungal biomineralization in the formation of early Carboniferous soil fabrics. Sedimentology 33, 831–838.

Wright, V.P., 1990. Syngenetic formation of grainstones and pisolites from fenestral carbonates in peritidal settings: discussion. Journal of Sedimentary Petrology 60, 309–310.

Wright, V.P., Alonso-Zarza, A.M., Sanz, M.E., Calvo, J.P., 1997. Diagenesis of Late Miocene micritic lacustrine carbonates, Madrid Basin, Spain. Sedimentary Geology 114, 81–95.

Wright, V.P., Azeredo, A.C., 2006. How relevant is the role of macrophyte vegetation in controlling peritidal carbonate facies? Clues from the Jurassic of Portugal. Sedimentary Geology 186, 147–156.

Wright, V.P., Marriott, S.B., 1993. The sequence stratigraphy of fluvial depositional systems: the role of floodplain sediment storage. Sedimentary Geology 86, 203–210.

Wright, V.P., Platt, N.H., 1995. Seasonal wetland carbonate sequences and dynamic catenas: a reappraisal. Sedimentary Geology 99, 65–71.

Wright, V.P., Platt, N.H., Wimbledon, W., 1988. Biogenic laminar calcretes: evidence of calcified root mat horizons in palaeosols. Sedimentology 35, 603–620.

Wright, V.P., Sandler, A., 1994. A hydrogeological model for the early diagenesis of Late Triassic alluvial sediments. Journal of the Geological Society 151, 897–900.

Table 1 Principal characteristics of the main facies formed in fluvial and associated carbonate systems.

Facies	Geometry of deposits	Textural characteristics and components	Sedimentary structures	Identifiable biological content	Common associated facies	Sedimentary environmental context
Intraclastic limestones Li	Lenticular and tabular, cm to dm thick and m to 10s of m in lateral extent.	Rudstones, packstones and grainstones of intraclasts. Rounded to angular, mainly carbonate clasts, mm to dm in diameter; without size evolution, coarsening or fining-upward.	Non-organized or with planar or trough cross-stratification.	Associated phytoclasts.	Packstones of bioclasts, mudstones of phytoclasts and oncoids.	Accumulation from rare floods that cause erosion of previous deposits.
Phytoclastic limestones Lph	Tabular and lenticular, up to several m thick and variable lateral extent.	Rudstones of phytoclasts. Coated stems, cm–dm, exceptionally m, long and up to 20–25 cm in diameter, and leaf moulds/imprints; with or without micrite matrix, and spar cement among them; interior cavities empty or filled with spar or micrite calcite. Phytoclast size may fine upward; long axes of stems can be parallel to stratification or to palaeoflow.	Non-organized or with cross-stratification.	Fragments of stems and leaf imprints.	Boundstones of stems (Lst 1); almost all facies.	Ubiquitous; agitated and slow-flowing areas; barrages and palustrine areas.

Microbioclastic limestones and sands Lmb Sb	Lenticular and tabular, up to 2–3 m thick and m to 10s of m in lateral extent.	Packstones and sands. 'Calcarenite' and sand of bioclasts and intraclasts with scatter small oncoids.	Layers mm–cm thick alternating with layers cm–dm thick of larger phytoclasts. Horizontal and low angle lamination, ripples.	Fragments of charophytes and molluscs.	Boundstones and rudstones of phytoclasts; floatstones of gastropods; coal.	Dammed areas, channels and bars.
Oncolitic limestones Lo	Lenticular, channel-like and tabular, commonly cm up to several m thick and m to 10s of m in lateral extent.	Rudstones of oncoids. Oncoids, cylindrical or elliptic to spheric in shape, mm to several m long. Nuclei of any type of grains, very common of plant stems and tufa fragments. Coatings of light and dark, micrite to spar calcite laminae µm to mm thick.	Non-organized or with planar or through cross-stratification.	Cyanobacteria and algae. Insect tubes.	Packstones of bioclasts. Rudstones of phytoclasts. Stromatolites.	Channel fills. Point, transverse and longitudinal bars. Overbank deposits. Lake shores.
Stromatolites Ls	Ls 1: Tabular, gentle stepped and hemidomic deposits, dm to m thick and m to 10s of m in lateral extent. Ls 2: Planar and undulatory layers; bioherms. Very variable thickness and lateral extent.	Boundstones of microbial communities. Alternating layers µm to 1–1.5 cm thick of different colour, thickness, porosity, crystal size (micrite and spar calcite) and/or microbial components. Planar, undulatory, domed and columnar growth.	Horizontal and undulating lamination.	Cyanobacteria, bacteria and algae.	Boundstones of bryophytes and stems (Lst 1, Lst 2); rudstones of phytoclasts.	Ls 1: Fast-flowing areas on the river bed (low slope zones and cascades). Ls 2: Calm fluvial and lacustrine areas. Inactive areas of waterfalls and caves.

Table 1. *(Continued)*

Facies	Geometry of deposits	Textural characteristics and components	Sedimentary structures	Identifiable biological content	Common associated facies	Sedimentary environmental context
Speleothem limestones Lsp	Variable shape and dimensions, generally m to 10s of m.	Cavities with stalactites, stalagmites, laminated crusts. Micrometre to mm thick laminae of micrite and spar calcite; parallel, sometimes concentric lamination.		Cyanobacteria, bacteria and bryophytes.	Boundstones of hanging stems	Caves and small cavities associated with waterfalls and cascades.
Carbonate breccias B	Commonly irregular masses m to 10s of m wide and long.	Clast-supported, carbonate sandy matrix and calcareous cement. Highly variable clast size; angular blocks, pebbles and clasts, mostly of tufa and associated facies.	Non-organized.	Those of the tufa clasts.	Rudstones of phytoclasts; boundstones of hanging stems and bryophytes; almost all.	Collapse of overlying tufas down into caves; commonly, bedrock is not affected.
Conglomerates and sandstones G and S	Commonly lenticular, channel-shaped, dm to several m thick; locally flat-convex. Metres to 10s of m in lateral extent.	Mainly clast-supported, sandy matrix and calcareous cement; clasts from drainage basin and nearby bedrock, as well as from older tufas; highly variable clast size. Siliceous sandstones, fine to coarse grained; common fining upward.	Structureless or with cross-stratification, horizontal and cross-lamination, ripples.	Phytoclasts.	Rudstones of phytoclasts; mudstones.	Channels and bars of different fluvial style.
Mudstones Fm Fl	Tabular, cm to m thick.	Ochre, red and grey mudstones.	Massive; rare parallel lamination.	Phytoclasts.	Conglomerates, sandstones, microbialite and tufa facies.	Settle-out of fine siliciclastics over floodplains.

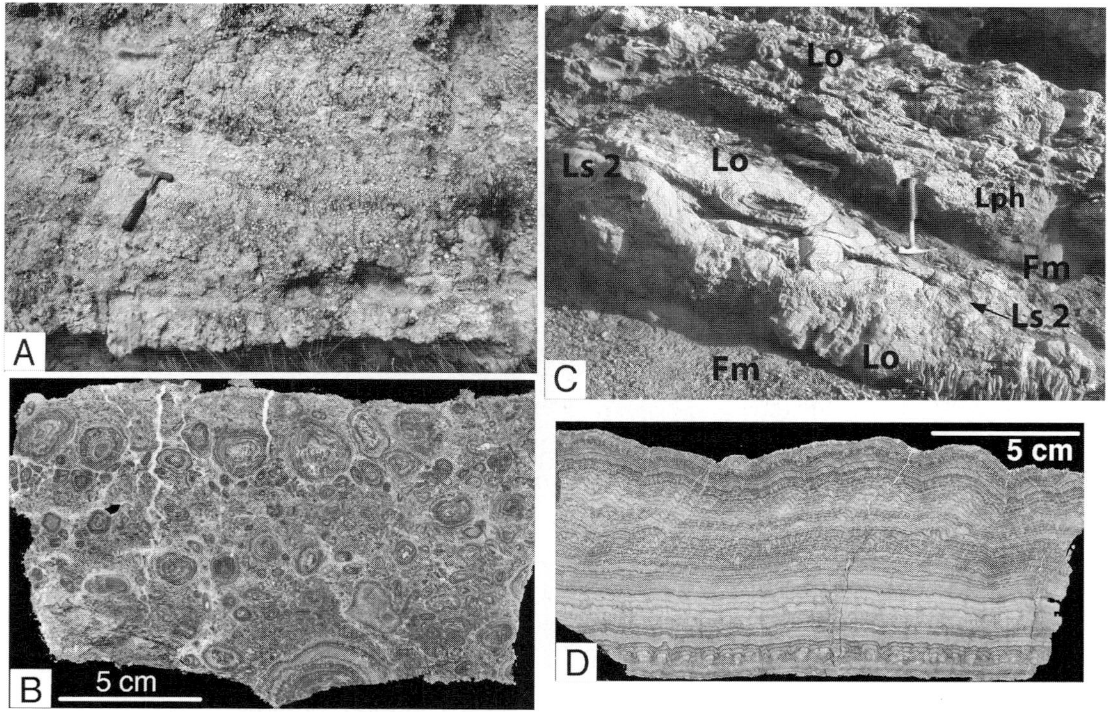

Figure 1 Field views of the most typical fluvial carbonate facies. Facies and sequences indicated are those of Table 1 and Figure 3. Hammer for scale in A, C, G, J and K is 32 cm long. A: Rudstone of oncoids with gentle cross-stratification. B: Rudstone of oncoids (polished section). C: Stromatolites (Ls 2) and large oncolites formed in ponded, calm areas, after phytoclast deposition over mudstone floodplains (sequence c6). D: Polished section of planar stromatolites formed in ponded, calm areas. E: Rudstone of phytoclasts. F: Stromatolitic facies originated in fast flow zones (Ls 1) associated with phytoclastic limestones (sequence a1). G: Boundstones of stems growing upward (Lst 1). H: Boundstone of hanging stems (Lst 2) alternating with boundstone of mosses (Lbr) originated in a waterfall over lacustrine zones with bioclastic limestones. I: Moss mound (Lbr) built as a result of progradation of barrage-waterfall deposits (flow inwards). Person is about 1.6 m tall. J: Detail of stacked moss layers in a prograding barrage. K: Boundstone of charophytes (Lch) originated in ponded, calm zones. The undulating topography is compensated by deposition of bioclastic facies (Lbg, Lbch) (sequence c3). L: Bioclastic limestones and carbonate sands with gastropods alternating with dark, rich-organic layers (facies C). Lens cap is 6 cm in diameter.

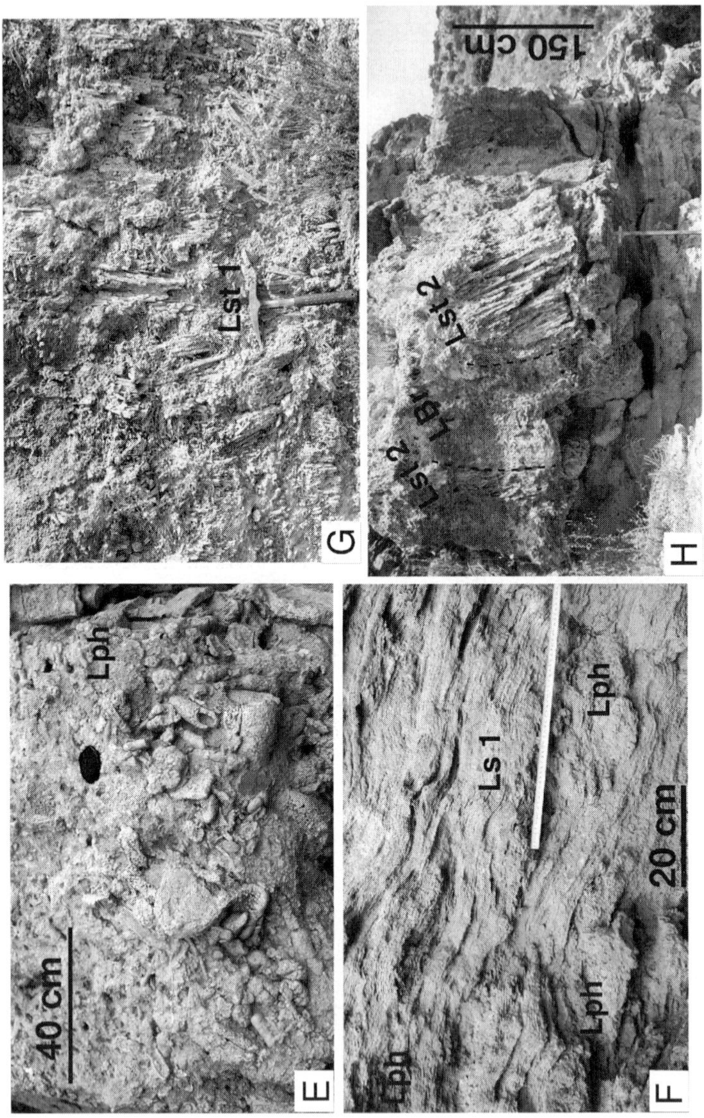

Figure 1 (Continued)

Figure 1 (Continued)

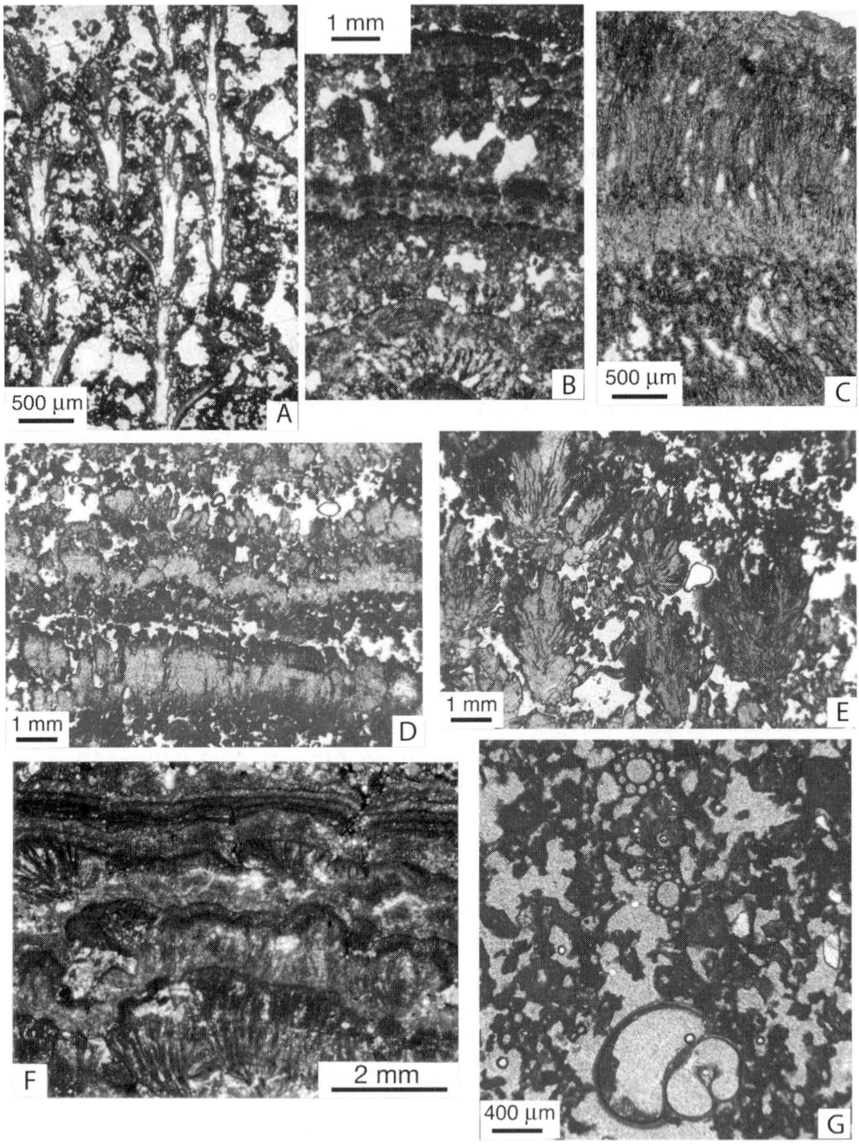

Figure 2 Photomicrographs (optical microscope) of some fluvial carbonate and associated facies. A: Boundstone of mosses. B: Oncoid coating. Note the alternating dense, domed and columnar zones rich in microbial remains, and porous grumulose layers. C: Coating of a stem with filamentous microbial remains within micrite and spar calcite. D: Alternating micrite and spar calcite laminae with highly porous zones, formed in zones of fast-flowing streams (Ls 1). E: Laminae with arborescent spar forms with enclosed bush-like filamentous microbes formed in zones of fast-flowing streams (Ls 1). F: Alternating light (spar) and dark (micrite) laminae of stromatolites deposited in ponded calm areas (Ls 2). Wavy lamination due to bush-like, microbial bodies. G: Bioclastic limestone (mudstone) of charophytes and gastropods.

(e.g. Jones and Renaut, 2009). In some cases, calcite contains Mg, Sr, Ba and other trace elements, and some minerals (e.g. Fe, Mn oxyhydroxides and Ba-sulphates) precipitated during or after deposition may be present in small amounts (Irion and Müller, 1968; Chafetz et al., 1998; Arenas et al., 2000), primarily depending on the characteristics of the related geological formations and on environmental conditions (Matsuoka et al., 2001; Ihlenfeld et al., 2003). Small amounts of quartz, phyllosilicates and dolomite can exist, usually as detrital, allochthonous grains.

Both micrite and sparry calcite with a wide array of crystals forms are commonly present (Rainey and Jones, 2007; Jones and Renaut, 2009). Both the depositional morphology and the size of calcite crystals vary generally in relation to the saturation index of the water and the hydrodynamic conditions (Jacobson and Usdowski, 1975; Braithwaite, 1979; Chafetz et al. 1991a). Chafetz et al. (1991a) found that, near springs, a low saturation index leads to generally well-developed rhombohedra of biologically induced calcite; in contrast, downstream of springs, a high saturation index produces a wider variety of forms (e.g. skeletal and dendritic crystals). Detailed work has been published by Chafetz and Folk (1984), Golubic et al. (1993), Chafetz et al. (1994) and Violante et al. (1994), in which the role of micro-organisms in calcite textural characteristics and lithification are emphasized. In many calcite rhombohedra, the centre is occupied by microbial bodies (preserved as empty holes), which indicates a biotic influence. In addition, some bladed spar crystals encompass upright cyanobacterial filaments. On the other hand, a variety of microbial organisms bore into the spar calcite and micritize it, which is a common diagenetic process (Kahle, 1977; Chafetz et al., 1994).

Table 1 shows the most important characteristics of the main carbonate and associated facies formed in fluvial and related carbonate systems, with the indication of their environmental context. Figure 1 shows some carbonate facies in the field. Facies such as oncoids, stromatolites, coatings of macrophyte stems (Figure 1A–C, G, F) display several micrometre to 1–1.5 cm thick layers distinguished by colour, crystal size, thickness, porosity and/or biological components (Figure 2B–F). The total thickness of the laminated deposits is very variable. Thickness variations of laminae may reflect short-term climate-related changes, as seasonal cycles (Matsuoka et al., 2001; Arenas et al., 2004; Andrews and Brasier, 2005; Kano et al., 2007; Vázquez-Urbez et al., 2009). These facies present a great variety of micro-organism remains that can form characteristic patterns or growth arrangements (Figure 2B–F). Most of these are due to cyanobacterial and algal growth subperpendicular to laminae in the form of isolated or loosely clustered filaments or grouped filaments forming palisades, bush-like or fan-like bodies (Anadón and Zamarreño, 1981; Golubic et al., 1993; Casanova, 1994; Arenas et al., 2007). In many cases, preserved calcite tubes evoke the filamentous micro-organisms that acted as templates. Among the most common cyanobacteria and algae are *Phormidium*, *Calothrix*, *Dichothrix*, *Schizothrix*, *Rivularia*, *Gongrosira*, *Cladophora*, *Oocardium* (Pentecost, 1978; Riding, 1991; Freytet and Verrecchia, 1998; Arp et al., 2001; Hägele et al., 2006). Other micro-organisms like bacteria, diatoms and fungi also contribute to form laminated carbonate deposits.

Diatoms and calcified bacteria are associated with almost all facies. Diatoms can be coated by a fine calcite film. However, their abundance in the fossil record is lower than in modern sediments, possibly due to a low preservation potential.

Boundstones of bryophytes, mostly mosses, can constitute spectacular buildups (Figure 1I) that consist of stacked layers millimetres to centimetres thick (Figure 1J). The coated moss plants are perpendicular to oblique to the accumulation surface, reflecting in some cases the current direction. The plants themselves are not preserved and their spaces are either empty (Figure 2A) or filled with micrite or, more commonly, sparry calcite. In some cases, the coatings present several laminae. Frequently encountered mosses in fluvial carbonate systems are *Bryum*, *Palustriela commutata* (*Cratoneuron*), *Rhynchostegium*, *Eurhynchium*, *Fissidens* (Pentecost, 2005).

The porosity of fluvial carbonate deposits is mainly fabric-selective, for instance due to intergranular, framework, mouldic or burrowing voids. Cementation, aggrading neomorphism and sparmicritization are widespread processes in many fluvial and associated carbonate deposits and can occur very early after deposition; these processes commonly take place by the influence of waters with the same chemical composition as that of the primary precipitates. Sparmicritization involves the presence of microbial organisms that bore into the spar and micritize it (Kahle, 1977; Chafetz et al., 1994). Dissolution and reprecipitation processes can be present depending on the prevailing conditions (see Jones and Renaut, 2009).

4. Facies Associations and Depositional Models

As compared with other carbonate depositional environments, fluvial carbonate systems develop rapidly, as proved by sedimentation rates calculated for recent and Quaternary records. The mean sedimentation rates in several present-day fluvial carbonate systems calculated from thicknesses accumulated on several artificial substrates vary between a few millimetres and 7.5 mm/year (Drysdale and Gillieson, 1997; Arenas et al., 2004; Vázquez-Urbez et al., 2009). Calculations from mass-balance methods in modern systems also indicate high precipitation rates (Emeis et al., 1987; Liu et al., 1995; Merz-Preiß and Riding, 1999; Vázquez-Urbez et al., 2009). High sedimentation rates favour rapid lateral facies shifts, which lead to generally complex vertical arrangements of facies. The resulting successions represent the superposition of laterally related subenvironments due to progradation, aggradation, retrogradation and lateral migration processes. In addition, erosion is a common process that occurs at different scales. In some fluvial systems with geographically restricted basin surfaces or with small transversal development, both lateral changes of facies and erosion events can be of higher magnitude than in larger fluvial systems.

There are few published reports in which vertical sequences of facies are analysed to represent the sedimentary processes that take place in a particular area and their time variations (e.g. Sanz Rubio et al., 1996; Meléndez and Gómez-Fernández, 2000; Vázquez-Urbez et al., 2002; Arenas et al., 2000, 2007). Instead, most contributions express such relationships by means of schemes that include geomorphological features. Pedley (1990), Violante et al. (1994), Ford and Pedley (1996), Carthew et al. (2003, 2006) and Pedley (2009) described a number of sedimentary models based on sedimentological, petrological and geomorphological characteristics. However, as they pointed out, most natural settings are present in more than one model system, so that most models share some facies and depositional environmental settings.

4.1. Vertical sequences of facies

Figure 3 shows some of the most common vertical associations of facies that include fluvial, palustrine and lacustrine deposits. These are actual associations that have been observed in various ancient and recent fluvial and associated carbonate systems, and represent cycles of variation of water level and/or flow conditions (hydrodynamics) and migration of active parts of fluvial systems. Many variations of these sequences exist based on the lack of one or more facies.

On the other hand, some comparable characteristics may be present in sequences formed in different environments. For instance, coarse detrital deposits (conglomerates and/or sandstones) can be present at the base of fluvial, fluvio-lacustrine and fluvio-palustrine carbonate sequences. They imply erosion phases of previous deposits (including bedrock) and generally represent high-energy events that may be associated with incision periods of the fluvial systems (sequences a1, b1, d1, e1).

4.1.1. Sequences a: fluvial s.s. without barrages

Three types of carbonate facies sequences (a1–a3) have been distinguished in fluvial environments without barrages.

The facies succession of sequence a1 represents a channel fill with initial formation and accumulation of oncoids as braided bar deposits, followed by accumulation of phytoclasts from breakage of coated hygrophytes of palustrine areas. This is followed by stromatolite deposits formed in high-energy stretches of the river bed during periods of low phytoclast or

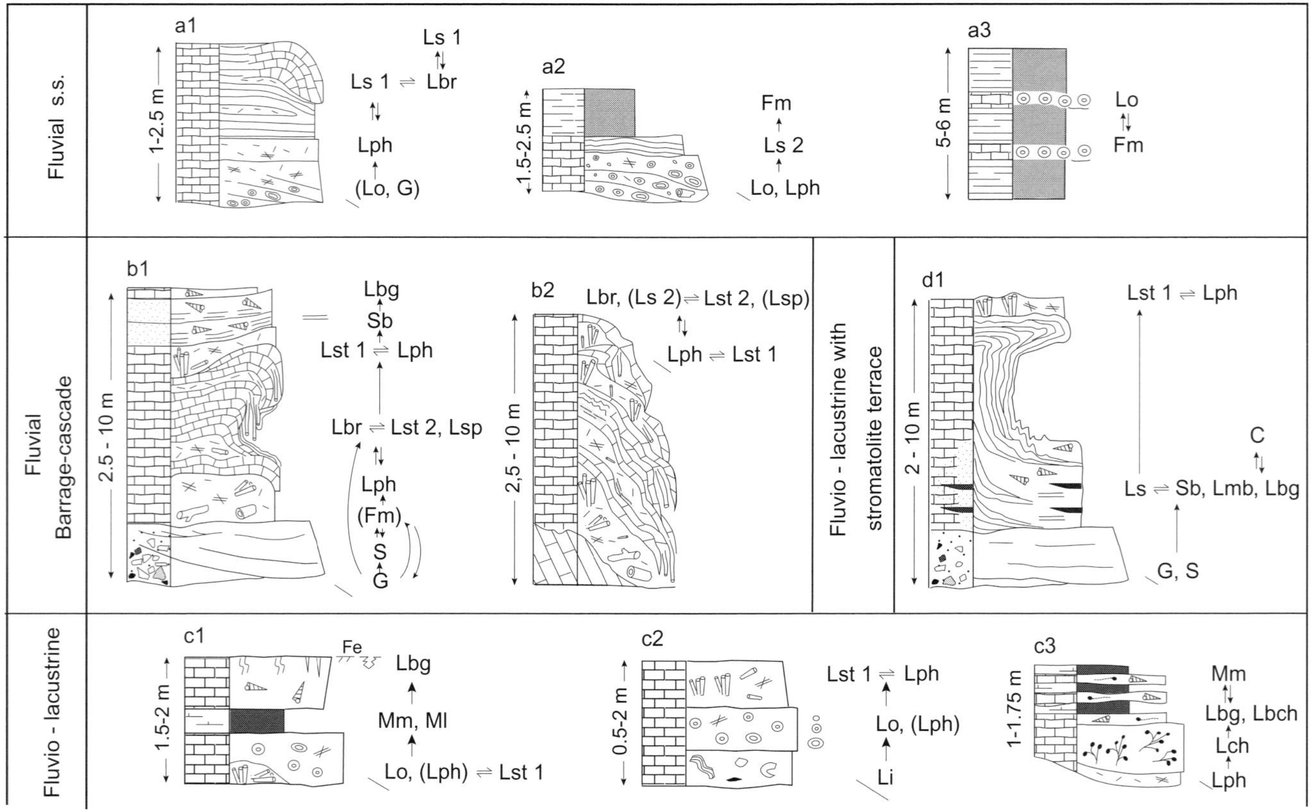

Figure 3 Some of the most characteristic facies sequences: a1, b1, b2, c3, e1, e3 and e4 from Quaternary deposits of the Piedra, Mesa and Añamaza Rivers (NE Spain). Sequences a2, a3, c5 and c6 are from Oligocene deposits in Mallorca (Arenas et al., 2007). Sequences c1 and c2 are from Late Miocene deposits of the central western Ebro Basin (Vázquez-Urbez et al., 2002) and c4 from the central Ebro Basin (modified from Arenas et al., 2000). Sequence d1 is from present-day and Holocene deposits of the Ruidera pool system (interpreted from information compiled from Pedley et al., 1996 and Ordóñez et al., 2005). Sequence e2 is from Late Miocene deposits of the Calatayud Basin, NE Spain (Sanz Rubio et al., 1996). Sequences published with permission of publisher and authors.

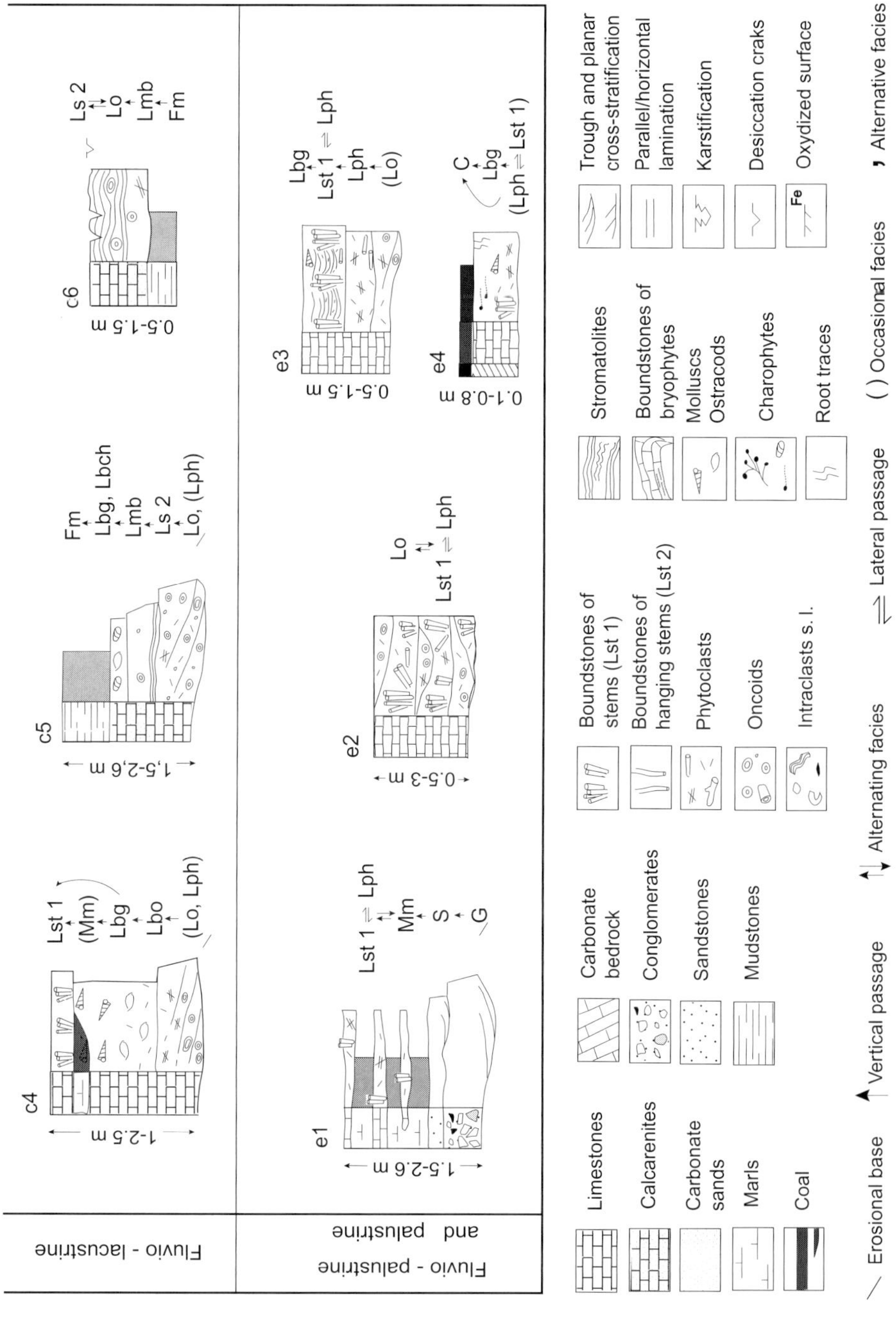

Figure 3 (*Continued*).

in ponded areas in which charophytes, ostracods and/or gastropods dwell (sequences c3, c4 and c5). During high-discharge events, these sediments can be reworked and accumulate prior to the establishment of calm pond conditions during which marls may be deposited (sequences c3 and c4). In fluvial systems with well-developed floodplains, fine siliciclastics are deposited after channel fills (sequence c5).

Dammed fluvial channels and floodplain ponds can be filled with oncoids and stromatolites and then become desiccated (sequence c6).

Sequence c3 has been recognized in Quaternary deposits of the Piedra River, NE Spain (Vázquez-Urbez, 2008). Sequence c4 has been recognized in Late Miocene? fluvio-lacustrine deposits of the central Ebro Basin (Arenas et al., 2000). Sequences c5 and c6 are common in Oligocene deposits of Mallorca, Spain (Arenas et al., 2007).

4.1.4. Sequences d: fluvio-lacustrine with barrages

This sequence type has been interpreted from information compiled from logs and schemes provided by Pedley et al. (1996) and Ordóñez et al. (2005) for some pools of the Ruidera system (Spain), and illustrates the prolonged lacustrine sedimentation that follows an initial deposit of coarse detrital sediments. Lacustrine sedimentation is represented by packstones, carbonate sands and lime mud with charophytes and aquatic gastropods, as well as sapropel. The lacustrine conditions are established as a result of barrier construction across the river valley. The barrages are not similar to other bryophyte-dominated buildups due to the development of stromatolite facies on their vertical upstream edges. This kind of stromatolite facies also occurs on all the lake walls, so that in the sequence this facies refers to both situations. Progradation of the stromatolite margins towards the lake leads to their colonization by plants and somehow is controlled by water-level changes.

4.1.5. Sequences e: fluvio-palustrine and palustrine

Four types of carbonate facies sequences (e1–e4) have been distinguished in fluvio-palustrine and palustrine environments.

Sequence e1 represents a channel fill with coarse detrital sediment during high-discharge events, channel abandonment and finally, establishment of floodplain sedimentation. This includes marl deposition and the development of hygrophytes the stems of which are coated by calcite. Flood stages introduce fine siliciclastic and carbonate sediment, and cause the breakage of coated plants that are deposited in nearby areas. This sequence has been recognized from Quaternary deposits of the Mesa River in NE Spain (Vázquez-Urbez, 2008).

In other situations, very shallow channels with oncoids run through the vegetated floodplain and give rise to sequence e2, which conforms with well-drained palustrine areas. Sequence e2 has been described by Sanz Rubio et al. (1996) from Late Miocene–Pliocene deposits of the Calatayud Basin (NE Spain).

In other cases, the draining channels are filled with phytoclast deposits that present crude stratification; later fill stages can include shallow-pond carbonate sedimentation (sequence e3). Stagnation of water on palustrine floodplains and abandoned channels may cause the spread of microphytes and establishment of anoxic conditions in which sapropelic coal can originate (sequence e4). Floodplains are also loci for peat formation and generation of humic coal. Sequences e3 and e4 have been observed in Quaternary deposits of the Piedra and Mesa Rivers in NE Spain (Vázquez-Urbez, 2008). The features and processes that build these palustrine sequences have been described by many authors from ancient palustrine environments (Buccino et al., 1978; Pedley, 1990; Sanz Rubio et al., 1996; Pedley et al., 2003).

4.2. Sedimentary facies models

The various depositional sedimentary contexts in which fluvial and associated carbonate deposits originate can be grouped into two broad situations: (1) low-gradient, non-stepped fluvial and fluvio-lacustrine conditions, generally with great development of oncoid and palustrine facies and (2) high-gradient and stepped fluvial conditions with typical laminated fluvial and lacustrine facies and moss buildups, and variable development of barrages, waterfalls and dammed areas. Both situations can occur in a single system, although one is usually dominant. Palustrine conditions, present in the two situations, can be the dominant environmental context in the first case (paludal model of Pedley, 1990; Pedley et al., 2003). Deposits close to springs, composed of stromatolites, boundstones of mosses and other macrophytes and bioclastic limestones formed in ponds, can be present in both situations.

4.2.1. Low-gradient, non-stepped fluvial and fluvio-lacustrine conditions

These occur in gently sloping areas without abrupt topographic changes in the longitudinal profile and commonly with a wide transverse extent. Typically these include low-sinuosity and meandering river styles, presenting wide floodplains. Oncoidal facies can be one of the most abundant to form and accumulate as bedforms that give rise to longitudinal, diagonal, transverse and point-bar deposits. Massive oncoid channel fills are also present. Stromatolites appear as crusts and bioherms in shallow conditions, in some cases during shallowing stages of abandoned channels and ponds, and in general represent low-energy conditions and low detrital input.

Palustrine conditions exist on the channel banks, in interchannel areas and on wide floodplains that can be inundated during high-discharge events. In these areas, the submerged parts of plants are coated with calcite and will be transformed into boundstones of up-growing stems. Rudstones of phytoclasts are abundant; these originate from breakage of coated hygrophytes and accumulation of the broken stems in channels and nearby palustrine areas. Plant fragments, coated and uncoated, and molluscs constitute common nuclei for oncoid formation.

Ponds of variable size, isolated or interconnected and/or connected with the channels may exist on the floodplains and are sites for accumulation of lime mud, with the gastropods, pelecypods, ostracods, etc. that live there. Organic-rich sediments can also accumulate in ponds. Boundstones of mosses can form on the river banks in proximity to small local positive relief forms or around springs, but typically are not abundant facies. Some systems present well-developed, extensive floodplains in which fine siliciclastic sediment is deposited and preserved.

Present-day examples of low-gradient, non-stepped carbonate fluvial streams with oncoid formation have been studied in northern Vietnam and Poland (Glazek, 1965), France (Casanova and Lafont, 1985; Verrecchia et al., 1997) and southern Germany (Hägele et al., 2006).

Several ancient deposits that can be interpreted as low-gradient, non-stepped systems have been described from some Cenozoic basins in France (Casanova and Nury, 1989) and in Spain: the Ebro Basin (Zamarreño et al., 1997; Arenas et al., 2000; Vázquez-Urbez et al., 2002), the Madrid Basin (Ordóñez and García del Cura, 1983), the Calatayud Basin (Sanz Rubio et al., 1996) and the island of Mallorca (Arenas et al., 2007). Mäcker (1997) described oncoidal and stromatolitic deposits in fluvial to tidal-influenced deltaic complexes from the Garumnian of the southern Pyrenees that evolved through time to alluvial plains with meandering channels. Among the very few Mesozoic examples are those described by Leinfelder (1985) and Meléndez and Gómez-Fernández (2000). The latter described Cretaceous fluvial systems in the Iberian Chain (NE Spain) in which oncoidal facies accumulated in both meandering and braided channels.

Quaternary examples of low-gradient systems with dominant palustrine conditions have been described from Italy (Buccino et al., 1978) and central Spain (Pedley et al., 2003), in the latter case developed as valley-floor deposits. In this situation, boundstones of stems surrounded by laminated microbial pelmicrites (phytoherm cushions *sensu* Pedley, 1990) pass laterally into microbial laminite facies in peripheral areas (Pedley et al., 2003).

Figure 4 presents a sedimentary facies model for the low-gradient and non-stepped fluvial carbonate system that existed at the end of the Miocene in the Ebro Basin (Vázquez-Urbez et al., 2002; Vázquez-Urbez, 2008).

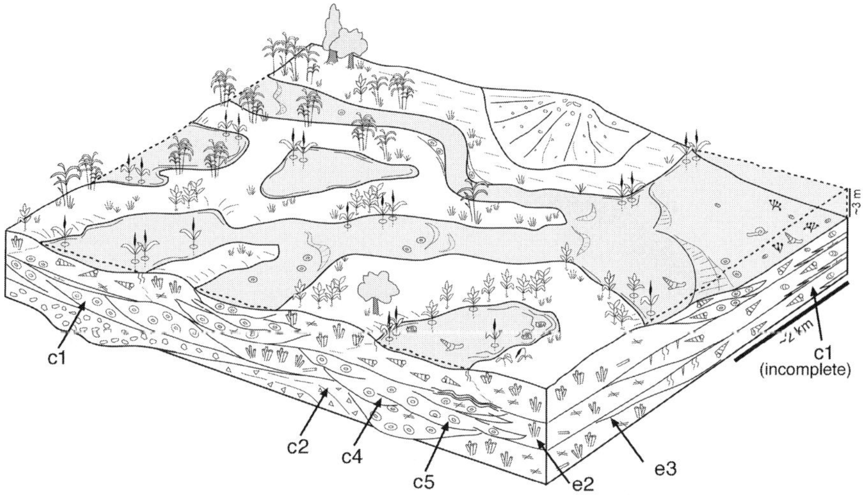

Figure 4 Sedimentary facies model for low-gradient, non-stepped conditions with strong development of oncolitic and palustrine facies. Facies sequences are indicated. Based on Late Miocene deposits of the central-western Ebro Basin (from Vázquez-Urbez, 2008). For legend see Figure 3. Published with permission of author.

This model corresponds to a low-sinuosity multichannel system with laterally related lacustrine and palustrine environments. Oncoidal facies are dominant in channel deposits. Locally some high-sinuosity river stretches originated. This system was connected downward to a permanent shallow lake with facies Mm, Lbg, Lbch and associated Lo on the lake margins (see Table 1 for facies codes). The scheme also shows the simple vertical sequences of facies that predominate in each area. One of the most significant characteristics is the rapid and persistent lateral shift of subenvironments. The system is interpreted to have been developed in a relative depression controlled by local tectonics and probably under humid temperate conditions.

Another example of a low-gradient and non-stepped fluvial system with carbonate deposits is that of the Oligocene on Mallorca, Spain (Arenas et al., 2007) (Figure 5), developed under subtropical conditions. In contrast to the example from the Ebro Basin presented above (Figure 4), the deposits on Mallorca are characterized by the presence of wide floodplains with fine siliciclastic sedimentation, limited development of lacustrine, gastropod-bearing deposits on the floodplains and poor preservation of boundstones of stems. Apart from oncoids of very different size, carbonate fragments (intraclasts) are ubiquitous components of channel fills. Together, this gives evidence for fluvial systems with prolonged intervals of distal (fine-grained) alluvial sedimentation and with a slow lateral mobility of the channels.

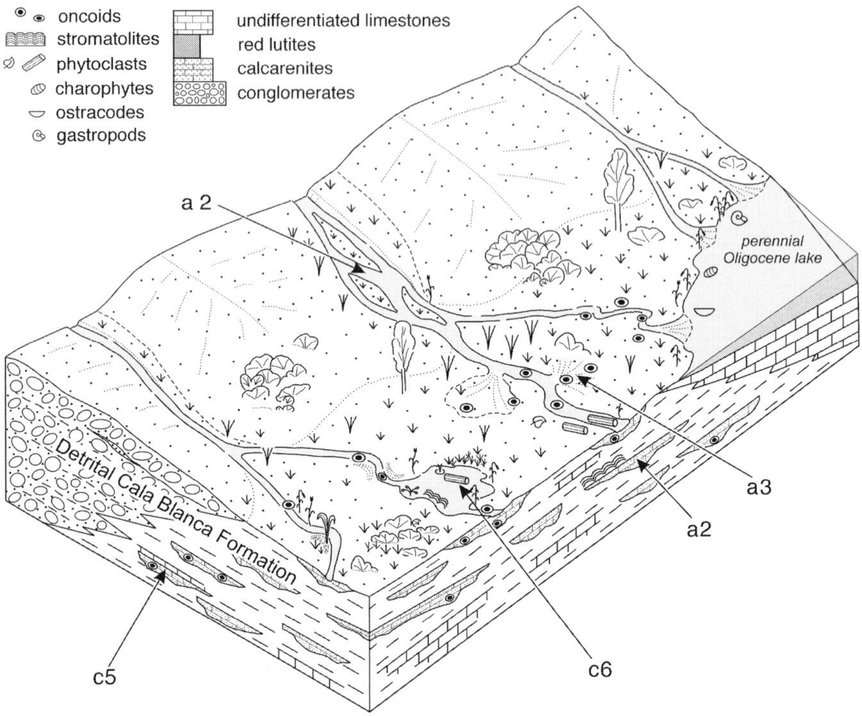

Figure 5 Sedimentary facies model for low-gradient, non-stepped conditions with strong development of oncolitic and stromatolitic facies and mudstone floodplains. Facies sequences are indicated. Based on Oligocene deposits in Mallorca, Spain (modified from Arenas et al., 2007). Published with permission of Elsevier.

4.2.2. High-gradient and stepped fluvial conditions

These conditions occur in some parts of fluvial basins with high slopes or with several topographic breaks along the longitudinal profiles of the rivers, so that a stepped fluvial morphology usually exists. Generally, these conditions occur in restricted basins or valleys with a number of lateral tributaries that present transverse profiles of reduced dimensions which, in part, controls the limited extent of floodplains. Bedrock characteristics (controlled by lithology and/or structure) may determine the origin of some topographic breaks; in other cases, breaks in the longitudinal profile form as a result of aggradation of some facies, e.g. those related to waterfalls and cascades created by dams or barrages.

In high-gradient conditions, deposits of generally reduced extent originate on valley sides from spring waters that run downstream onto steep profiles. Basically, this situation fits the 'perched springline tufa model' of Pedley (1990) and Ford and Pedley (1996). The slope system of Violante et al. (1994) and travertine fans and cliff-drape tufas of Chafetz and Folk

(1984) compare well with this model. Ponded areas with carbonate sands and bioclastic facies and with variable development of palustrine facies exist beside the springs. The outer peripheral parts of these perched deposits commonly consist of steep surfaces that result from cascade and waterfall development. Typically these are formed of bryophyte, phytoclast and phytoherm limestones of hanging stems. Barrages, terraced cascades and lakes can originate downstream. Some Quaternary examples have been studied in Italy, the United Kingdom and Belgium (Violante et al., 1994; Ford and Pedley, 1996), in Spain (Martín Algarra et al., 2003; Pedley et al., 2003) and in NW Australia (Viles and Goudie, 1990b).

Within stepped fluvial carbonate situations, several depositional environmental settings are present, mostly dependent on the topography and biofacies, giving rise to deposits with distinct facies and geometry. These subenvironments are: vertical waterfalls, cascades (gently sloped or stepped waterfalls), barrages, caves behind the overhangs, dammed areas or lakes, gently sloped, low-sinuosity river stretches and palustrine zones.

Energy conditions are controlled by the general slope between successive topographic breaks that determine two situations. The first situation is characterized by the presence of flowing water courses with limited standing water areas. Examples are the Honey and Falls creeks in the Arbuckle Mountains, USA (Chafetz et al., 1991b), the Louie Creek in northern Australia (Drysdale and Gillieson, 1997), the Caerwys and Alport sites in southern United Kingdom (Pedley et al., 1996), the middle course of the Piedra River in NE Spain (Arenas et al., 2004) and Big Hill Springs in Canada (Turner and Jones, 2005). Stromatolitic deposits are typical facies of the fast-flowing stretches between topographic breaks. The second situation is characterized by the development of large, shallow to deep pools or lakes, dammed by barrages. Examples are lakes in Plitvice Natural Park, Croatia (Emeis et al., 1987; Horvatinčić et al., 2000), the Ruidera pools in southeast central Spain (Pedley et al., 1996; Ordóñez et al., 2005), pools and waterholes along some rivers of the Barkly karst in northern Australia (Drysdale et al., 2002; Taylor et al., 2004) and pools along the Havasu Creek, Arizona, USA (O'Brien et al., 2006). Lacustrine facies (carbonate sands, packstones, mudstones, floatstones and marls, in some cases with organic-rich fine sediments) are dominant in areas between barrages.

Boundstones of mosses are among the most abundant and characteristic deposits formed on steep floors. The most spectacular boundstones of mosses are those formed in waterfalls and cascades. These typically form steeply inclined stacked moss layers that may alternate with layers of boundstones of hanging plants and rudstones of phytoclasts. Stromatolites develop in calm situations, such as protected river margins, spray zones and lacustrine margins, and also in fast-flowing fluvial zones, such as gently sloped, low-sinuosity river stretches and cascades. The latter conditions usually result in more

An example of a stepped fluvial system with barrages and several large and deep dammed areas is that of the Ruidera pools in south-central Spain (Pedley et al., 1996; Ordóñez et al., 2005), an active system in a semi-arid region. Up to 15 pools of variable size and depth (12–20 m) exist between the barrages. The barrages are mostly formed of bryophyte and stromatolite layers. The most distinctive feature of this system is the growth of stromatolites on the vertical walls of some pools and the presence of stromatolite terraces. Conical stromatolites dominate the lower part, while laterally accreting mammillary forms appear on the middle and upper walls (see sequence d1 in Figure 3). Marginal hydrophilous vegetation is present on the stromatolite benches. Several cores retrieved Holocene lime mud alternating with organic-rich, clayey lime deposits, which gives evidence of sedimentation in a deep lake. In some Holocene pools, rare charophyte-rich lime muds indicate shallow-lake conditions. Deep-lake conditions existed at least during 4.5 ka at Redondilla Lake, which required the simultaneous development of a barrage to keep water depth. This fluvial system is considered to represent long-term morphological stability (Pedley et al., 1996) under warm, semi-arid conditions.

5. Stable Isotopes of Carbon and Oxygen

The isotopic ratios ($\delta^{13}C$ and $\delta^{18}O$) of fluvial and associated carbonate deposits help in paleoenvironmental interpretation by providing information concerning temperature, humidity, soil and vegetation development, CO_2 source in carbonates and characteristics of the catchment areas (see Andrews, 2006; Deocampo, 2009).

Most fluvial and associated carbonate deposits present a wide range in $\delta^{13}C$ and a much narrower range in $\delta^{18}O$ (Table 2), with exceptions due to particular sedimentary and diagenetic conditions. The covariance between $\delta^{13}C$ and $\delta^{18}O$ is therefore low, particularly in facies originating in river and stream settings, in which $\delta^{18}O$ mostly reflects the ratio between oxygen isotopes in the inflowing meteoric water. Better covariant trends record the effects of evaporation due to longer residence time of water, e.g. in ponds and lakes. Except for that and some laminated facies, there are in general little or no significant differences in $\delta^{18}O$ among the several facies within a fluvial carbonate system. Common $\delta^{18}O$ values are between −8 and −5‰ PDB. The carbon-isotopic composition generally ranges between −4 and −12‰ PDB, although higher values have been reported (Andrews et al., 1997).

The isotopic ratios of fluvial and associated carbonate facies depend on the combination and relative contribution of a number of factors, which are discussed in Sections 5.1 and 5.2.

Table 2 Stable-isotope composition of some ancient and modern fluvial carbonate deposits.

Site and authors	Age/climate	Facies/environment	$\delta^{13}C$ ‰ Pee Dee Belemnitella	$\delta^{18}O$ ‰ Pee Dee Belemnitella
Creeks in Arbuckle Mountains (USA), Chafetz et al. (1991b)	Recent Semi-arid	Sediment on test objects Creeks Stromatolites Creeks	Warm period: −8.21 to −4.99 Mean: −7.0 Cool period: −8.86 to −6.92 Mean: −7.7 Spar lam.: −7.25 to −5.76 Mean: −6.5 Micrite lam.: −7.59 to −6.34 Mean: −7.0	Warm period: −7.48 to −5.47 Mean: −6.5 Cool period: −5.59 to −4.17 Mean: −4.9 Spar lam.: −6.87 to −5.75 Mean: −6.4 Micrite lam.: −5.44 to −4.92 Mean: −5.2
Streams in British Isles (UK), Andrews et al. (1993)	Recent Cool temperate	Stromatolites and oncoids Streams, waterfalls, lakes	Streams: −12.0 to −5.0 Lakes: −6.4 to +3.0	Mean: −5.8 Mean: −4.5 (high standard deviation)
River Piedra at Monasterio de Piedra (Spain), Vázquez-Urbez et al. (2005)	Recent Semi-arid	Stromatolites, mosses, loose sediment Streams and cascades	Warm period: −8.77 to −7.3 Mean: −7.9 Cool period: −8.6 to −7.04 Mean −8.1	Warm period: −9.24 to −8.4 Mean: −8.8 Cool period: −8.4 to −7.15 Mean: −7.75

Table 2. *(Continued)*

Site and authors	Age/climate	Facies/environment	$\delta^{13}C$ ‰ Pee Dee Belemnitella	$\delta^{18}O$ ‰ Pee Dee Belemnitella
Western central Ebro Basin (Spain), Vázquez-Urbez et al. (2002)	Late Miocene Humid temperate?	Oncoids, coated stems, mud-wack of bioclasts, marls Fluvial, palustrine and lacustrine	−8.94 to −7.17	−7.74 to −6.13
Central Ebro Basin (Spain), Arenas et al. (2000)	Late Miocene? Warm and humid?	Boundstones of mosses, coated stems, stromatolites, oncoids; mud-wack of bioclasts Fluvial and lacustrine	−6.81 to −6.06	−7.99 to −7.01 Biomicrites (mean):−5.69 Carbonate marls (mean): −9.24
Mallorca Island (Spain), Arenas et al. (2007)	Middle Eocene and Oligocene Tropical to subtropical	Stromatolites, oncoids, tufas (phytoclasts, bryophytes), biomicrites Fluvial and lacustrine	Eocene: −7.81 to −4.46 Mean: −5.74 Oligocene: −8.65 to −4.87 Mean: −6.71 Eocene tufas: −5.57 to −4.46 Eocene oncoids: −7.81 to −5.70	Eocene: −5.96 to −4.16 Mean: −4.77 Oligocene: −5.87 to −4.87 Mean: −5.24 Eocene tufas: −5.96 to −4.46 Eocene oncoids: −4.46 to −4.32
Eastern Ebro Basin (Spain), Zamarreño et al. (1997)	Late Palaeocene–Eocene Tropical	Stromatolites and oncoids Fluvial and lacustrine	−11.3 (Late Palaeocene) to −4.4 (Middle–Late Eocene) No significant difference between alternating light and dark laminae	−8.5 to −5.2

5.1. Stable carbon isotopes

Catchment areas containing marine carbonate give rise to ^{13}C-enriched HCO_3^- water, whereas organic-matter-bearing formations result in ^{13}C-depleted CO_2 derived from decay of organic matter (Andrews et al., 1993). Variations in $\delta^{13}C$ may also inform about changes in configuration of the drainage areas (i.e. changes in the topography and altitude that control the development of vegetated areas: Zamarreño et al., 1997). In addition, the $\delta^{13}C$ value of fluvial and associated carbonate deposits may reflect changes in the residence time of aquifer water related to water recharge as a function of climate (Garnett et al., 2004; Andrews, 2006). During arid intervals, carbonate deposits may record an increase in $\delta^{13}C$ due to the influence of aquifer water that is isotopically more enriched, resulting from a longer contact and dissolution of aquifer marine carbonates, than during intervals of higher water recharge. Carbon dioxide degassing and in-aquifer calcite precipitation are also favoured during low-recharge intervals, which cause $\delta^{13}C$, Mg/Ca and Sr/Ca to increase in the aquifer water (Ihlenfeld et al., 2003; Garnett et al., 2004).

The isotope composition of the dissolved inorganic carbon (DIC) is controlled by equilibrium reactions between the DIC and the atmospheric CO_2, which lead to a ^{13}C enrichment of the HCO_3^-, by calcite precipitation and by biological activity.

In fluvial carbonate systems, the effects of intense CO_2 degassing near springs cause rapid calcite precipitation and disequilibrium of the carbon-isotope system (Usdowski et al., 1979). The progressive downstream CO_2 degassing, enhanced by high-velocity current conditions, implies a downstream increase of $\delta^{13}C$ values of DIC and carbonate deposits (Pentecost and Spiro, 1990). However, there are several variables that may interfere with this trend along the stream and that may cause the calcite $\delta^{13}C$ composition to lack regular downstream variations (Chafetz et al., 1991b; Chafetz and Lawrence, 1994; Lojen et al., 2004). These are additional surface or underground water recharge along the stream, the presence of areas of different current conditions, and hence variable rates of calcite precipitation, different types of precipitates, changes in soil-CO_2 contribution and photosynthesis. In areas with rapid and turbulent current conditions (e.g. rapids, waterfalls, lake shores), mechanical CO_2 outgassing is enhanced, so that the correlative deposits should reflect slightly higher $\delta^{13}C$ values with respect to areas with slower currents (Arenas et al., 2007).

Photosynthesis and most enzymatic reactions cause organic matter to be ^{12}C-depleted, leading to a ^{13}C enrichment of the remaining dissolved CO_2, and hence of the precipitated carbonates. In addition, the $\delta^{13}C$ composition of plants varies according to their metabolism for atmospheric-CO_2 fixation (see Deocampo, 2009). While most systems with C3 plants of temperate climate have calcite $\delta^{13}C$ values below −8‰ PDB, those with C4 plants in arid and semi-arid regions have calcite $\delta^{13}C$ values between −6 and +2‰ PDB (Smith et al., 2004).

The magnitude of the photosynthetic effects on the $\delta^{13}C$ composition of calcite is still a matter of discussion. The preferential $^{12}CO_2$ uptake through photosynthesis produces an increase of $\delta^{13}C$ values of the DIC and carbonate deposits that generally is considered very small and dependent on environmental conditions (e.g. duration of day) and the type of plants. The effects are more significant in micro-environments dominated by cyanobacteria and some algae (Pentecost and Spiro, 1990; Arp et al., 2001). Pentecost (1987) reported an increase in $\delta^{13}C$ of up to 0.7‰ in the photosynthetic calcites produced during summer by *Rivularia*, relative to bulk calcite. However, Shiraishi et al. (2008) indicated that photosynthesis is not always traceable in the carbon-isotope ratio, and the absence of a photosynthetic signature in carbon isotopes does not imply inorganic precipitation.

Humidity controls the development of flora and soils, so that $\delta^{13}C$ values of fluvial carbonates and associated deposits may reflect changes in this parameter through space and time. Moisture availability affects the rate of soil respiration and induces changes in plant communities, i.e. biomass type and abundance (Andrews et al., 2000; Makhnach et al., 2004; O'Brien et al., 2006). Low $\delta^{13}C$ values are commonly related to a strong development of plants (primarily C3 plants) and/or a high soil-derived CO_2 input during low-aridity periods. This is particularly well recorded by sedimentary systems with dominant palustrine conditions. High $\delta^{13}C$ values in calcite can be associated with either a change towards C4 plants or deeper introduction of atmospheric CO_2 into the soil during arid stages, when soil respiration diminished, leading to an increase in ^{13}C of the soil CO_2 (Andrews, 2006).

Unlike $\delta^{18}O$, $\delta^{13}C$ values of some laminated deposits do not always show a cyclic, seasonal pattern, probably because the carbon isotopes rather reflect a complex combination of the different carbon sources and fractionation processes (Vázquez-Urbez et al., 2005), in which temperature influence may be masked. However, seasonal $\delta^{13}C$ variability in sub-recent fluvial carbonate deposits has been reported by Matsuoka et al. (2001). The good covariance between $\delta^{13}C$ and $\delta^{18}O$ values led them to interpret the seasonal change in $\delta^{13}C$ as linked to temperature. They considered that CO_2 degassing of the groundwater is enhanced in winter due to the higher temperature of the air in the subsurface than in the atmosphere, causing an increase in $\delta^{13}C$ of the groundwater.

5.2. Stable oxygen isotopes

The $\delta^{18}O$ composition of meteoric water varies primarily with latitude and altitude; other factors include intense precipitation in low-latitude areas and seasonal variability in polar and continental settings (Andrews, 2006; Deocampo, 2009). In most fluvial and associated carbonate systems, the

$\delta^{18}O$ water values are quite constant and reflect the stability of groundwater composition (Chafetz et al., 1991a,b; Matsuoka et al., 2001).

Variations in the isotopic composition of meteoric water due to changes in altitude of the catchment areas have been reported to explain $\delta^{18}O$ variations through time. Zamarreño et al. (1997) explained the $\delta^{18}O$ lowering of 2.5‰ PDB from Late Palaeocene to Middle–Late Eocene stromatolites of the Ebro Basin by an increase in altitude of the catchment area of about 1,000 m, which agreed with tectono-geomorphological calculations. These variations also affected the development of vegetation in the catchment areas, which was reflected by a decrease of soil influence on $\delta^{13}C$ through time.

The $\delta^{18}O$ composition of calcite may also reflect different moisture sources that contribute to the $\delta^{18}O$ of the meteoric water recharge (Makhnach et al., 2000, 2004; Andrews, 2006). All together, these factors affecting changes in the $\delta^{18}O$ value of the meteoric water must be taken into account when mid- to long-term variations in water temperature are estimated from the $\delta^{18}O$ composition in calcite, as water temperature may not be the only variable that causes changes in $\delta^{18}O$ values through space and/or time.

Isotopic fractionation effects due to temperature are more significant for oxygen than for carbon, and appear to be the principal factor for short-term $\delta^{18}O$ variations. Some laminated carbonate deposits (e.g. stromatolites) record periodic variations in $\delta^{18}O$ that are related to seasonal changes in temperature (Chafetz et al., 1991b; Matsuoka et al., 2001; Andrews and Brasier, 2005), with the lower values dating from warm intervals (Table 2). In modern systems with a semi-arid climate, sediments of warm and cool seasons present $\delta^{18}O$ differences of about 1–1.5‰ PDB (Chafetz et al., 1991b; Vázquez-Urbez et al., 2005). In subtropical and tropical climates, these seasonal differences in $\delta^{18}O$ are less obvious or even absent (Zamarreño et al., 1997).

Sedimentary environmental conditions, i.e. hydrological, hydrodynamic and biological characteristics, determine the relative influence of factors such as the residence time of the water and hence variable effects of evaporation, rate of CO_2 degassing and influence of soil-derived CO_2 (the two latter were discussed above).

Typically, most fluvial and associated carbonate environments correspond to hydrologically open systems, and therefore have a low residence time of the water with minimal effects of evaporation, so that the $\delta^{18}O$ composition generally reflects the meteoric inflow. This causes the low covariance with $\delta^{13}C$ (Arenas et al., 2000; Leng and Marshall, 2004; O'Brien et al., 2006). However, some ponds on the alluvial plains, abandoned channels and terraced pools along the river profile may exist, at least temporarily, as hydrologically closed depressions; in such situations, the carbonate sediments may record a correlative enrichment in

ACKNOWLEDGEMENTS

This contribution benefited from research carried out for several projects of the Spanish Government and the European Regional Development (PB97-0882-C03-02, REN2002-03575/CLI, CGL2006-05063/BTE) and the University of Zaragoza. The authors thank N. Meléndez, R. Shapiro, E. Ramos, L. Cabrera, C. Osácar, L. Auqué, M.A. García del Cura and S. Ordóñez for their comments and information, as well as Dr. M. Pedley, the editors of this volume and the series editor for their criticism to improve the manuscript.

REFERENCES

Anadón, P., Zamarreño, I., 1981. Paleogene nonmarine algal deposits of the Ebro Basin, Northeastern Spain. In: Monty, C. (Ed.), Phanerozoic Stromatolites. Springer-Verlag, Berlin, pp. 140–154.

Andrews, J.E., 2006. Palaeoclimatic records from stable isotopes in riverine tufas: synthesis and review. Earth-Science Reviews 75, 85–104.

Andrews, J.E., Brasier, A.T., 2005. Seasonal records of climate change in annually laminated tufas: short review and future prospects. Journal of Quaternary Science 20, 411–421.

Andrews, J.E., Pedley, H.M., Dennis, P., 2000. Palaeoenvironmental records in Holocene Spanish tufas: stable isotope approach in search of reliable climatic archives. Sedimentology 47, 961–978.

Andrews, J.E., Riding, R., Dennis, P.F., 1993. Stable isotopic compositions of recent freshwater cyanobacterial carbonates from the British Isles: local and regional environmental controls. Sedimentology 40, 303–314.

Andrews, J.E., Riding, R., Dennis, P.F., 1997. The stable isotope record of environmental and climatic signals in modern terrestrial microbial carbonates from Europe. Palaeogeography, Palaeoclimatolology, Palaeoecololgy 129, 171–189.

Arenas, C., Auqué, L.F., Elías, M.J., Longares, L.A., Osácar, M.C., Pardo, G., Peña, J.L., Sancho, C., Vázquez-Urbez, M., 2004. Depósitos tobáceos actuales y fósiles del río Piedra en el entorno del Monasterio de Piedra (provincia de Zaragoza). Geo-Guías, 1: Itinerarios Geológicos por Aragón. Sociedad Geológica de España, pp. 11–31.

Arenas, C., Cabrera, L., Ramos, E., 2007. Sedimentology of tufa facies and continental microbialites from the Palaeogene of Mallorca Island (Spain). Sedimentary Geology 197, 1–27.

Arenas, C., Gutiérrez, F., Osácar, C., Sancho, C., 2000. Sedimentology and geochemistry of fluvio-lacustrine tufa deposits controlled by evaporite solution subsidence in the central Ebro Depression, NE Spain. Sedimentology 47, 883–909.

Arp, G., Wedemeyer, N., Reitner, J., 2001. Fluvial tufa formation in a hard-water creek (Deinschwanger Bach, Franconian Alb, Germany). Facies 44, 1–22.

Bhaya, D., Schwarz, R., Grossman, A.R., 2000. Molecular responses to environmental stress. In: Whitton, B.A., Potts, M. (Eds), The Ecology of Cyanobacteria. Kluwer Academic Publishers, Dordrecht, The Netherlands, pp. 397–442.

Braithwaite, C.J.R., 1979. Crystal textures of recent fluvial pisolites and laminated crystalline crusts in Dyfed, South Wales. Journal of Sedimentary Petrology 49, 181–194.

Buccino, G., D'Argenio, B., Ferreri, V., Brancaccio, L., Panichi, C., Stanzione, D., 1978. I travertini della bassa Valle del Tanagrio (Campania): studio geomorphologico, sedimentologico e geochimico. Bulletino Società Geológica Italiana 97, 617–646.

Carthew, K.D., Taylor, M.P., Drysdale, R.N., 2003. Are current models of tufa sedimentary environments applicable to tropical systems? A case study from the Gregory River. Sedimentary Geology 162, 199–218.

Carthew, K.D., Taylor, M.P., Drysdale, R.N., 2006. An environmental model of fluvial tufas in the monsoonal tropics, Barkly karst, northern Australia. Geomorphology 73, 78–100.

Casanova, J., 1994. Stromatolites from the East African Rift: a synopsis. In: Bertrand-Sarfati, J., Monty, C. (Eds), Phanerozoic Stromatolites II. Kluwer Academic Publishers, Dordrecht, The Netherlands, pp. 193–226.

Casanova, J., Lafont, R., 1985. Les cyanophycées encroûtantes du Var (France). Verhandlungen der Internationalischen Vereinigung für Theoretische und Angewandte Limnologie 22, 2805–2810.

Casanova, J., Nury, D., 1989. Biosédimentologie des stromatolites fluvio-lacustres du Fossé Oligocène de Marseille. Bulletin de la Societé Géologique de France 5, 1173–1184.

Chafetz, H.S., Akdim, B., Julià, R., Reid, A., 1998. Mn-and-Fe-rich black travertine shrubs: bacterially (and nannobacterially) induced precipitates. Journal of Sedimentary Petrology 68, 404–412.

Chafetz, H.S., Folk, R.L., 1984. Travertines: depositional morphology and the bacterially constructed constituents. Journal of Sedimentary Petrology 54, 289–316.

Chafetz, H.S., Lawrence, J.R., 1994. Stable isotopic variability within modern travertines. Géographie Physique et Quaternaire 48, 257–273.

Chafetz, H.S., Rush, P.F., Utech, N.M., 1991a. Microenvironmental controls on mineralogy and habit of $CaCO_3$ precipitates: an example from an active travertine system. Sedimentology 38, 107–126.

Chafetz, H.S., Srdoc, D., Horvatincic, N., 1994. Early diagenesis of Plitvice Lakes waterfall and barrier travertine deposits. Géographie Physique et Quaternaire 48, 247–255.

Chafetz, H.S., Utech, N.M., Fitzmaurice, S.P., 1991b. Differences in the $\delta^{18}O$ and $\delta^{13}C$ signatures of seasonal laminae comprising travertine stromatolites. Journal of Sedimentary Petrology 61, 1015–1028.

Chen, J., Zhang, D.D., Wang, S., Xiao, T., Huang, R., 2004. Factors controlling tufa deposition in natural waters at waterfall sites. Sedimentary Geology 166, 353–366.

Deocampo, D.M., 2010. On the geochemistry of continental carbonates. In: Alonso-Zarza, A.M., Tanner, L.H. (Eds), Carbonates in Continental Settings: Geochemistry, Diagenesis and Applications. Developments in Sedimentology, Vol. 62. Elsevier, Amsterdam, pp. 1–60.

Drysdale, R.N., Gillieson, D., 1997. Micro-erosion meter measurements of travertine deposition rates: a case study from Louie Creek, northwest Queensland, Australia. Earth Surface Processes and Landforms 22, 1037–1051.

Drysdale, R.N., Taylor, M.P., Ihlenfeld, C., 2002. Factors controlling the chemical evolution of travertine-depositing rivers of the Barkly karst, northern Australia. Hydrological Processes 16, 2941–2962.

Dunham, R.J., 1962. Classification of carbonate rocks according to depositional texture. In: Ham, W.E. (Ed.), Classification of Carbonate Rocks. American Association of Petroleum Geologists Memoir 1, pp. 108–121.

Durán, J.J., 1989. Geocronología de los depósitos asociados al karst en España. In: Durán, J.J., López-Martínez, J. (Eds), El karst en España. Sociedad Española de Geomorfología, Monograph 4, pp. 243–256.

Embry, A.F., Klovan, E., 1971. A late Devonian reef tract on northeastern Banks Island, N.W.T. Canadian Society of Petroleum Geologists Bulletin 19, 730–781.

Emeis, K.-C., Richnow, H.H., Kempe, S., 1987. Travertine formation in Plitvice National Park, Yugoslavia: chemical versus biological control. Sedimentology 34, 595–609.

Evans, J.E., 1999. Recognition and implications of Eocene tufas and travertines in the Chadron Formation, White River Group, Badlands of South Dakota. Sedimentology 46, 771–789.

Ford, T.D., Pedley, H.M., 1996. A review of tufa and travertine deposits of the world. Earth-Science Reviews 41, 117–175.

Freytet, P., Verrecchia, E., 1998. Freshwater organisms that build stromatolites: a synopsis of biocrystallization by prokaryotic and eukaryotic algae. Sedimentology 45, 535–563.

Garnett, E.R., Andrews, J.E., Preece, R.C., Dennis, P.F., 2004. Climatic change recorded by stable isotopes and trace elements in a British Holocene tufa. Journal of Quaternary Science 19, 251–262.

Glazek, J., 1965. Recent oncolites in streams of North Vietnam and the Polish Tatra Mountains. Rocznik Polskiego Towarzystwa Geologicznego 35, 221–242.

Glover, C., Robertson, A.H.F., 2003. Origin of tufa (cool-water carbonate) and related terraces in the Antalya area, SW Turkey. Geological Journal 38, 329–358.

Golubic, S, Violante, C, Ferreri, V., D'Argenio, B., 1993. Algal control and early diagenesis in Quaternary travertine formation (Rochetta a Volturno, Central Apennines). In: Barattolo, F. (Ed.), Studies on Fossil Algae. Bolletino Società Paleontológica Italiana, Special Volume 1, pp. 231–247.

Hägele, D., Leinfelder, R., Grau, J., Burmeister, E.-G., Struck, U., 2006. Oncoids from the river Alz (southern Germany): tiny ecosystems in a phosphorus-limited environment. Palaeogeography, Palaeoclimatolology, Palaeoecology 237, 378–395.

Heimann, A., Sass, E., 1989. Travertines of the northern Hula Valley, Israel. Sedimentology 36, 95–108.

Henning, G.J., Grün, R., Brunnacker, K., 1983. Speleothems, travertines and paleoclimates. Quaternary Research 20, 1–29.

Horvatinčić, N., Calic, R., Geyh, M.A., 2000. Interglacial growth of tufa in Croatia. Quaternary Research 53, 185–195.

Ihlenfeld, C., Norman, M.D., Gagan, M.K., Drysdale, R.N., Maas, R., Webb, J., 2003. Climatic significance of seasonal trace element and stable isotope variations in a modern freshwater tufa. Geochimica et Cosmochimica Acta 67, 2341–2357.

Irion, G., Müller, G., 1968. Mineralogy, petrology and chemical composition of some calcareous tufa from the Schwäbische Alb, Germany. In: Müller, G., Friedman, G.M. (Eds), Recent Developments in Carbonate Sedimentology in Central Europe. Springer-Verlag, Berlin, pp. 157–171.

Jacobson, R.L., Usdowski, E., 1975. Geochemical controls on a calcite precipitating spring. Contributions to Mineralogy and Petrology 51, 65–74.

Janssen, A., Swennen, R., Podoor, N., Keppens, E., 1999. Biological and diagenetic influence in recent and fossil tufa deposits from Belgium. Sedimentary Geology 126, 75–95.

Jones, B., Renaut, R., 2010. Calcareous spring deposits in continental settings. In: Alonso-Zarza, A.M., Tanner, L.H. (Eds), Continental Settings: Facies, Environments and Processes. Developments in Sedimentology, Vol. 61. Elsevier, Amsterdam, pp. 177–224.

Kahle, C.F., 1977. Origin of subaerial Holocene calcareous crusts: role of algae, fungi and sparmicritisation. Sedimentology 24, 413–435.

Kano, A., Hagiwara, R., Kawai, T., Hori, M., Matsuoka, J., 2007. Climatic conditions and hydrological change recorded in a high-resolution stable-isotope profile of a recent laminated tufa on a subtropical island, southern Japan. Journal of Sedimentary Research 77, 59–67.

Leinfelder, R.R., 1985. Cyanophyte calcification morphotypes and depositional environments (Alenquer Oncolite Upper Kimmeridgian? Portugal). Facies 12, 253–274.

Leng, M.J., Marshall, J.D., 2004. Palaeoclimate interpretation of stable isotope data from lake sediment archives. Quaternary Science Reviews 23, 811–831.

Liu, Z., Svensson, U., Dreybrodt, W., Daoxian, Y., Buhmann, D., 1995. Hydrodynamic control of inorganic calcite precipitation in Huanglong Ravine, China: field measurements and theoretical prediction of deposition rates. Geochimica et Cosmochimica Acta 59, 3087–3097.

Lojen, S., Dolenec, T., Vokal, B., Cukrow, N., Mihelèiae, G., Papesch, W., 2004. C and O stable isotope variability in recent freshwater carbonates (River Krka, Croatia). Sedimentology 51, 361–375.

Mäcker, H., 1997. Upper Cretaceous and Lower Tertiary brackish to freshwater oncoids and stromatolites of the Garumnian facies in the Àger Basin (Central Southern Pyrenees, Spain). Facies 36, 246–249.

Makhnach, N., Zernitskaja, V., Kolosov, I., Demeneva, O., Simakova, G., 2000. $\delta^{18}O$ and $\delta^{13}C$ in calcite of freshwater carbonate deposits as indicators of climatic and hydrological changes in the Late-Glacial and Holocene in Belarus. Journal of Geochemical Exploration 69/70, 435–439.

Makhnach, N., Zernitskaja, V., Kolosov, I., Simakova, G., 2004. Stable oxygen and carbon isotopes in late Glacial-Holocene freshwater carbonates from Belarus and their palaeoclimatic implications. Palaeogeography, Palaeoclimatology, Palaeoecology 209, 73–101.

Malusa, J., Overby, S.T., Parnell, R.A., 2003. Potential for travertine formation: Fossil Creek, Arizona. Applied Geochemistry 18, 1081–1093.

Martín Algarra, A., Martín-Martín, M., Andreo, B., Julià, R., González-Gómez, C., 2003. Sedimentary patterns in perched spring travertines near Granada (Spain) as indicators of the paleohydrological and paleoclimatological evolution of a karst massif. Sedimentary Geology 161, 217–228.

Martínez Tudela, A., Cuenca, F., Santisteban, C., Grun, R., Hentzsch, B., 1986. Los travertinos del río Matarraña, Beceite (Teruel) como indicadores paleoclimáticos del Cuaternario. In: López-Vera, F. (Ed.), Quaternary Climate in Western Mediterranean. Universidad Autónoma de Madrid, Madrid, pp. 307–324.

Matsuoka, J., Kano, A., Oba, T., Watanabe, T., Sakai, S., Seto, K., 2001. Seasonal variation of stable isotopic compositions recorded in a laminated tufa, SW Japan. Earth and Planetary Science Letters 192, 31–44.

Meléndez, N., Gómez-Fernández, J.C., 2000. Continental deposits of the Eastern Cameros Basin (Northern Spain) during Tithonian-Berriasian time. In: Gierlowski-Kordesch, E.H., Kelts, K.R. (Eds), Lake Basins Through Space and Time. American Association of Petroleum Geologists, Studies in Geology 46, pp. 263–278.

Merz-Preiβ, M., Riding, R., 1999. Cyanobacterial tufa calcification in two freshwater streams: ambient environment, chemical thresholds and biological processes. Sedimentary Geology 126, 103–124.

Nickel, E., 1983. Environmental significance of freshwater oncoids, Eocene Guara Formation, Southern Pyrenees, Spain. In: Peryt, T.M. (Ed.), Coated Grains. Springer, Berlin, pp. 308–329.

O'Brien, G.R., Kaufman, D.S., Sharp, W.D., Atudorei, V., Parnell, R.A., Crossey, L.J., 2006. Oxygen isotope composition of annually banded modern and mid-Holocene travertine and evidence of paleomonsoon floods, Grand Canyon, Arizona, USA. Quaternary Research 65, 366–379.

Ordóñez, S., García del Cura, M.A., 1983. Recent and Tertiary fluvial carbonates in Central Spain. In: Collinson, J.D., Lewin, J. (Eds), Ancient and Modern Fluvial Systems. International Association of Sedimentologists Special Publication 6, pp. 485–497.

Ordóñez, S., González Martín, J.A., García del Cura, M.A., Pedley, H.M., 2005. Temperate and semi-arid tufas in Pleistocene to Recent fluvial barrage system in the Mediterranean area: The Ruidera Lakes Natural Park (Central Spain). Geomorphology 69, 332–350.

Pazdur, A., Drobrowolski, R., Durakiewicz, T., Piotrowska, N., Mohanti, M., Das, S., 2002. $\delta^{13}C$ and $\delta^{18}O$ time record and palaeoclimatic implications of the Holocene

calcareous tufa from south-eastern Poland and eastern India (Orissa). Geochronometria 21, 97–108.
Pazdur, A., Pazdur, M.F., Starkel, L., Szulc, J., 1988. Stable isotopes of Holocene calcareous tufa in southern Poland as palaeoclimatic indicators. Quaternary Research 30, 177–189.
Pedley, H.M., 1990. Classification and environmental models of cool freshwater tufas. Sedimentary Geology 68, 143–154.
Pedley, H.M., 2009. Tufas and travertines of the Mediterranean region: a testing ground for freshwater carbonate concepts and developments. Sedimentology 56, 221–246.
Pedley, H.M., Andrews, J., Ordóñez, S., García del Cura, M.A., Gónzalez-Martín, J.A., Taylor, D., 1996. Does climate control the morphological fabric of freshwater carbonates? A comparative study of Holocene barrage tufas from Spain and Britain. Palaeogeography, Paleoclimatology, Palaeoecology 121, 239–257.
Pedley, H.M., González-Martín, J.A., Ordóñez-Delgado, S., García del Cura, M.A., 2003. Sedimentology of Quaternary perched springline and paludad tufas: criteria for recognition, with examples from Guadalajara Province, Spain. Sedimentology 50, 23–44.
Peña, J.L., Sancho, C., Lozano, M.V., 2000. Climatic and tectonic significance of Pleistocene and Holocene tufa deposits in the Mijares River canyon, eastern Iberian range, Northeasten Spain. Earth Surface Processes and Landforms 25, 1403–1417.
Pentecost, A., 1978. Blue green algae and freshwater carbonates. Proceedings of the Royal Society of London, Series B 200, 43–62.
Pentecost, A., 1987. Growth and calcification of the freshwater cyanobacterium *Rivularia haematites*. Proceedings of the Royal Society of London, Series B 232, 125–136.
Pentecost, A., 1995. The Quaternary travertine deposits of Europe and Asia Minor. Quaternary Science Reviews 14, 1005–1028.
Pentecost, A., 1996. Moss growth and travertine deposition: the significance of photosynthesis, evaporation and degassing carbon dioxide. Journal of Bryology 19, 229–234.
Pentecost, A., 2005. Travertine. Springer-Verlag, Berlin.
Pentecost, A., Spiro, B., 1990. Stable carbon and oxygen isotope composition of calcites associated with modern freshwater cyanobacteria and algae. Geomicrobiology Journal 8, 17–26.
Pentecost, A., Viles, H.A., 1994. A review and reassessment of travertine classification. Géographie Physique et Quaternaire 48, 305–314.
Pentecost, A., Zhang, Z., 2002. Bryophytes from some travertine-depositing sites in France and the U.K.: relationships with climate and water chemistry. Journal of Bryology 24, 233–241.
Rainey, D.K., Jones, B., 2007. Rapid cold water formation and recrystallization of relict bryophyte tufa at the Fall Creek cold springs, Alberta, Canada. Canadian Journal of Earth Sciences 44, 889–909.
Riding, R., 1991. Calcareous Algae and Stromatolites. Springer-Verlag, Berlin.
Roiron, P., 1997. Apport des flores des travertins à la reconstitution des paléoenvironnements néogènes et quaternaires. Études Géographie Physique, suppl. XXVI, Colloque hispano-français: Milieux carbonatés continentaux, pp. 39–42.
Sanz Rubio, E., Hoyos, M., Cañaveras, J.C., Sánchez-Moral, S., Calvo, J.P., 1996. Caracterización sedimentológica de los sistemas fluviolacustres y tobáceos del Mioceno Superior-Plioceno de la Cuenca de Calatayud (Zaragoza). Geogaceta 20, 277–280.
Shackleton, N.J., Opdyke, N.D., 1973. Oxygen isotope and palaeomagnetic stratigraphy of Equatorial Pacific core V28–238: oxygen isotope and ice volume on a 10^5 years and 10^6 years scale. Quaternary Research 3, 39–55.
Shapiro, R.S., Fox, K., Maxson, J., 2005. Rolling stones and dino bones: oncoids of the Cedar Mountain Formation, Lowe Cretaceous, Utah. Abstracts with Programs, Geological Society America 37 (4), 76.

Shiraishi, F., Reimer, A., Bisset, A., De Beer, D., Arp, G., 2008. Microbial effects on biofilm calcification, ambient water chemistry and stable isotope records in a highly supersaturated setting (Westerhöfer Bach, Germany). Palaeogeography, Palaeoclimatology, Palaeoecology 262, 91–106.
Smith, J.R., Giegengack, R., Schwarcz, H.P., 2004. Constraints on Pleistocene pluvial climates through stable-isotope analysis of fossil-spring tufas and associated gastropods, Kharga Oasis, Egypt. Palaeogeography, Palaeoclimatology, Palaeoecology 206, 157–175.
Taylor, M.P., Drysdale, R.N., Carthew, K.D., 2004. The formation and significance of calcite rafts in tropical tufa-depositing rivers of the Barkly karst, northern Australia. Sedimentology 51, 1089–1101.
Turner, E.C., Jones, B., 2005. Microscopic calcite dendrites in cold-water tufa: implications for nucleation of micrite and cement. Sedimentology 52, 1043–1066.
Usdowski, E., Hoeffs, J., Menschel, G., 1979. Relationship between ^{13}C and ^{18}O fractionation and changes in major element composition in a recent calcite-depositing spring – a model of chemical variations with inorganic $CaCO_3$ precipitation. Earth and Planetary Science Letters 42, 267–276.
Vázquez-Urbez, M., 2008. Caracterización y significado ambiental de depósitos tobáceos del Neógeno de la Cuenca del Ebro. Comparación con ambientes cuaternarios. Ph.D. Thesis Universidad de Zaragoza, Spain.
Vázquez-Urbez, M., Arenas, C., Pardo, G., 2002. Facies fluvio-lacustres de la unidad superior de la Muela de Borja (Cuenca del Ebro): modelo sedimentario. Revista de la Sociedad Geológica de España 15, 41–54.
Vázquez-Urbez, M., Arenas, C., Sancho, C., Osácar, C., Auqué, L., Pardo, G., 2009. Factors controlling present-day tufa dynamics in the Monasterio de Piedra Natural Park (Iberian Ranges, Spain): depositional environmental settings, sedimentation rates and hydrochemistry. International Journal of Earth Sciences, in press.
Vázquez-Urbez, M., Osácar, C., Arenas, C., Sancho, C., Auqué, L., 2005. Variabilidad de la señal isotópica ($\delta^{13}C$ y $\delta^{18}O$) del sistema tobáceo actual del Parque del Monasterio de Piedra (provincia de Zaragoza). Geo-Temas 8, 119–123.
Verrecchia, E.P., Freytet, P., Julien, J., Baltzer, F., 1997. The unusual hydrodynamic behaviour of freshwater oncolites. Sedimentary Geology 113, 225–243.
Viles, H.A., Goudie, A.S., 1990a. Tufas, travertines and allied carbonate deposits. Progress in Physical Geography 14, 19–41.
Viles, H.A., Goudie, A.S., 1990b. Reconnaissance studies of the tufa deposits of the Napier Range, NW Australia. Earth Surface Processes and Landforms 15, 425–443.
Viles, H.A., Pentecost, A., 2007. Tufa and travertine. In: Nash, D., McLaren, S. (Eds), Geochemical Sediments and Landscapes. Blackwell Publishing, Oxford, pp. 173–199.
Viles, H.A., Taylor, M.P., Nicoll, K., Neumann, S., 2007. Facies evidence of hydrodynamic regime shifts in tufa depositional sequences from the arid Naukluft Mountains, Namibia. Sedimentary Geology 195, 39–53.
Violante, C., Ferreri, V., D'Argenio, B., Golubic, S., 1994. Quaternary travertines at Rochetta a Volturno (Isernia, Central Italy). Facies analysis and sedimentary model of an organogenic carbonate system. In: PreMeeting Fieldtrip Guidebook, A1, International Association of Sedimentologists, Ischia'94, 15th Regional Meeting, Italy, pp. 3–23.
Williams, D.F., Thunell, R.C., Tappa, E., Rio, D., Raffi, I., 1988. Chronology of the Pleistocene oxygen isotope record: 0–1.88 m.y. B.P. Palaeogeography, Palaeoclimatology, Palaeoecology 64, 221–240.
Zamarreño, I., Anadón, P., Utrilla, R., 1997. Sedimentology and isotopic composition of Upper Palaeocene to Eocene non-marine stromatolites, eastern Ebro Basin, NE Spain. Sedimentology 44, 159–176.

Pentecost, 1995, 2005; Ford and Pedley, 1996), and they commonly precipitate spectacular deposits of calcite and/or aragonite that reflect, in their textures, the complex interplay between the physical, chemical, and biological factors that governed their formation (Figure 1). Some subaerial, calcareous spring deposits are confined to small, local discharge aprons below the spring vent (Bonny and Jones, 2003; Rainey and Jones, 2007) whereas others, such as the Antalya Travertines of Turkey, are thick (245 m, Glover and Robertson, 2003) and cover an extensive area ($\sim 630\,km^2$, Koşun et al., 2005). Impressive subaerial travertine deposits like those at Pamukkale in Turkey (Figures 1A, B and 2) and Mammoth Hot Springs in Wyoming (USA) are world-famous and visited by many thousands of tourists annually (e.g., Ekmekci et al., 1995). Others discharge subaqueously from vents on lake floors, and remain poorly known until exposed when lake level drops, as for example at Mono Lake (California) and Pyramid Lake (Nevada) in the southwestern USA.

The literature on springs is vast (Waring, 1965, e.g., listed 3,773 references pertaining to springs of all types). Among this literature are historical accounts that provide valuable records of spring behaviour over hundreds of years. Lamoreaux (2005), for example, made note of a Chinese book, "*Annotation on Water Scripture*" by Li Daoyuan, which was published in the second century AD, that described hot springs and recorded their use in 1134 BC. Similarly, the first record of activity of Geysir (eponym for Geyser), located in southwest Iceland, was in 1294 when the chronicle "Oddaverjaannáll" described the effect that earthquakes had on springs in the area (Thoroddsen, 1899; Einarsson, 1967; Storm, 1977).

Modern springs are natural laboratories for assessing the chemical, physical, and biological factors that control the precipitation of calcite or aragonite around their vents and downstream (Figures 1 and 2). Springs are geologically and biologically important because they provide the data and evidence needed to assess: (1) the role of water temperature and pCO_2 conditions in calcite or aragonite precipitation, (2) the factors that govern development of the broad array of crystal morphologies found in spring deposits, (3) the possible roles that plants, microbes, and animals play in calcite and aragonite precipitation, and (4) the early diagenetic processes that may transform the primary fabric of the rock.

Springs are multifaceted systems and their deposits originate through the interaction of many intrinsic and extrinsic factors. Calcareous deposits can develop from waters at many temperatures. Some, including examples in the Kenya Rift Valley (Figure 1E), precipitate from boiling water (e.g., Renaut and Jones, 1997), whereas others, like those around Antalya in southern Turkey (e.g., Koşun et al., 2005), developed from relatively cool waters. Bacteria, fungi, and plants, ranging from microphytes (e.g., eukaryotic algae: Figures 3 and 4) to macrophytes (e.g., grasses,

Figure 1 Calcareous spring deposits. A: Upper part of spring deposit at Pamukkale, Turkey; large, quiet-water terrace pools behind rimstone dams. Note people in upper central part of image for scale. B: Upper part of spring deposit at Pamukkale, Turkey; large terrace pool with no water. Note sediment on floor of pool (Sed), rimstone dam (Rim), and calcareous stalactite curtain (Stal) on front of overhanging rimstone dame. Pool is ~6 m wide. C: Güney Falls, Turkey, ~30 m high, surface covered with growing moss (green) that is being actively calcified. Water temperature ~10°C. D: Modern spring at Lýsuhóll, western Iceland, with discharge apron covered with shallow terrace pools. Spring pool, with water at 45°C in upper central part of image with artificial containing wall. E: Travertine mound around perpetually spouting hot spring (97°C), Loburu, shores of Lake Bogoria, Kenya Rift Valley. F: Calcite rafts (large rafts ~20 cm long) on surface of spring pool, Satono-yu Spa, Japan. G: Calcified gas bubbles (largest bubble is ~1 mm diameter) lying on surface of microbial mat, spring discharge area, Daimaru Ryokan, Japan.

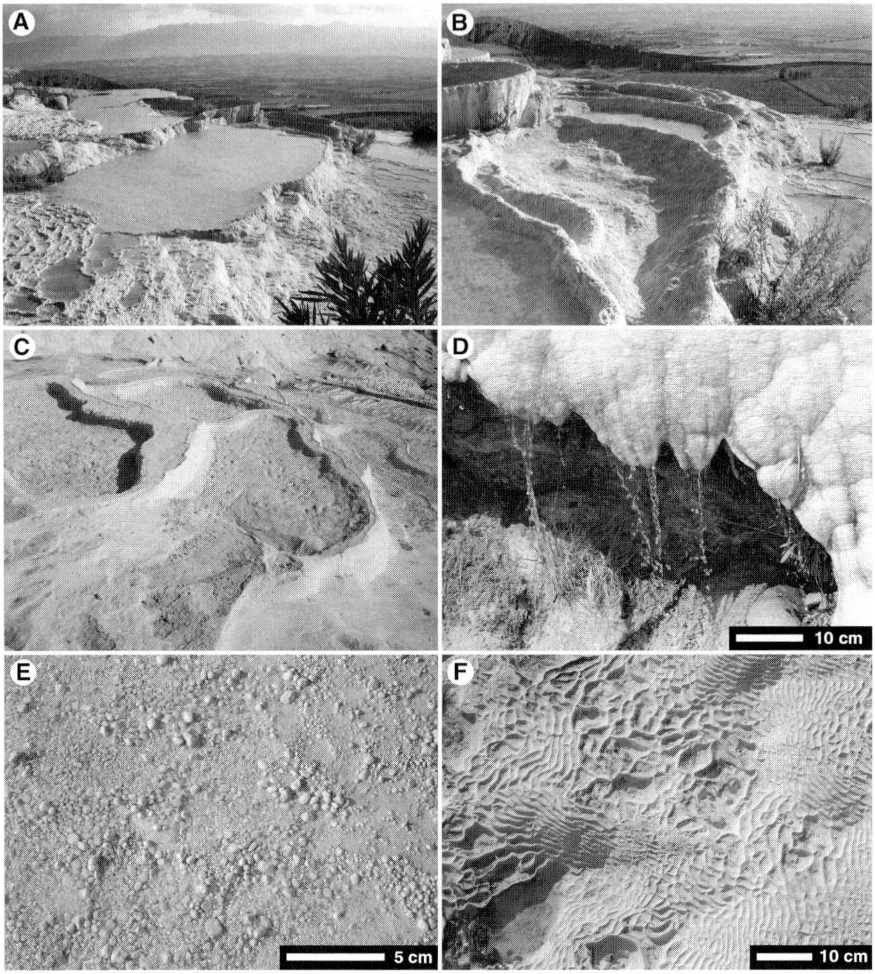

Figure 2 Features of travertine deposits at Pamukkale, Turkey. A: Rimstone dams with pools (main pool ~2 m wide). B: Rimstone dams with empty pools (~1.5 m wide). C: Rimstone dams and pools (~2.0 m wide) with sediment on floors, but no water. D: Water flowing off stalactites growing on terrace front. Note merger of stalactites to form curtain and cave formed behind curtain. E: Coated grains (pisoids or oncoids) on pavement. F: Surface covered with small rimstone dams and small terrace pools.

mosses: Figures 5 and 6) may induce $CaCO_3$ precipitation through their metabolic processes or by acting as templates for abiotic crystal growth.

The present contribution deals with the factors that control carbonate formation in spring settings, comments on the classification of spring systems, provides examples of common facies in calcareous spring systems, and discusses the factors that govern the formation of tufa and travertine in

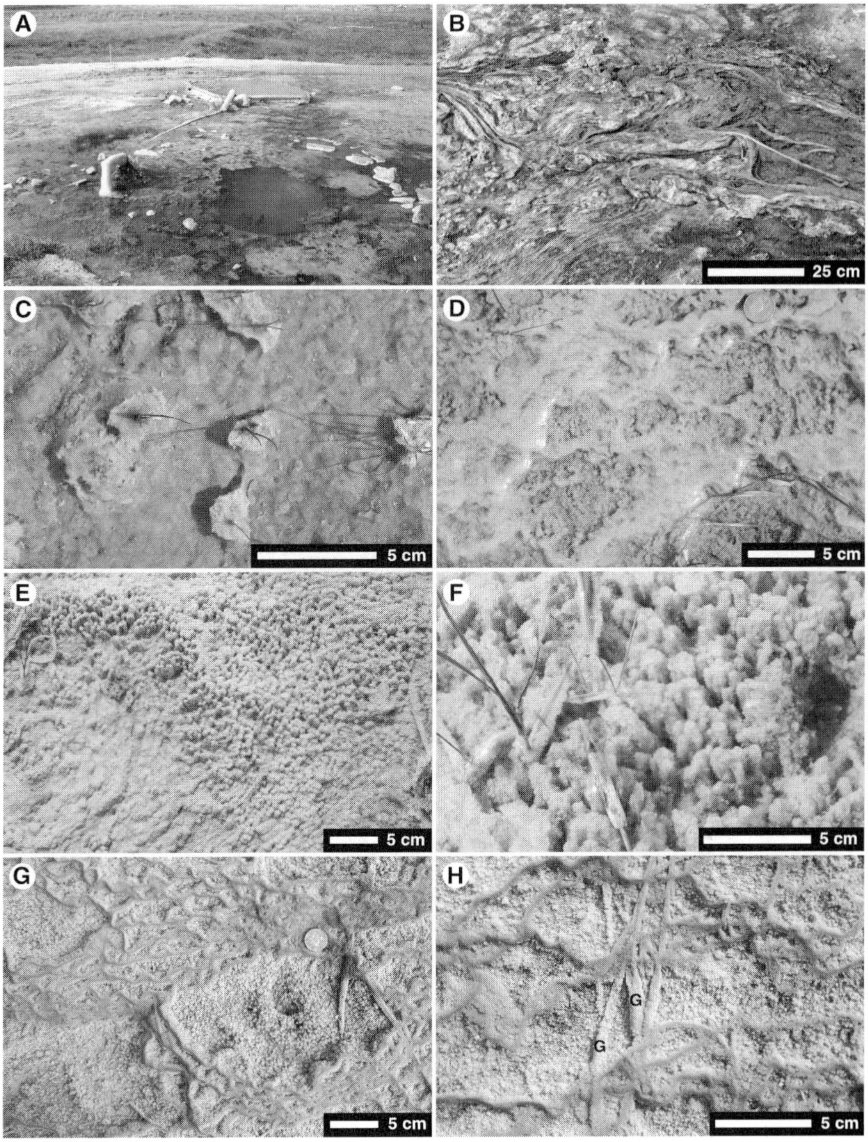

Figure 3 Calcite spring deposits at Lýsuhóll (Snæfellsnes, Iceland). A: Spring pool (temperature of 45°C) and discharge apron. B: Microbial mat formed of long, contorted streamers on upper part of discharge apron. C: Marginal area on upper part of discharge apron with tufts of actively growing grass and surface covered with microbial mat. D: Area slightly downstream of area shown in Figure 3C, showing calcite precipitates partly covered with microbial mat. E: Shallow (<3 cm deep) pool with clusters of calcite dendrite crystals growing on floor. F: Enlarged view of calcite dendrites growing on pool floor. G: Lower part of discharge apron showing low rimstone dams with shallow terrace pools. H: Terrace pools, rimstone dams, and pieces of grass (G) encrusted with calcite.

Figure 4 Spring deposits along the Ichi River, Japan. A: Deposit formed from spring waters flowing down the riverbank (bank ~4 m high). Dark colours due to microbial mats covering surface of calcite precipitates. B: Leaf, microbial mat, and calcite-encrusted gas bubbles on surface of spring deposits. C: Microbial mat and calcite-encrusted gas bubbles on surface of spring deposit.

the diverse spring environments. This overview also highlights some areas in need of further research.

2. Classification of Springs by Water Temperature or Source

Modern active springs have been classified according to their water temperature or the geological source of the spring water.

Classification by water temperature is attractive because it (1) is easily measured in most active springs, (2) is independent of factors such as water source and spring geomorphology, (3) exerts a strong control on the distribution of the biota that inhabit a spring system, and (4) may influence the type of precipitates and facies that form at the vent and on the discharge apron. Terms such as "cold springs", "warm springs", "hot springs", "boiling springs", and "thermal springs" have routinely been used. Nevertheless, such terms have been largely debased through poor usage or lack of precise definition (cf. Pentecost et al., 2003). This debate has existed since Fuchs (1865) discussed the difference between hot springs defined in relative and absolute terms. Much of this debate, as noted by

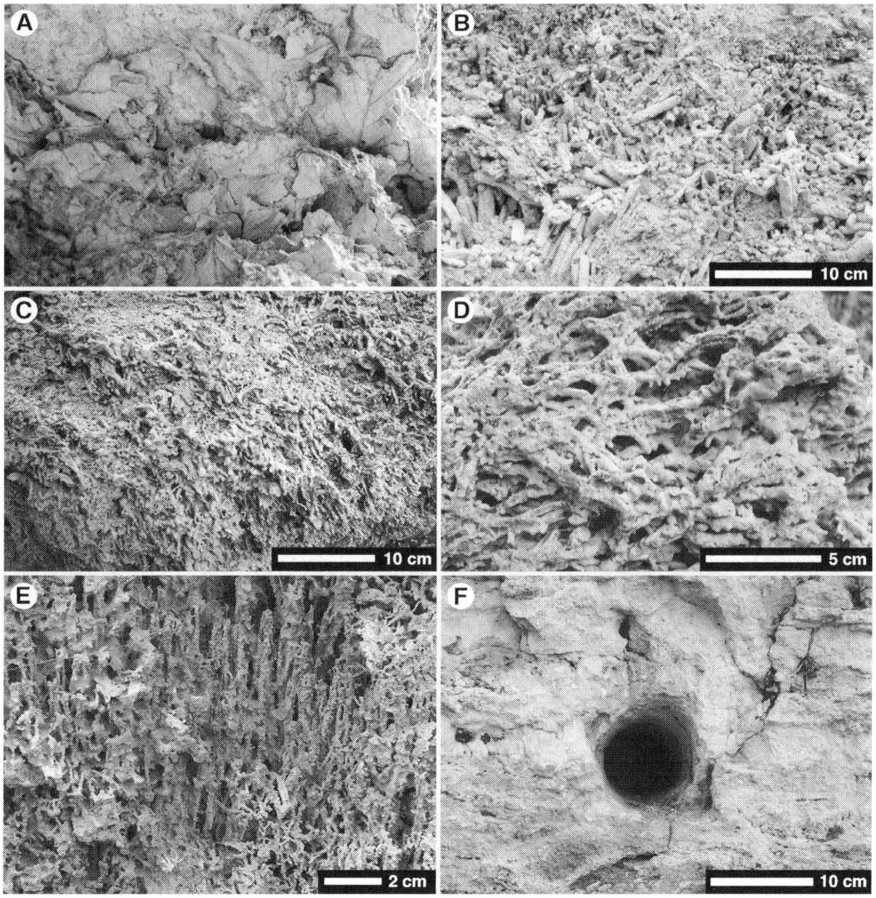

Figure 5 Plants preserved in calcareous tufa. A: Imprint of leaves (~5 cm wide), Caerwys, North Wales. B: Reed stems, Caerwys, North Wales. C–D: Bryophytes preserved by calcite encrustation, Guney Falls, Turkey. E: Bryophyte stems preserved by calcite encrustation, Big Hills tufa, near Calgary, Canada. F: Cast of tree, note deflection of overlying tufa beds over tree, tufa at Fairmont Hot Springs, British Columbia, Canada.

Tuxen (1944), revolves around the use of the absolute temperature of the spring water or the temperature of the spring water relative to the local ambient air temperature. Waring (1965), for example, noted that definitions of a thermal (i.e., hot) spring included: (1) average water temperature notably higher than the local mean annual air temperature, (2) water temperature >20°C, (3) water temperature >15°C above that of the local ambient air, and (4) spring water that does not freeze during winter in cold climates. A further complication is that the location where the water temperature was measured is not always specified (Pentecost

tufa, travertine-tufa, calc tufa, calcareous tufa, plant-tufa, thermal tufa, petrified moss, moss tufa, algal tufa, Oscillatoriaceae tufa, chironomid tufa, sinter, spring-sinter, calcic-sinter, calc-sinter, and sinter crust (e.g., Irion and Muller, 1968; Julià, 1983). Although Pia (1933) advocated the use of the term "sinter" for all (assumed) abiotic deposits that are denser and less porous than tufa, such usage should be avoided. In sedimentology, sinter is usually restricted for spring deposits formed of silica (cf. Ford, 1989) that are precipitated from high-temperature (i.e., high enthalpy) hot springs and geysers. To avoid unnecessary confusion, the use of the term "sinter" for calcareous deposits of all types (including speleothems) is not recommended.

Much of the problem with the usage of tufa and travertine is rooted in the philosophical basis of their definition. These terms have been variously defined to include aspects of water temperature (e.g., Glover and Robertson, 2003), depositional settings (e.g., Emig, 1917; Fouke et al., 2000), the presence or absence of microorganisms (e.g., Riding, 1991), or the hardness and age of the deposits (e.g., Ford, 1989). Koban and Schweigert (1993), for example, suggested that tufa can "... be distinguished from travertine by their framework of incrusted ... plant remains." Ford and Pedley (1996) suggested that "tufa" is a product of $CaCO_3$ precipitation under cool water (near ambient temperature) regimes that typically contain remains of micro- and macrophytes, invertebrates, and bacteria. In contrast, they used the term "travertine" for hard, crystalline deposits that lack macrophytes or invertebrates and formed mainly from hydrothermal waters. Similarly, Pedley et al. (2003) applied the term "tufa" to any carbonate deposit, regardless of crystallinity, which developed through biomediation and/or physiochemical processes under ambient temperature conditions associated with fresh-water discharge. Inherent to this view is the notion that travertine is precipitated from thermal waters. In contrast, Ford (1989) used tufa for all "... the deposits of calcium carbonate in fresh-water streams and lakes ...", whereas travertine "...may be used for the harder, generally stratigraphically older material, used for ornamental stone." Tufa is also commonly used for calcareous precipitates of sublacustrine springs and groundwater seepage zones, including those precipitated around submerged hot springs.

At present, there is little consensus on usage of the terms "tufa" and "travertine" as advocated by Pedley et al. (2003). Andrews (2006), for example, used "tufa" even though Julià (1983) and Pentecost (2005) argued that this term was not needed because the term "travertine" should be applied to *all* continental limestones that have formed around groundwater seepages, springs, and along streams and rivers. Chafetz and Folk (1984) applied the term "travertine" to all of the Italian spring deposits that they considered in their classic paper and further noted that, in the western USA, travertine-depositing springs have a temperature range of 5–95 °C with most being between 10 and 30°C. Similarly, Heimann and Sass (1989) argued that travertine formed from water with a temperature of 5–95°C, but with most being between 10 and 30°C.

Application of the terms "tufa" and "travertine" as advocated by Ford and Pedley (1996) and Pedley et al. (2003) is perhaps reasonable where deposits from cool (<20°C) water deposits are being compared with hot (>80°C) water deposits. Nevertheless, the lack of a definitive temperature division poses problems, especially if that definition also involves comparison with local ambient air temperatures. The problem is accentuated when dealing with older deposits that were produced by springs that are no longer active. In those cases, temperatures can only be inferred from the crystal fabrics and/or preserved biota in the spring precipitates and sediments. Hence, the distinction between tufa and travertine becomes a matter of interpretation. In some respects, these issues parallel the problems that are encountered when classifying springs by water temperature or water source.

4. Aragonite and Calcite Precipitation

Calcareous spring deposits, which are variably formed of aragonite, calcite, and in rare cases dolomite, are characterized by an amazing array of crystal morphologies. The mineralogy and crystallography of these deposits are controlled by a complex set of parameters that may act alone or in unison. Interpretations are commonly confounded by the apparent lack of correlation between mineral species, crystal form, and the physiochemical attributes of the spring water. Part of this problem, however, may be due to the fallacy of assuming that the water currently flowing over a spring deposit is the parent water from which it is formed. In some cases, the two may not be genetically linked. The precipitates might have formed many years before, when the physiochemical attributes of the spring water were significantly different. For example, the ionic composition of the fluids may remain essentially constant, but the pCO_2 may vary with time such that calcite or aragonite only precipitates when CO_2 briefly declines or the water is seasonally warmed. Alternatively, the apparent lack of correlation between the composition of the spring water and the spring deposits may also reflect the complex interplay of environmental factors, some as yet unresolved, that control carbonate precipitation from spring waters.

The reason(s) why calcite precipitates from spring fluids instead of aragonite is still unclear, despite many hypotheses involving temperature control, growth inhibitors, degree of supersaturation with respect to $CaCO_3$ as caused by CO_2 degassing rates and/or evaporation, and impact of plant or bacterial growth (e.g., Kitano and Kawasaki, 1958; Kitano, 1962a,b; Kitano et al., 1962; Busenburg and Plummer, 1986; Renaut and Jones, 1997; Pentecost, 2005). Many suggestions have originated from laboratory experiments where individual parameters can be varied in the absence of others. In natural settings, however, the causes are commonly masked because all factors may operate simultaneously and are difficult to segregate.

4.1. Mineralogy

4.1.1. Water composition

Precipitation of calcium carbonate in spring systems usually occurs when CO_2 is degassed from CO_2-rich groundwater that contains $>2\,mmol\,L^{-1}$ ($\sim 80\,ppm$) Ca^{2+} (Pentecost, 2005). This is not universally true, however, because some springs with extensive carbonate deposits are fed by water with a very low Ca^{2+} concentration. For example, modern calcite is precipitating at Loburu hot springs at Lake Bogoria, Kenya, from fluids with $<2\,mg\,L^{-1}$ Ca^{2+} (Jones and Renaut, 1995, their Table 1), and at Waikite Springs on North Island, New Zealand (Jones et al. 1996, their Table 1), even though their discharged waters contain $<10\,ppm$ Ca. At Loburu, modern precipitation has been confirmed by encrustation of recent artefacts such as beer-bottle caps.

4.1.2. Water temperature

Aragonite precipitation has commonly been attributed to precipitation from waters with temperatures of $>40-45°C$ (e.g., Moore, 1956; Siegel, 1965; Folk, 1994). Kitano (1962a) also favoured a temperature control, but suggested that aragonite would form only if the temperature was $>60°C$. Artificially adding other ions to the water, however, can modify this apparent temperature control (Kitano, 1962b; Kitano et al., 1962). Similarly, agitation of the water, irrespective of its composition and temperature, will increase the rate of CO_2 exsolution that, in turn, may affect the $CaCO_3$ saturation state of the water (Kitano, 1962a). These considerations indicate that the temperature/polymorph relationship may only apply in situations where the parent waters are not strongly agitated (rare in many springs) and do not contain the concentrations of ions or trace elements that appear to influence precipitation of the polymorphs. Calcite precipitates directly from waters with temperatures of $>90°C$ in Kenya (Jones and Renaut, 1995) and New Zealand (Jones et al., 1996), so the assumption that aragonite is favoured by temperatures of $>40-45°C$ does not always apply.

4.1.3. Growth inhibitors

Various elements, when added to water, may promote or inhibit calcite and/or aragonite precipitation. Alkali chlorides, for example, may inhibit aragonite precipitation (Kitano, 1962b), whereas the addition of Sr or Mg can apparently enhance aragonite precipitation (Kitano and Kawasaki, 1958;

Kitano, 1962a,b). Precipitation of aragonite as opposed to calcite has commonly been linked to the Mg/Ca ratio of the parent water. Kitano et al. (1962), for example, were able to induce aragonite precipitation in the laboratory by adding $MgCl_2$ to the parent solution. Similarly, the precipitation of aragonite instead of calcite in lakes and cool springs has commonly been attributed to a high Mg/Ca ratio (Müller et al., 1972; Kelts and Hsu, 1978; Popp and Wilkinson, 1983). The Mg/Ca ratio needed to trigger aragonite rather than calcite precipitation is debatable. Folk (1994) suggested that aragonite will precipitate from any water that has an Mg/Ca ratio $>1:1$, irrespective of water temperature. In springs with very high Mg/Ca ratios (e.g., >20), Mg-carbonates such as hydromagnesite and nesquehonite may precipitate around the vent (Stamatakis et al., 2007).

4.1.4. CO_2 Degassing and saturation levels

Laboratory experiments and interpretations of natural precipitates have shown that changes in supersaturation caused by CO_2 degassing and/or evaporation commonly influence calcite or aragonite precipitation (Branner, 1901; Holland et al., 1964; Folk, 1974; Ishigami and Suzuki, 1977; Cabrol and Coudray, 1982; Chafetz et al., 1991a). The rate of CO_2-degassing from spring water is controlled by: (1) the $p CO_2$ of the spring water when discharged at its vent (Herman and Lorah, 1987, 1988; Chafetz and Lawrence, 1994), (2) the area and depth of water on the discharge apron (Dandurand et al., 1982), and (3) the amount of water agitation and turbulence (Jacobson and Usdowski, 1975; Herman and Lorah, 1988), and (4) aeration, jet-flow, and low-pressure effects (Zhang et al., 2001). Rapid degassing of CO_2 from Ca-bearing spring waters with high $p CO_2$ can produce waters strongly supersaturated with respect to $CaCO_3$, leading to precipitation around the vent. Turbulent flow and agitation over the lips of pools on discharge aprons, for example, commonly causes increased CO_2 degassing that elevates the saturation levels and thereby promotes $CaCO_3$ precipitation at the pool margins. Aragonite precipitation, as opposed to calcite, has generally been associated with waters that have attained very high levels of supersaturation with respect to $CaCO_3$ (Kitano, 1962a; Holland et al., 1964; White and Gundy, 1974; Ishigami and Suzuki, 1977; Arnórsson, 1989).

4.1.5. Microbial growth

Buczynski and Chafetz (1991) suggested that microbial exopolymers (mucus) might dictate which $CaCO_3$ polymorph forms, irrespective of water temperature and composition, because they may inhibit the transfer of ions to developing nucleation centres. Similarly, Guo and Riding (1992) argued that aragonitic laminae in the Rapolano Terme travertines in Italy formed from waters with a temperature $<45°C$ because of associated microbial activity.

4.2. Crystal morphology

$CaCO_3$ precipitation in spring systems produces a vast array of crystal forms that range from minute crystals in micrite to equant spar and microspar mosaics, to large, columnar (palisade) crystal beds, to morphologically complex dendrite and skeletal crystals. Although the term "micrite" has caused considerable debate (Folk, 1959; Friedman and Sanders, 1978; Friedman, 1985a,b; Milliman et al., 1985), it is best used in a non-genetic sense for calcite particles $<4\,\mu m$ long (Milliman et al., 1985; Jones, 1989a), and therefore includes crystals that formed as a sediment (precipitated in the fluid or detrital), as a cement in existing carbonates, and as the breakdown particles of older components (cf. Jones and Kahle, 1995). Varieties of equant spar, which commonly dominate many calcareous spring deposits, are commonly constructed of numerous subcrystals (Sandberg, 1985; Jones, 1989b; Jones and Renaut, 1996a; Jones et al., 2005). Columnar crystals aligned normally to the substrate are common on terraces and sites of runoff, and commonly form beds several centimetres thick, separated in places by discontinuities. Some columnar crystal beds are metres thick (Folk et al., 1985).

Many calcite crystals in travertines show bizarre morphologies, including skeletal crystals (Folk et al., 1985; Jones and Renaut, 1996b). Among the more common forms are dendrites, named for their tree-like morphology (Keith and Padden, 1964; Lofgren, 1974; Jones and Kahle, 1986, 1993; Jones et al., 2000) with multiple levels of branching (Jones et al., 2005; Jones and Renaut, 2008) (Figure 7). With crystals locally >10 cm long, these complex crystals are divided into crystalline and non-crystalline forms. Crystalline calcite dendrites (branches follow crystallographic structures) have been described by Kirchmayer (1987) and Jones and Kahle (1986, 1993), whereas non-crystallographic dendrites (branches do not follow crystallographic structures) have been documented by Jones and Renaut (1995) and Jones et al. (2000). Although dendrite crystals can grow rapidly (Jones and Kahle, 1986), they also develop in spring systems through recurrent annual growth cycles (Jones et al., 2005; Jones and Renaut, 2008). Curiously, the basic architectural design of the dendrite crystals seems to vary from locality to locality, with (to date) no two locales having identical crystal architecture (Jones and Renaut, 2008).

In general, aragonite is less common than calcite in calcareous spring deposits except in hot springs. Typically, acicular (typically $<1\,\mu m$ wide, $200\,\mu m$ long) aragonite crystals are found as sprays, sheaths (dumbbells of Folk et al., 1985), and spherulites (Guo and Riding, 1992; Pentecost, 2005). Needle crystal shrubs have also been reported from some deposits (Pentecost, 1990). Although acicular crystals seem to be the norm, large pseudohexagonal (up to 4 cm long, 4 mm wide) crystals formed of complex nested skeletal crystals, are found in hot-spring travertines at Chemurkeu on

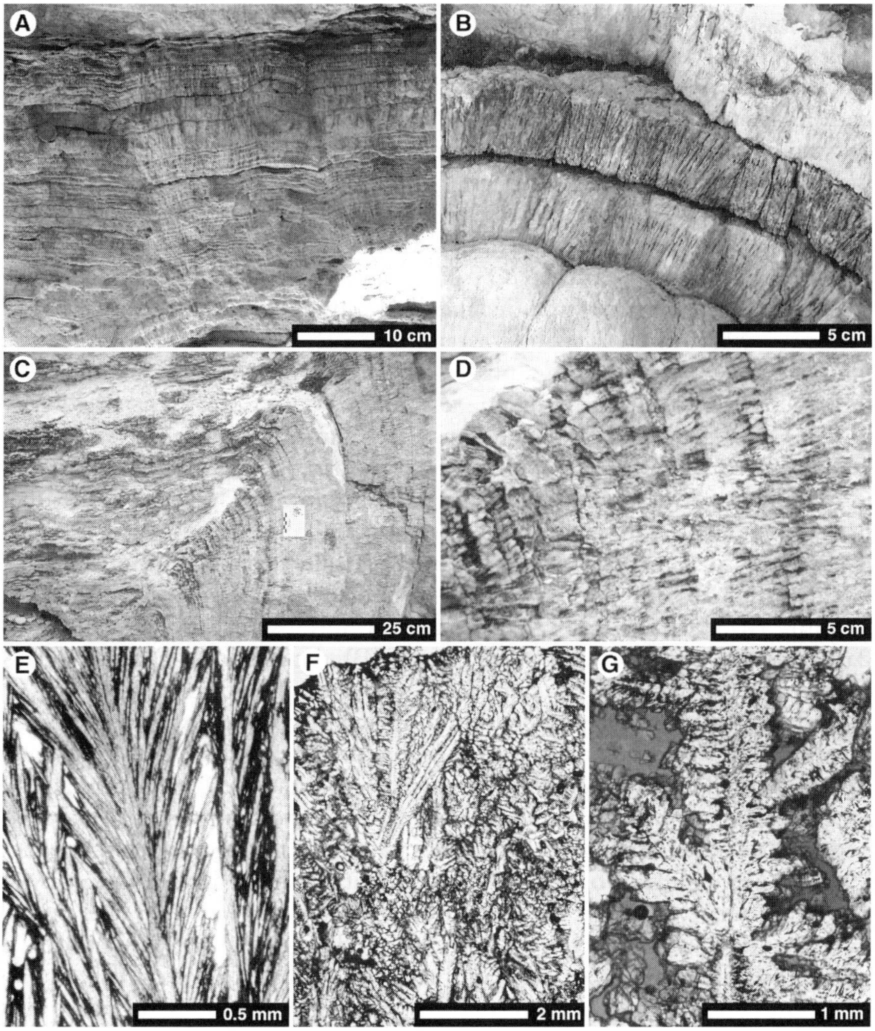

Figure 7 Crystalline calcite travertine deposits. A: Beds formed of calcite dendrite crystals. Fairmont Hot Springs, British Columbia, Canada. B: Beds formed of calcite dendrite crystals, upper part of "roll-over" structure, Fairmont Hot Springs, British Columbia, Canada. C: Vertical section through a travertine deposit showing vertical beds of dendrites crystals on the front of a rimstone dam that pass laterally (to left) into horizontal beds that formed in a pool behind that dam. Clinton travertine, British Columbia. D: Enlarged view of the beds of dendrite crystals from Figure 7C. E–G: Thin-section photomicrographs showing examples of feather dendrite crystals (E) from Loburu hot springs, Lake Bogoria, Kenya, and complex dendrites formed of trilete subcrystals from Lýsuhóll (Snæfellsnes, Iceland) (F, G).

The distribution of these biotic elements throughout a spring system is closely related to environmental parameters such as water temperature, pH, oxygen status (Eh), and/or toxins in the water. Water temperature, for example, is of fundamental importance (Figure 8). Only heterotrophic bacteria and archaea can thrive in hyperthermal spring waters with temperature >75°C, whereas all of the plant groups, bacteria and fungi can survive in areas where the water temperature is <40°C (Figure 4). Cyanobacteria and some eukaryotes can tolerate waters with temperatures of 40–75°C. The distribution of different plants and prokaryotes in a hot-spring system is generally related to the cooling gradient of the spring water as it flows downstream.

Assessing the flora that may have influenced growth and development of spring carbonates is greatly hindered by the fact that plants generally have a low preservation potential, largely because of a lack of hard parts. Nevertheless, the presence of plants is reflected in many different ways, including imprints of leaves (Figure 5A) and hollow tubes (moulds) that formed through calcite being precipitated around plant material that subsequently decayed (Figure 5B–F). Calcification of plant tissue is rare. Some plants, including bryophytes and reeds, actively grow in the spring

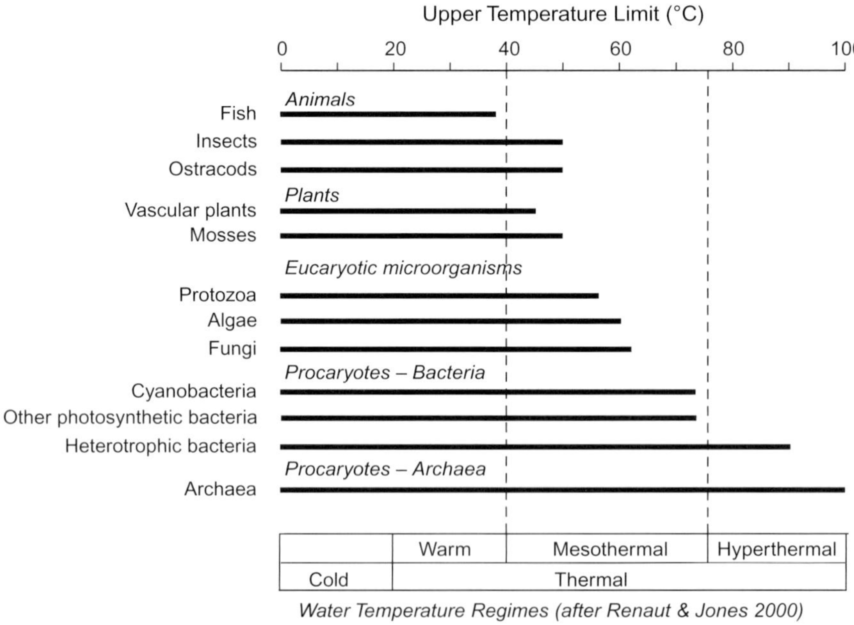

Figure 8 Upper temperature limits for organisms that inhabit spring systems (adapted from Brock, 1994, p. 11) compared to the classification of springs based on the maximum water temperature as measured in the spring vent (from Renaut and Jones, 2000).

waters, whereas other plant material, including leaves and wood fragments, may be washed or blown into the spring waters, or fall from adjacent areas. Tree trunks and branches have commonly had a major influence on spring deposition because they are large enough to influence water flow and hence, the large-scale pattern of carbonate deposits (Figure 5F). Subsequent loss of the organic matter commonly produces elongate round pores that will remain open until filled with internal sediment and/or cement (Figure 5F). Plant material (e.g., wood), if preserved, may provide a means of obtaining ^{14}C dates for young spring deposits, although oxidation and organic decay are rapid in many spring environments.

Our knowledge of the biota in spring systems varies from group to group. The cyanobacteria that inhabit spring systems have been extensively studied, with many taxa having a worldwide distribution (e.g., Pentecost, 2005, pp. 155–166). Spring bryophytes, for example, are quite well known from many different studies (e.g., Gams, 1932; Privittera and Guidice, 1986; Couderc, 1977). Diatom floras are known from both hot and cool springs (e.g., Owen et al., 2004, 2008). Other groups such as fungi are less well known.

Assessment of the bacteria in spring systems is fraught with problems as it depends largely on the approach that is taken to their taxonomic classification. Bacteria have limited morphological variation (e.g., rod-shaped, coccoid, filamentous). Conversely, classifications based on their DNA reveal a far greater diversity. Fouke et al. (2003), for example, used DNA analyses to show that the bacterial assemblages in Mammoth Hot Springs, Wyoming, contained 221 taxa, many of which were unnamed. They argued that a distinct bacterial assemblage characterized each facies in the spring system. This actualistic approach poses significant problems, however, when assessing lithified tufas, because the preserved microbes (if present) will probably display only few morphological features (e.g., size, shape) that will be of limited taxonomic value in terms of extant taxa, and not yield DNA data that can be used in the manner of Fouke et al. (2003). This issue is compounded by the fact that most newly defined microbes are now based largely on their DNA characteristics with only passing attention being given to their morphology. Merging the microbiological assessment of modern living bacterial assemblages with the geological assessments of fossilized bacteria found in tufa or travertine may be impossible to overcome at present simply because the methodologies involved are mutually exclusive.

Tufa and travertine formation has been attributed to both physicochemical (abiotic) precipitation and biogenic precipitation (e.g., Adolphe, 1981, 1991; Adolphe et al., 1989; Pedley, 1992, 1994). The issue of whether or not plants play an active or passive (i.e., indirect) role in calcite precipitation in spring systems has been widely debated (e.g., Ford and Pedley, 1996; Pedley et al., 2003; Kepčija et al., 2006). Active involvement

7. Classification of Spring Systems

No single classification covers the full gamut of spring deposits and none consider those associated with sublacustrine springs or the separation of spring tufa/travertine from fluvial tufa/travertine. Existing classifications tend to treat the tufa of the cool-water realm and travertine of the hot-water realm as separate entities, even though they commonly overlap in sites where hot waters cool downstream. Although many classifications have been proposed for tufa, comparatively few exist for the deposits that have formed in hotter waters. Such classifications are commonly rooted in the spring systems on which they were based and/or the underlying theme of the study for which they were developed.

Most classifications presume that the lithofacies in a tufa reflect factors related to the underlying substrate (Ford and Pedley, 1996). Building on studies by Pedley (1990), Pentecost and Viles (1994), Violante et al. (1994), and Ford and Pedley (1996) developed a scheme with four end members:

- fluvial tufa that encompasses deposits formed in shallow braided streams and the development of arcuate, downstream-facing phytoherm dams that extend from one valley margin to the other;
- perched spring-line tufa with proximal lobate deposits that thicken downslope from a spring on a valley side and with distal deposits formed of sheet-like deposits inclined at a low angle;
- lacustrine tufa formed in large, fresh-water lakes;
- paludal (marsh) tufa that formed on poorly drained terrains.

This classification system, focused entirely on tufas as defined by Ford and Pedley (1996), excludes spring systems where deposits (travertine) formed from hot-water springs and those discharging subaqueously in saline lakes. Discharge aprons with their spectacular rimstone dams and pools (Figures 1A–D and 2A, B), for example, cannot be classified under this scheme. It could be argued, however, that this was not the intent of Ford and Pedley (1996) for they were only classifying tufa, which by their definition would exclude the deposits formed on the discharge aprons below hot springs. Although a viable argument, this fundamental division between tufa and travertine, itself open to debate, precludes an overarching system that incorporates all calcareous spring deposits.

Recently, it has also been noted that the classification system proposed by Ford and Pedley (1996) cannot be applied to all tufa and travertine deposits, because it was modelled on deposits formed in cool temperate climes and does not account for variances in tufas that develop under seasonally humid tropical (Carthew et al., 2003) or hot arid (Viles et al., 2007) climates. Carthew et al. (2003) noted that tufa systems in the humid tropical climate of the Gregory River area of Queensland (Australia) differ from those forming in cool temperate zones because (1) bryophytes grow on inactive dam surfaces rather than on actively accreting surfaces, (2) aquatic insect larvae play a significant role in tufa formation because they cover large areas in the flow zones, (3) oncoids are absent, (4) calcite rafts are abundant, (5) tufa domes develop below dam spillways, and (6) upstream-dipping ramps formed of wedge-shaped masses of tufa-encrusted woody debris and gravel that accumulate behind the tufa dams. Similarly, Viles et al. (2007) argued that tufa systems which formed in hot arid climates like those associated with the tufa of the Naukluft Mountains in Namibia differ from other tufas because they are characterized by downstream-prograding barrage forms with steeply dipping fronts that commonly include cemented gravel facies.

There are no well-established classification schemes for calcareous deposits that formed from hot spring waters. Mesothermal and hyperthermal spring waters will cool downstream and may either become the source for streams or flow into other streams. Yet, there are no established criteria for separating a spring travertine/tufa from a fluvial tufa other than geography and hence, by default, proximity to the spring vent. Thus, the question of how to distinguish between a spring deposit and a fluvial tufa is open to debate. Possible contrasts that may enable distinction and classification include:

- spring deposits form on discharge aprons from point-sourced springs with minimal input from rainfall and runoff, whereas fluvial tufa forms almost entirely from water that comes mainly from rainwater, runoff, and groundwater;
- fluvial tufa forms from water that is generally confined to channels, whereas spring deposits on a discharge apron may form from water that is flowing in shallow channels, as sheetflow, and/or residing in pools, some of which are self-built.

Deposits found in hot- and warm-spring systems have been classified in various ways. In this context, the words "hot" and "warm" refer to water temperature at the spring vent. Such usage, however, may disguise the fact that many deposits formed downslope under much cooler regimes. Deposits in and around the vent might be true hot-water deposits with the characteristics of travertine, whereas the distal deposits might have formed from water with a temperature of $<20^{\circ}C$ and, hence, in the range generally ascribed to tufa formation. This underscores the problems of using the temperature of the water in the vent to describe a spring deposit. Notwithstanding this issue, Chafetz and Folk (1984) classified Italian travertines on the basis of geomorphology into: (1) waterfall or cascade, (2) lake-fill, (3) steeply sloping mounds, fans, or cones, (4) terraced mounds, and (5) fissure ridges. Similarly, Ford and Pedley (1996) made a distinction between fissure ridges and terraces. Fissure ridges form where waters, supersaturated with respect to $CaCO_3$, are brought to the surface along a fault plane or linear fracture, and the resultant precipitates form a linear ridge that follows the trace of the permeable structure (Renaut et al., 2002; Brogi et al., 2005, their Figures 5 and 7; Yanik et al., 2005, their Figures 3 and 4, Table 1). Many discharge aprons associated with warm and hot springs that discharge on valley-side slopes are characterized by successions of pools that formed behind the arcuate rimstone dams (Figures 1A, D and 2A–C).

8. Typical Facies

8.1. Overview

Calcareous springs encompass a wide array of environmental niches that give rise to many different lithofacies and biofacies. Although all are $CaCO_3$ in composition, their fabrics vary with factors such as their mineralogy, crystal morphology, the influence of microorganisms and macrophytes, flow patterns across discharge aprons, streams and in pools, and the hydrological regime – i.e., whether springs are continuously flowing or periodically dry up. Many of these factors are under climatic control. Climate also influences the levels of $CaCO_3$ saturation by controlling temperature, fluid dilution, vegetation, and other factors.

Facies classifications in tufa systems have been based on the processes involved in their formation (e.g., Julià, 1983), the different types of fossil plants that they contain (e.g., Irion and Muller, 1968; Pentecost and Lord, 1988; Pentecost and Viles, 1994), sedimentological criteria (Ordoñez and García del Cura, 1980; Pedley, 1990), and petrographic features (Buccino et al., 1978; Ferreri, 1985; Violante et al., 1994). Ford and Pedley (1996, their Figure 15), however, integrated various elements from each of these schemes into a single classification system that included their division into allochthonous (allogenic) and autochthonous (autogenic) deposits (Table 1). Framed on Dunham's (1962) textural classification of carbonate rocks, the scheme proposed by Ford and Pedley (1996) is flexible enough to allow designation of most facies found in tufa. This scheme also allows the introduction of different facies from different areas while maintaining the fundamental division between allochthonous and autochthonous deposits (e.g., Koşun et al., 2005, their Table 1). Arenas-Abad et al. (2010, this volume) also present a classification of fluvial carbonates based on texture, components, and geochemical aspects.

The classification of facies in travertine deposits, especially those that formed in mesothermal and hyperthermal regimes, has received less attention, and no unifying scheme has yet been proposed. The designation

Table 1 Classification of facies in tufa deposits.

Allochthonous		Autochthonous
Micro detrital tufa	Macro detrital tufa	Phytoherm tufa
Micrite tufa	Oncoid and cyanolith tufa	(A) Boundstone sheets of micrite and peloids (stromatolith-like bacterioherms)
Peloidal tufa	Intraclast tufa	(B) Microherm shrubby framework of bacterial colonies
Sapropelitic tufa (organic rich)	Phytoclast tufa	(C) Framestone true "reef" framework of macrophytic coated with mixed micritic and sparry calcite fringes
Lithoclast tufa (inorganic rich)	Lithoclast tufa	

Source: From Ford and Pedley (1996, their Figure 15).

of a mesothermal or hyperthermal spring refers to the maximum-recorded water temperature at the vent. As that water cools downslope, it may eventually cool below 20–25°C and thus give rise to tufas that are typically associated with cooler waters. At Miette Hot Springs in the Rocky Mountains of Alberta, Canada, for example, relict tufa deposits formed from spring waters that had a temperature of 50–65°C in their vents (Bonny and Jones, 2003). In places downslope, where the waters had cooled sufficiently, stromatolitic tufa formed and, in the cooler more distal parts, bryophyte tufa also developed.

Classification schemes for facies in warm- and hot-spring systems have generally been designed to enable description and discussion in accord with the theme of the study: hence, different schemes use different criteria. Guo and Riding (1998), for example, classified Late Pleistocene travertines at Rapolano Terme (Italy), using a tripartite system that involved lithotypes (e.g., shrub travertine, pisoid travertine), depositional systems (e.g., slope depositional system), and facies (e.g., terrace slope facies, smooth slope facies, waterfall facies). In some cases, the facies were further subdivided; for example, the "Terrace Slope Facies" in the "Slope Depositional System" was divided into terrace walls, terrace pools, and terrace rims. Fouke et al. (2000, 2003), in their analysis of Angel Terrace in Mammoth Hot Springs, Wyoming, defined five facies based on their fabrics and distance from the vent (e.g., vent facies with aragonite botryoids, distal slope facies with dendrite calcite crystals).

8.2. Common facies in calcareous spring systems

8.2.1. Plant and microbial tufa/travertine

Plants and prokaryotes are common in many spring systems, providing the water temperature does not exceed their maximum tolerance levels. Where water supply is maintained with more or less constant temperature and composition, archaea, bacteria, cyanobacteria, green algae, diatoms, fungi, lichens, bryophytes, liverworts, vascular plants (including ferns, herbs, trees) commonly thrive (Pentecost, 2005). Such plants, fungi, bacteria, and archaea play varied roles in the development of spring deposits.

Plants that act as substrates for $CaCO_3$ precipitation commonly give rise to highly porous tufa deposits with primary pores of variable size and shape being generated through their growth patterns or formed once the plants had decayed. Calcification (permineralization) of the plant material seems to be rare. Tufas of this type, which are commonly named after their formative plant, include, for example, bryophytic tufa (Figures 1C and 5C, D), moss tufa (e.g., Emig, 1918; Irion and Muller, 1968; Weijermars et al., 1986; Zhang and Pentecost, 2000; Pavlović et al., 2002; Rainey and Jones, 2007), and reed travertine (Guo and Riding, 1998; Viles et al., 2007).

Consortia of microbes (e.g., cyanobacteria, bacteria, fungi, green algae) with their associated biofilms (Figures 3B–D and 4B, C) commonly mediate stromatolite growth by providing sites for calcite or aragonite nucleation and/or trapping and binding $CaCO_3$ grains to their growth surfaces (Love and Chafetz, 1990; Pedley, 1993; Pedley et al., 1996; Das and Mohanti, 1997). Some stromatolites in calcareous springs have been named after their dominant microbe; Irion and Muller (1968), for example, used the terms Oscillatoriaceae tufa and *Schizothrix* tufa.

8.2.2. Insect tufa/travertine

Many insects inhabit spring systems (Pentecost, 2005). Some insects, like the chironomids, create feeding burrows that are commonly preserved in carbonate deposits (Pentecost, 2005). The chironomid larvae (typically ~5 mm long, 0.3 mm wide) spin a tubular netting that soon becomes calcite encrusted. In some places, large numbers of these tubes (up to 15 mm long, 0.3–0.6 mm wide) become concentrated in small areas and thereby form highly porous tufa (Pentecost, 2005).

In seasonally wet, humid tropical areas, such as those along the Gogol River in northern Papua New Guinea and the Gregory River in Queensland, Australia, larvae play a significant role in tufa formation (Humphreys et al., 1995; Drysdale and Gale, 1997; Drysdale, 1998, 1999, 2003; Carthew et al., 2003). Formed primarily by the caddis fly larvae (Trichoptera), the insect-constructed tubes, 1–2 mm in diameter with a wall up to 1 mm thick, are formed of discontinuous micrite layers

(Humphreys et al., 1995). Hydropsychid caddis fly are a major biotic component of travertine formation in Venezuela (Paprocki and Holzenthal, 2003). Similarly, Kepčija et al. (2006) showed that the silk pads produced by blackfly provide substrates for calcite precipitation on tufa barriers in the Plitvice Lakes of Croatia.

8.2.3. Shrub tufa/travertine

Calcitic shrubs, generally < 3 cm high, are characterized by an upward-expanding, irregular dendrite morphology (Chafetz and Meredith, 1983; Chafetz and Folk, 1984; Guo and Riding, 1994, 1998; Chafetz and Guidry, 1999) and typically grow in terrace pools and depressions (Guo and Riding, 1998), especially in sulphurous hot springs. Such shrubs, formed of micrite and spar-rhombs or needle crystals (Guo and Riding, 1994, 1998), have been attributed to abiogenic (Pentecost, 1990) and biogenic (Chafetz and Meredith, 1983; Chafetz and Folk, 1984; Chafetz and Guidry, 1999) processes. Mn- and Fe-rich black shrubs have been found in hot-water travertines in southern Morocco (Chafetz et al., 1998).

8.2.4. Coated grains

Coated grains (Figure 2E), characterized by a nucleus encased by cortical laminae, have been variously classified in terms of their size (e.g., ooids < 2 mm diameter vs. pisoids > 2 mm in diameter) and their biogenicity (e.g., ooids and pisoids being abiogenic vs. oncoids and rhodoids being biogenic) (Peryt, 1983a, his Table 2). Nevertheless, the terms applied to coated grains formed in spring systems do not always conform to such distinction. For example, coated grains up to 5 cm in diameter have been called "ooids" (Richter and Besenecker, 1983) and the term "pisoid" has been applied to coated grains to which a biogenic origin was ascribed (e.g., stromatolitic mammillated pisoids of Risacher and Eugster, 1979; bacterial pisoliths of Guo and Riding, 1998). In cases where the origins of coated grains are debatable, several terms could be used. Risacher and Eugster (1979), for example, applied the term "pisolith" to the large, coated grains found in hot-spring pools at Laguna Pastos Grandes (Bolivia), whereas Jones and Renaut (1994) noted that they could equally be called "ooids", "vadoids", or "oncoids", depending on the importance placed on their biogenicity and depositional setting. This variable application of terms must be taken into account when assessing the distribution and origin of coated grains in spring systems.

Ooids are known from cool and warm (Peryt, 1983b; Richter and Besenecker, 1983) as well as from mesothermal and hyperthermal springs (Jones et al., 1996; Alçiçek et al., 2005). The ooids described by Richter and Besenecker (1983), formed of aragonite and up to 5 cm in diameter,

and Gandin, 2005). These millimetre-thick rafts, also termed "hot water ice" (Allen and Day, 1935) and "calcite ice" (Bargar, 1978), are akin to those in cave pools (González and Lohmann, 1988; Chafetz et al., 1991a).

The upper surfaces of the rafts are flat (Weed, 1889), whereas the lower surfaces are dentate because the crystals grow downwards into the water. Disturbance of the surface water commonly breaks the delicate calcite precipitates into rafts of various sizes and shapes. With ever-increasing density due to preferential crystal growth on their lower surface, the rafts eventually sink to the bottom of the pool (e.g., Jones, 1989a; Taylor et al., 2004). There, the rafts settle at variable angles, depending on how they sank through the water column and on the consistency of the substrate.

Calcite rafts are common in the tufas found along rivers in the seasonally wet, humid climes of northern Queensland (Taylor et al., 2004). Filamentous cyanobacteria and diatoms, which commonly colonize those rafts, are typically calcified and/or entombed by continued growth of the calcite crystals. Loss of the cyanobacteria and/or diatoms through decay produces distinctive elongate biomoulds.

8.2.9. Coated bubble tufa/travertine

Gas bubbles associated with spring waters are commonly encased with precipitated calcite and/or aragonite (Kitano, 1963; Schreiber et al., 1981; Chafetz and Folk, 1984; Chafetz et al., 1991a; Koban and Schweigert, 1993; Guo and Riding, 1998). The gas bubbles are produced by microbial activity in the mats or sediment on the floors of spring pools (Figures 1G and 4B, C) or, in some hot springs by CO_2 or steam produced by pressure release or shallow boiling. These extremely delicate coated bubbles are easily broken and will be preserved only if they are trapped and protected in some way. Spherical gas bubbles, typically 1–4 mm in diameter, may be rapidly coated with a micrite layer that is encased by a thin layer of euhedral aragonite crystals (Reis, 1926; Chafetz et al., 1991a).

Rocks formed of masses of coated bubbles have been termed "bubble limestone" (Chafetz and Folk, 1984), "lithified bubbles", "honeycombe rock", or "foam rock" (Chafetz and Folk, 1984). Tubes formed of amalgamated, vertically elongated bubbles are commonly found in sediments that accumulated on pool floors (Guo and Riding, 1992, 1998).

8.2.10. Crystalline travertine

Many spring deposits are characterized by large areas of crystalline calcite or aragonite that formed through rapid precipitation from fast flowing waters (Guo and Riding, 1992, 1998). The attitude of these layers conforms to the underlying topography and therefore ranges from (sub) horizontal on terraces to vertical or overhanging on terrace fronts (Figure 7A–D).

The elongate crystals in these crusts typically have their long axes at right angles to the growth surface. Younger crusts are commonly soft and easily crushed, whereas older crusts are typically very hard and resistant (Chafetz and Folk, 1984).

The crystalline calcite layers are formed of a morphologically diverse array of crystals, including ray crystals (Folk et al., 1985; Guo and Riding, 1992), feather crystals (Koban and Schweigert, 1993, their plate 48 and Figure 2; Pentecost, 1994; Jones and Renaut, 1995), scandulitic dendrites (Jones et al., 2000), trigonal dendrite crystals (Jones et al., 2000), complex dendrites formed of trilete subcrystals (Jones et al., 2005) or subcrystals of variable morphology (Jones and Renaut, 2008). Although commonly associated with hot springs ($>80^{\circ}C$), dendrites also form in cold-water spring systems (Turner and Jones, 2005). Jones and Renaut (1995) suggested that the morphology of the dendrite crystals is related to a disequilibrium factor (Figure 9) that depends on various parameters, including the $CaCO_3$ saturation level. Many dendrite crystals appear to be the product of rapid precipitation from waters that are supersaturated with respect to calcite because of rapid CO_2 degassing (Jones et al., 2005). Nevertheless, evidence from some localities indicates that seasonal changes in the CO_2 degassing rates causes dendrite growth to temporarily cease, thereby allowing the development of thin opal-A laminae (Jones et al., 2005) or growth lines (Jones and Renaut, 2008).

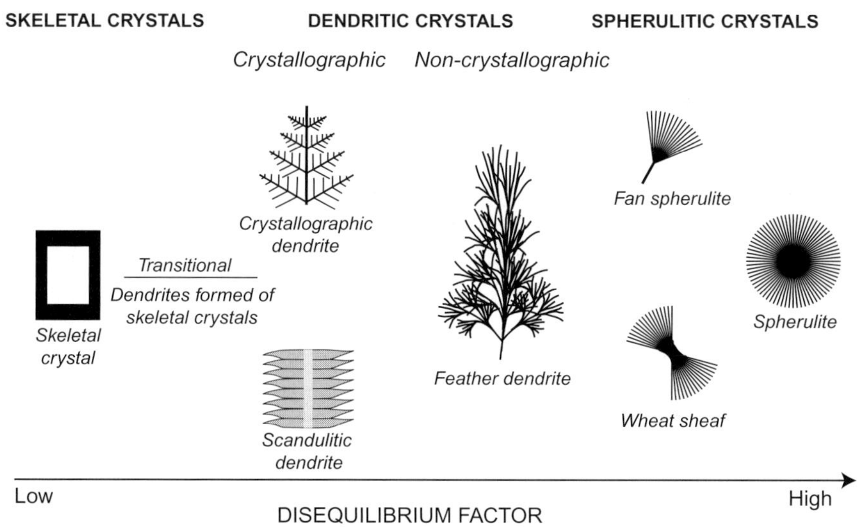

Figure 9 Possible relationship between the calcite crystal morphology and the disequilibrium factor, which is a general term that encompasses all factors that control disequilibrium precipitation. Modified from Jones and Renaut (1995, their Figure 14).

Aragonite is generally a minor component of these crystalline successions, commonly forming only thin micritic laminae (Guo and Riding, 1992). In some spring deposits, however, beds formed of complex pseudohexagonal aragonite crystals comprised of nested skeletal crystals are intercalated with beds formed of calcite dendrite crystals (Jones and Renaut, 1996a; Renaut and Jones, 1997).

9. Diagenesis

Flowing surface water is the most obvious feature of most springs. In many systems, however, large volumes of water are diverted into and through the older, highly porous and permeable deposits. Such waters, depending on their chemical attributes, can mediate a range of diagenetic processes that will modify the older deposits. These changes, including any permutation of dissolution, recrystallization, cementation, oxidation of organic matter, compaction, microbial alteration, or precipitation of authigenic minerals (cf. Tucker and Wright, 1990; Pentecost, 2005), may begin immediately after precipitation of the calcite or aragonite deposits. The pattern of diagenetic alteration, strongly dependent on subsurface flow patterns, will vary spatially and temporally as some permeability pathways become blocked (by cementation or compaction) while others are created (by dissolution or fracturing). Early diagenesis in calcareous spring precipitates is, however, poorly known because relatively little attention has been paid to this aspect of the deposits.

Given the continuous delivery of flowing waters at many springs, even in dry climates, calcareous spring deposits are susceptible to diagenetic processes such as aragonite inversion, cementation, and recrystallization. Although a poorly understood process, the time required for the aragonite-to-calcite inversion is highly variable, as it is with marine aragonite (e.g., Rehman et al., 1994). Aragonite in Bagni San Filippo in Italy inverts to calcite in less than 15 days (Malesani and Vannucchi, 1975), whereas it can be totally dissolved in the Pleistocene Rapolano Terme deposits (Guo and Riding, 1992), but it is still preserved in Miocene travertine (Gregor, 1982). Aragonite is replaced by dolomite in tufa mounds at the Great Salt Lake, Utah (Pedone and Dickson, 2000).

Calcite crystals in spring deposits are commonly composite crystals formed as the original crystals were progressively enlarged by successive generations of syntaxial calcite cement precipitated from water that flowed around them. The precipitation of calcite cements in cavities found in moss tufa (e.g., Irion and Muller, 1968), for example, reduces their porosity while making them less friable (Pentecost, 2005). Emeis et al. (1987) noted that hard lithified surfaces were produced in Plitvice lake dams (Croatia) as calcite cements were precipitated in older deposits. In a recent study of large dendrite crystals from the Clinton travertine (British Columbia, Canada), Jones and Renaut (2008) demonstrated that many features of the original crystals had been disguised by later precipitation of calcite cement. Original fabrics in Quaternary tufa from the Kharga Oasis (Egypt) have been extensively altered through recrystallization, cementation, sparmicritization, and neomorphism (Nicoll et al., 1999). Cementation of the once porous tufas included gravitational, meniscus, isopachous, and random mosaic cements that formed under vadose and phreatic conditions. Smith et al. (2004) also noted similar diagenetic fabrics in four, aerially extensive fossil-spring tufas in the area around the Kharga Oasis. Periodic slippage and slope failures are common on some terraced travertines producing open fractures that are commonly filled by coarse calcite cements.

There are relatively few documented examples of recrystallization in calcareous spring deposits. Chafetz et al. (1998) suggested that the acicular to bladed morphology of much of the calcite in Quaternary hot-spring travertines in southern Morocco was evidence that it had been diagenetically derived from aragonite. Recently precipitated travertine in the Arbuckle Mountains of Oklahoma is formed of layers composed of filamentous cyanobacterial bushes that are encased by one or more elongate spar calcite crystals that alternate with layers formed of equidimensional crystals that surround individual cyanobacterial filaments (Love, 1985; Love and Chafetz, 1988, 1990). Older parts of the same travertine, however, are formed of dense, laminated crusts composed almost entirely of columnar calcite crystals (up to 1 cm long) that display little affinity to the modern laminae (Love and Chafetz, 1988). Etching and staining of those crystals, however, revealed the remains of the microbial bushes akin to those in the modern laminae. The loss of the well-defined microbial bushes and laminae couplets was attributed to aggrading neomorphism (Love and Chafetz, 1988). Similarly, Rainey and Jones (2007) showed that many original fabrics in bryophyte tufa from Fall Creek (Alberta, Canada) had been diagenetically masked even though the deposit was probably $<5,000$ years old. These three examples clearly illustrate the susceptibility of young calcareous spring deposits to recrystallization.

The calcite in some tufas and travertines can be significantly altered by sparmicritization (e.g., Nicoll et al., 1999). This process, mediated by the activity of boring cyanobacteria, fungi and bacteria, involves the transformation of spar calcite crystals to micrite (e.g., Kahle, 1977; Jones, 1987; Chafetz et al., 1994). Such diagenetic alteration usually takes place when microbial infestations become established on exposed surfaces during periods when little or no calcite precipitation is taking place. Without careful examination, micrite laminae produced by this process could easily be confused with primary micrite.

10. Sublacustrine Calcareous Spring Deposits

Tall narrow chimneys and domal mounds of calcium carbonate (sublacustrine "tufa") are present on the floor of several lakes and most have been attributed to precipitation from sublacustrine springs and areas of groundwater seepage. Examples have been reported from modern and Pleistocene lakes in Turkey (Lake Van: Kempe et al., 1991), the western United States (Mono Lake, Pyramid Lake, Big Soda Lake, and others: Scholl and Taft, 1964; Shearman et al., 1989; Benson, 1994, 2004; Rosen et al., 2004), Pavilion Lake, British Columbia (Laval et al., 2000), Lac Abbe, Djibouti (Fontes and Pouchan, 1975), and several other lakes in Africa and Australia.

In many of these examples, dilute groundwater provides Ca^{2+}, whereas the lake water, which is often saline and alkaline, is the main source of HCO_3^- and CO_3^{2-}. An exception is Pavilion Lake, which is a fresh-water lake in limestone terrain. Water temperatures range from a few degrees to $>100°C$. The carbonate mineralogy is diverse. In addition to calcite and aragonite, monohydrocalcite, ikaite, and Mg-bearing carbonates including dolomite and hydromagnesite, have been reported from the mounds or, in the case of ikaite, inferred to have been present. Precipitation is commonly attributed to a combination of abiotic processes and biomediation by a range of cyanobacteria and heterotrophic bacteria (Kempe et al., 1991; Rosen et al., 2004; Benzerera et al., 2006). Many sublacustrine carbonates, such as the small aragonite chimneys in Lake Tanganyika (Tanganydro Group, 1992), are geothermally heated and some have associated sulphide mineralization. Precipitation at Lake Tanganyika has been attributed to mixing of hydrothermal fluids with dilute lake water (Stoffers and Botz, 1994).

Sublacustrine spring carbonates typically have domal mound or chimney-like morphologies (Figure 10A–C). They range from small localized deposits <1 m high and wide to tall chimneys such as those in Lake Van, Turkey, which rise up to 40 m above the lake floor (Kempe et al., 1991), and large domal mounds such as those at Pyramid Lake, Nevada (Benson, 2004). Most mounds lie near the lake shorelines. In some shallow playa lakes, the mounds provide evidence of periodic emergence and phases of subaerial discharge or inactivity during low lake levels (Figure 10D).

Unlike subaerial spring carbonates, sublacustrine tufa precipitation is normally restricted to the area directly around the vent(s), but this can be broad if the groundwater discharge zone is extensive. Distally from the vents, the spring fluids increasingly mix with lake water and although they may contribute to carbonate precipitation on the lake floor, they do not usually produce spring deposits analogous to the discharge apron of subaerial springs. The tufas range from massive to laminated (stromatolitic)

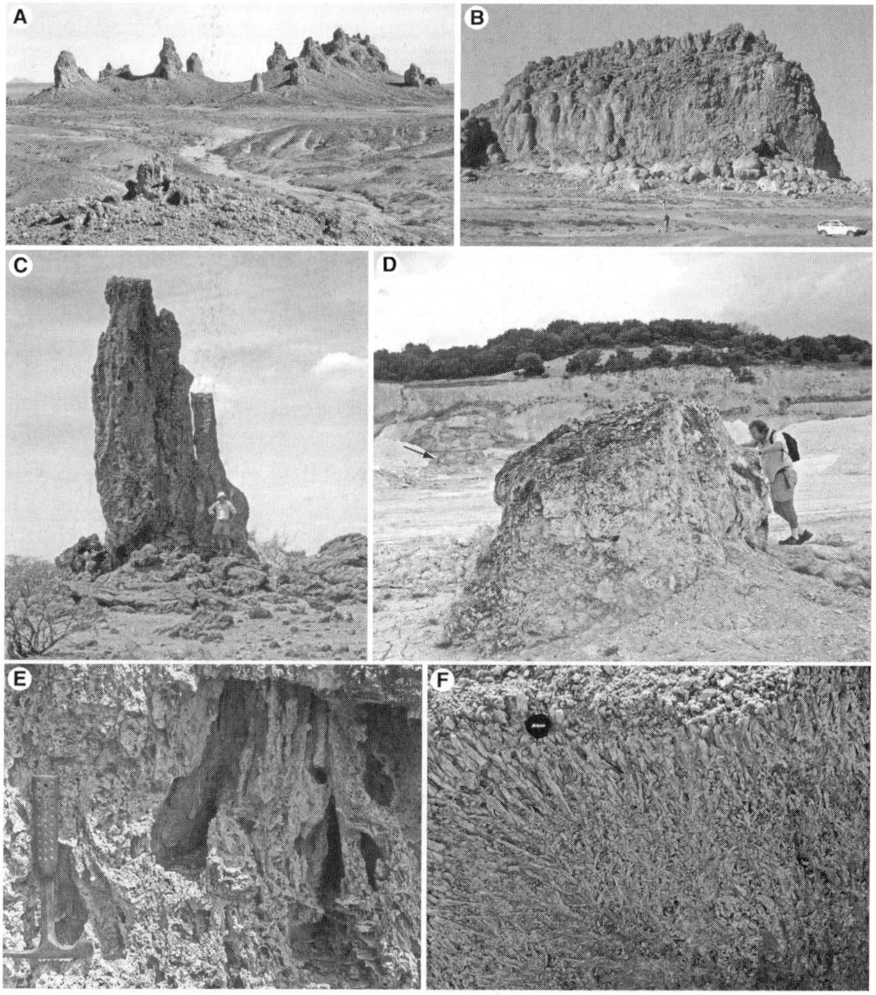

Figure 10 Sublacustrine carbonate mounds. A: Pleistocene sublacustrine tufa "pinnacles", Searles Lake, California. B: Large tufa mound, Pyramid Lake, Nevada. Subaerial hot springs discharge near the base of the mound. C: Pleistocene sublacustrine hydrothermal chimney (calcite), ~15 km northwest of Lake Magadi, Kenya. D: Small tufa mound (calcite) in a huntite-hydromagnesite quarry near Kozani, northern Greece. The mound (calcite) is the site of spring discharge into a shallow alkaline Mg-rich lake. A second mound (arrow) lies along the same fracture or fault. E: Porous (vertical) tubular tufa at the base of the mound near Magadi shown in C. F: Thinolite (calcite that has replaced ikaite) tufa from a large mound at Pyramid Lake, Nevada.

and banded. Others show clotted and globular textures and porous fenestral or tubular fabrics (Figure 10E). Several sublacustrine tufa mounds from the western United States show unusual calcite textures attributed to pseudomorphic replacement of ikaite (Figure 10F) (Shearman et al., 1989; Benson, 2004).

11. Discussion

Springs are complex depositional and biological systems that develop in response to processes operative at all scales (cf. Guo and Riding, 1994; Carthew et al., 2003). The aspect of scale has created many problems with classification schemes because it is difficult to design a multi-scaled system that encompasses everything from the microtextures created through microbial activity to the large-scale structures that reflect the interplay of water flow and geomorphology. Carthew et al. (2003) recognized this problem when they compared tufa formation in seasonally humid tropical settings (e.g., Gregory River of Queensland) with tufa formation in cooler temperate climes (e.g., Great Britain). Although similar at a large scale, the tufas developed in the different climates reflect significant differences caused by the different climatic settings. Viles et al. (2007, their Table 2) expanded on this problem when they compared tufa that formed in an arid tropical area with tufa from Great Britain and the Gregory River area of Australia.

Tufa and travertine formation in spring systems is the result of the combined action of large-, medium-, and small-scale processes. Spring systems, however, vary tremendously in scale with the smallest being triangular-shaped discharge aprons of limited extent (e.g., Lýsuhóll, Iceland – Figures 1D and 3) and the largest covering vast areas (e.g., Pamukkale, Turkey – Figures 1A, B and 2). From a process perspective, therefore, scale must be considered relative to the system being examined. Thus, a rimstone dam-pool will have the same fundamental elements even though their absolute sizes are vastly different. At Lýsuhóll, for example, a rimstone dam-pool setting is measured in centimetres (Figures 1D and 3G, H), whereas the same geomorphic feature at Pamukkale is measured in tens of metres (Figures 1A and 2A–C). Despite the vast size differences, both are characterized by a pool that forms behind a dam. Processes of formation may be similar, but the magnitude of their effects will differ with scale.

Large-scale processes control development of large-scale features (i.e., visible at the scale of a map or air photograph) in a tufa and travertine system. Vent location and the surrounding geomorphological setting largely control the structures at this scale. Accordingly, spring systems have been divided into: (1) cascade/waterfall (Figure 1C), (2) barrage (lacustrine) (Figure 6A), (3) perched (Figure 4A), (4) paludal, (5) fluvial, (6) fissure, and (7) mound (e.g., Ford and Pedley, 1996; Carthew et al., 2003, 2006;

Viles et al., 2007). This scheme can be applied irrespective of water temperatures, facies, biotas, or mineralogy because it focuses on the setting and general appearance and morphology of the deposits on a large scale.

At the medium scale, focus is on the deposits forming at any given location or segment in a spring system. For convenience, these locations can be referred to as proximal, intermediate or distal, relative to their position from the vent. Alternatively, specific geomorphological terms can be used. Guo and Riding (1998), for example, divided the depositional regimes found in the Rapolano Terme travertines into the (1) slope system, (2) depression system, and (3) mound system. Each regime encompasses its own depositional system that will be controlled by factors such as water-temperature regime (e.g., hot vs. cold), local climate conditions (e.g., arid vs. humid), water flow rate (e.g., fast flowing vs. stagnant), and $p\mathrm{CO_2}$ of the spring water and CO_2 degassing (e.g., slow vs. fast) that strongly influence $CaCO_3$ saturation levels and precipitation, and the biota (macrophytes vs. microphytes). Any combination of these parameters may act to control the mineralogy, crystallography, and textures of the calcareous deposits that are formed. Not all factors, however, will operate in all settings. In the hyperthermal regimes, for example, living macrophytes are absent because of high water temperature, and deposits in those regimes lack textures attributable to their activity. In the absence of macrophytes, saturation levels with respect to calcite or aragonite that are controlled primarily by CO_2 degassing may become the dominant control. In contrast, macrophytes may exert the primary control over textures found in cool-water tufa/travertine deposits.

At the micro-scale, focus is on the textures apparent in hand samples, in thin section, and on the scanning electron microscope. At these scales, the focus is on the individual grains and crystals in the tufa or travertine, with emphasis being on the manner in which they formed, the role that animals, plants, or microbes played in their formation, and the micro-environmental conditions that dictated their formation.

Spring discharge regimes are dynamic systems that are, in many respects, self-generating. Thus, changes at the micro-scale may engender changes at the medium scale that may, in turn, promote large-scale changes. Guo and Riding (1998) alluded to this notion in their discussion of the temporal evolution of sequence geometries in the Pleistocene travertines at Rapolano Terme (Italy). For example, they described depression systems that evolved upwards into slope deposition. Under this scenario, increased flow rates might flood a depression, and cause an increase in deposition that, in turn, may transform the area into a slope system. It must also be remembered that changes in one part of a system may well cause changes in another, downstream part of the system. Continual precipitation of calcite or aragonite in one area, for example, may cause the flowing spring water to be diverted to another area. Thus, an area once characterized by active precipitation may suddenly become an area where exposure, erosion, and early diagenesis is the rule. At the same time, however, areas where exposure and erosion dominated may suddenly be transformed into areas where active precipitation is taking place.

Today, springs are found on every continent with some producing thick successions that cover large areas. By comparison, ancient calcareous spring deposits are poorly known, apart from a few well documented examples such as those from the Paleoproterozoic Fennoscandian Shield (Melezhik and Fallick, 2001), the Upper Jurassic (Purbeck Formation) in southern England (Perry, 1994), the Eocene of South Dakota (Evans, 1999), the Neogene of northern Kenya (Renaut et al., 2002), and Pliocene strata in the Creede volcanic caldera, Colorado, USA (Steven and Ratte, 1965). This discrepancy can be attributed to various factors, including the fact that not every spring will produce deposits, springs commonly produce thin, aerially restricted spring deposits that may be difficult to locate in ancient successions, many calcareous spring deposits may be destroyed through erosion, weathering, and karstification before they have the chance to become preserved, and ancient calcareous spring deposits may pass unrecognized, especially if they have been extensively altered through diagenesis.

12. Conclusions

Spring systems are extremely complex environments where calcareous deposits commonly form under the influence of biotic and abiotic processes. This review has highlighted the following points:

- the classification of active springs according to water temperature and/or water source is difficult to apply to deposits in fossil systems where water is no longer flowing;
- the terms "tufa" and "travertine", commonly applied to calcareous spring deposits, have been largely debased because of their inconsistent usage in the literature;
- the classification of calcareous spring deposits is still in a state of flux; no single system encompasses all known spring facies;
- although calcite and aragonite are the main minerals precipitated in calcareous spring systems, some of the factors that control the precipitation of these two polymorphs are still not fully understood;
- many different types of plants, fungi, and prokaryotes inhabit calcareous spring systems; whether or not these organisms play an active or passive role in $CaCO_3$ precipitation is still open to debate;
- calcareous spring systems are characterized by a wide variety of facies; the factors that control their formation are, in some cases, poorly understood;

- early diagenesis is a neglected factor in many studies of travertine and tufa genesis; in particular, the role of spring water and runoff waters percolating through recently precipitated deposits merits detailed study.

Spring systems are remarkable natural laboratories where the development of calcareous deposits is closely linked to water temperature, water pH, water chemistry, and, in many situations, biogenic processes. As such, they offer the means for assessing the factors that control the myriad of facies that form in these environs.

ACKNOWLEDGEMENTS

The chapter is an outgrowth of the authors' research projects that have been funded by the Natural Sciences and Engineering Research Council of Canada (grants A6090 to Jones and GP0000629 to Renaut). We are indebted to Drs. C. Arenas, A.J. van Loon, and A.M. Alonso-Zarza who critically reviewed an earlier version of this manuscript.

REFERENCES

Adolphe, J.P., 1981. Exemples de contributions microorganiques dans les constructions carbonatées continentals. Bulletin de l'Association Géographique de France 479–480, 194–195.

Adolphe, J.P., 1991. Microoganisms and carbonated sediments in continental formations in North Africa. Journal of African Earth Sciences 12, 397–407.

Adolphe, J.P., Hourimeche, A., Loubiere, J.F., Paradas, J., Soleilhavoup, F., 1989. The carbonate formations of bacterial origin – continental formations from North Africa. Bulletin de la Société Géologique de France 5, 55–62.

Alçiçek, H., Varol, B., Özkul, M., 2005. Pisolith formation in self-built channel travertine, Pamukkale District, western Turkey. In: Özkul, M., Yagiz, S., Jones, B. (Eds), Proceedings of the 1st International Symposium on Travertine. Kozan Ofset Matbaacilik San. ve Tic. Ltd. Sti. Ankara, pp. 62–69.

Allen, E.T., Day, A.L., 1935. Hot springs of the Yellowstone National Park 466. Carnegie Institute Washington, Washington, 525 pp.

Andrews, J.E., 2006. Palaeoclimatic recorded from stable isotopes in riverine tufas: synthesis and review. Earth-Science Reviews 75, 85–104.

Andrews, J.E., Brasier, A.T., 2005. Seasonal records of climatic change in annually laminated tufas: short review and future prospects. Journal of Quaternary Science 20, 411–421.

Anzalone, E., Ferreri, V., Sprovieri, M., D'Argenio, B., 2007. Travertines as hydrologic archives: the case of the Pontecagnano deposits (southern Italy). Advances in Water Resources 30, 2159–2175.

Arenas, C., Cabrera, L., Ramos, E., 2007. Sedimentology of tufa facies and continental microbialites from the Palaeogene of Mallorca Island (Spain). Sedimentary Geology 197, 1–27.

Arnórsson, S., 1989. Deposition of calcium carbonate minerals from geothermal waters – theoretical considerations. Geothermics 18, 33–39.

Arenas-Abad, C., Vãzquez-Urbez, M., Pardo-Tirapu, G., Sancho-Marcén, C., 2010. Fluvial and associated carbonate deposits. In: Alonso-Zarza, A.M., Tanner, L.H. (Eds),

Folk, R.L., 1974. The natural history of crystalline calcium carbonate: effect of magnesium content and salinity. Journal of Sedimentary Petrology 44, 40–53.
Folk, R.L., 1994. Interaction between bacteria, nannobacteria and mineral precipitation in hot springs of central Italy. Géographie Physique et Quaternaire 48, 233–246.
Folk, R.L., Chafetz, H.S., 1983. Pisoliths (pisoids) in Quaternary travertines of Trivoli, Italy. In: Peryt, T.M. (Ed.), Coated Grains. Springer, Berlin, pp. 474–487.
Folk, R.L., Chafetz, H.S., Tiezzi, A., 1985. Bizarre forms of depositional and diagenetic calcite in hot-spring travertines, central Italy. In: Schneidermann, N., Harris, M. (Eds), Carbonate Cements. SEPM Special Publication 36, pp. 349–369.
Fontes, J-C., Pouchan, P., 1975. Les cheminées du lac Abhé (TFAI): stations hydro-climatiques de l'Holocène. Comptes Rendues de l'Academie des Sciences, Paris 280D, 383–386.
Ford, T.D., 1989. Tufa-the whole dam story. Cave Science 16, 39–49.
Ford, T.D., Pedley, H.M., 1996. A review of tufa and travertine deposits of the world. Earth-Science Reviews 41, 117–175.
Fouke, W.F., Bonheyo, G.T., Sanzenbacher, B., Frias-Lopez, J., 2003. Partitioning of bacterial communities between travertine depositional facies at Mammoth Hot Springs, Yellowstone National Park, U.S.A. Canadian Journal of Earth Sciences 40, 1531–1548.
Fouke, B.W., Farmer, J.D., Des Marais, D.J., Pratt, L., Sturchio, N.C., Burns, C., Discipulo, M.K., 2000. Depositional facies and aqueous-solid geochemistry of travertine-depositing hot springs (Angel Terrace, Mammoth Hot Springs, Yellowstone National Park, USA). Journal of Sedimentary Research 70, 565–585.
Freytet, P., Plet, A., 1991. Les formations stromatolitiques (tufs calcaires) récentes de la région de Tournus (Saone et Loire). Geobios 24, 123–139.
Freytet, P., Plet, A., 1996. Modern freshwater microbial carbonates: the Phormidium stromatolites (tufa-travertine) of southeastern Burgundy (Paris Basin, France). Facies 34, 219–238.
Friedman, G.M., 1985a. The problem of submarine cement in classifying reefrock: an experience in frustration. In: Schneidermann, N., Harris, M. (Eds), Carbonate Cements. SEPM Special Publication 36, pp. 117–121.
Friedman, G.M., 1985b. The term micrite or micrite cement is a contradiction. Discussion of micritic cements in microborings is not necessarily a shallow-water indicator. Journal of Sedimentary Petrology 55, 777.
Friedman, G.M., Sanders, J.E., 1978. Principles of Sedimentology. Wiley, New York, 792 pp.
Fuchs, C.W.C., 1865. Die vulkanischen Erscheinungen der Erde. Leipzig.
Gams, H., 1932. Bryo-cenology (Moss Societies). In: Verdoorn, F. (Ed.), Manual of Bryology. Martinus Nijhoff, The Hague, pp. 323–366.
Gierlowski-Kordesch, E.H., 2010. Lacustrine carbonates. In: Alonso-Zarza, A.M., Tanner, L.H. (Eds), Carbonates in Continental Settings: Facies, Environments and Processes. Developments in Sedimentology 61, pp. 1–102.
Glover, C., Robertson, A.F.H., 2003. Origin of tufa (cool water carbonate) and related terraces in the Antalya area, SW Turkey. Geological Journal 38, 329–358.
González, L.A., Lohmann, K.C., 1988. Controls on mineralogy and composition of spelean carbonates: Carlsbad Caverns, New Mexico. In: James, N.P., Choquette, W. (Eds), Paleokarst. Springer, New York, pp. 81–101.
Gregor, H.H., 1982. Die miozäne Flora aus dem Böttinger Thermalsinterkalk; eine Revision. Stuttgart Beiträge zur Naturkunde, Series B 1–15.
Guo, L., Riding, R., 1992. Aragonite laminae in hot water travertine crusts, Rapolano Terme, Italy. Sedimentology 39, 1067–1079.

Guo, L., Riding, R., 1994. Origin and diagenesis of Quaternary travertine shrub fabrics, Rapolano Terme, central Italy. Sedimentology 41, 499–520.

Guo, L., Riding, R., 1998. Hot-spring travertine facies and sequences, Late Pleistocene, Rapolano Terme, Italy. Sedimentology 45, 163–180.

Heimann, A., Sass, E., 1989. Travertines in the northern Hula Valley, Israel. Sedimentology 36, 95–108.

Herman, J., Lorah, M., 1987. CO_2 outgassing and calcite precipitation in Falling Spring Creek, Virginia, U.S.A. Chemical Geology 62, 251–262.

Herman, J., Lorah, M., 1988. Calcite-precipitation rates in the field: measurement and prediction for a travertine-depositing stream. Geochimica et Cosmochimica Acta 52, 2347–2355.

Holland, H.D., Kirsipu, T.V., Huebner, J.S., Oxburgh, U.M., 1964. On some aspects of the chemical evolution of cave waters. Journal of Geology 72, 36–67.

Humphreys, W.F., Awramik, S.M., Jebb, M.H.P., 1995. Freshwater biogenic tufa dams in Madang Province, Papua New Guinea. Journal of the Royal Society of Western Australia 78, 43–54.

Irion, G., Muller, G., 1968. Mineralogy, petrology and chemical composition of some calcareous tufa from the Schwabische Alb, Germany. In: Müller, G., Friedman, G.M. (Eds), Recent Developments in Carbonate Sedimentology in Central Europe. Springer, Berlin, pp. 157–171.

Ishigami, T., Suzuki, R., 1977. Factors affecting crystalline form of calcareous sinters. Chikyu Kagaku 11, 9–13.

Jacobson, R.L., Usdowski, E., 1975. Geochemical controls on a calcite precipitating spring. Contributions to Mineralogy and Petrology 51, 65–74.

Janssen, A., Swennen, R., Podoor, N., Keppens, E., 1999. Biological and diagenetic influence in recent and fossil tufa deposits from Belgium. Sedimentary Geology 126, 75–95.

Jones, B., 1987. The alteration of sparry calcite crystals in a vadose setting, Grand Cayman Island. Canadian Journal of Earth Sciences 24, 2292–2304.

Jones, B., 1989a. Calcite rafts, peloids and micrite in cave deposits from Cayman Brac, British West Indies. Canadian Journal of Earth Sciences 26, 654–664.

Jones, B., 1989b. Syntaxial overgrowths on dolomite crystals in the Bluff Formation, Grand Cayman, British West Indies. Journal of Sedimentary Petrology 59, 839–847.

Jones, B., Kahle, C.F., 1986. Dendritic calcite crystals formed by calcification of algal filaments in a vadose environment. Journal of Sedimentary Petrology 56, 217–227.

Jones, B., Kahle, C.F., 1993. Morphology, relationship and origin of fiber and dendrite calcite crystals. Journal of Sedimentary Petrology 63, 1018–1031.

Jones, B., Kahle, C.F., 1995. Origin of endogenetic micrite in karst terrains: a case study from the Cayman Islands. Journal of Sedimentary Research 65, 283–293.

Jones, B., Renaut, R.W., 1994. Crystal fabrics and microbiota in large pisoliths from Laguna Pastos Grandes, Bolivia. Sedimentology 41, 1171–1202.

Jones, B., Renaut, R.W., 1995. Noncrystallographic dendrites from hot-spring deposits at Lake Bogoria, Kenya. Journal of Sedimentary Research 65, 154–169.

Jones, B., Renaut, R.W., 1996a. Morphology and growth of aragonite crystals in hot-spring travertines at Lake Bogoria, Kenya Rift Valley. Sedimentology 43, 323–340.

Jones, B., Renaut, R.W., 1996b. Skeletal crystals of calcite and trona from hot-spring deposits in Kenya and New Zealand. Journal of Sedimentary Research 66, 265–274.

Jones, B., Renaut, R.W., 2008. Cyclic development of large, complex calcite dendrite crystals in the Clinton travertine, Interior British Columbia, Canada. Sedimentary Geology 203, 17–35.

Jones, B., Renaut, R.W., Owen, R.B., Torfason, H., 2005. Growth patterns and implications of complex dendrites in calcite travertines from Lýsuhóll, Snæfellsnes, Iceland. Sedimentology 52, 1277–1301.
Jones, B., Renaut, R.W., Rosen, M.R., 1996. High-temperature (>90°C) calcite precipitation at Waikite Hot Springs, North Island, New Zealand. Geological Society of London 153, 481–496.
Jones, B., Renaut, R.W., Rosen, M.R., 2000. Trigonal dendritic calcite crystals forming from hot spring waters at Waikite, North Island, New Zealand. Journal of Sedimentary Research 70, 586–603.
Julià, R., 1983. Travertines. In: Scholle, P.A., Bebout, D.G., Moore, C.H. (Eds), Carbonate Depositional Environments. American Association of Petroleum Geologists Memoir 33, pp. 64–72.
Kahle, C.F., 1977. Origin of subaerial Holocene calcareous crusts: role of algae, fungi and sparmicritization. Sedimentology 24, 413–435.
Kano, A., Matsuoka, J., Kojo, T., Fujii, H., 2003. Origin of annual laminations in tufa deposits, southwest Japan. Palaeogeography, Palaeoclimatology, Palaeoecology 191, 243–262.
Keith, H.D., Padden, F.J., 1964. Spherulitic crystallization from the melt. I. Fractionation and impurity segregation and their influence on crystalline morphology. Journal of Applied Physics 35, 1270–1285.
Kelts, K., Hsu, K.J., 1978. Freshwater carbonate sedimentation. In: Lerman, A. (Ed.), Lakes: Chemistry, Geology, Physics. Springer, Berlin, pp. 295–323.
Kempe, S., Kazmierczak, J., Landmann, G., Konuk, T., Reimer, A., Lipp, A., 1991. Largest known microbialites discovered in Lake Van, Turkey. Nature 349, 605–608.
Kepčija, R.M., Habdija, I., Primc-Habdija, B., Miliša, M., 2006. Simuliid silk pads enhance tufa deposition. Archives für Mikrobiologie 166, 387–409.
Kirchmayer, M., 1987. Universal stage grain fabrics statistics on dendrites of mine cave pearls. Carbonates and Evaporites 2, 59–65.
Kitano, Y., 1962a. A study of the polymorphic formation of calcium carbonate in thermal springs with an emphasis on the effect of temperature. Bulletin of the Chemical Society of Japan 35, 1980–1985.
Kitano, Y., 1962b. The behavior of various inorganic ions in the separation of calcium carbonate from a bicarbonate solution. Bulletin of the Chemical Society of Japan 35, 1973–1980.
Kitano, Y., 1963. Geochemistry of calcareous deposits found in hot springs. Journal of Earth Science, Nagoya University 11, 68–100.
Kitano, Y., Hood, D.W., 1965. The influence of organic matter on the polymorphic crystallization of calcium carbonate. Geochimica et Cosmochimica Acta 29, 29–41.
Kitano, Y., Kawasaki, N., 1958. Behavior of strontium ion in the process of calcium carbonate separation from bicarbonate solution. Journal of Earth Sciences, Nagoya University 6, 63–74.
Kitano, Y., Park, K., Hood, D., 1962. Pure aragonite synthesis. Journal of Geophysical Research 67, 4873–4874.
Koban, C.G., Schweigert, G., 1993. Microbial origin of travertine fabrics – two examples from southern Germany (Pleistocene Stuttgart Travertines and Miocene Riedöschingen Travertine). Facies 29, 251–264.
Koşun, E., Sarigül, A., Varol, B., 2005. Sedimentological investigation of Antalya tufas. In: Özkul, M., Yagiz, S., Jones, B. (Eds), Proceedings of the 1st International Symposium on Travertine. Kozan Ofset Matbaacilik San. ve Tic. Ltd. Sti. Ankara, pp. 50–61.
Lamoreaux, P.E., 2005. History and classifications of springs. Geological Society of America, Program with Abstracts 37, 324.
Laval, B., Cady, S.L., Pollack, J.C., McKay, C.P., Bird, J.S., Grotzinger, J.P., Ford, D.C., Bohm, H.R., 2000. Modern freshwater microbialite analogues for ancient dendritic reef structures. Nature 407, 626–629.

Liu, Z., Li, H., You, C., Wan, N., Sun, H., 2006. Thickness and stable isotope characteristics of modern seasonal climate-controlled sub-annual travertine lamina in a travertine-depositing stream at Baishuitai, SW China: implications for paleoclimate change. Environmental Geology 51, 257–265.
Lofgren, G., 1974. An experimental study of plagioclase crystal morphology: isothermal crystallization. American Journal of Science 274, 243–273.
Love, K.M., 1985. Petrology of Quaternary travertine deposits, Arbuckle Mountains, Oklahoma. Unpublished M.Sc. Thesis, University of Houston.
Love, K.M., Chafetz, H.S., 1988. Diagenesis of laminated travertine crusts, Arbuckle Mountains, Oklahoma. Journal of Sedimentary Petrology 58, 441–445.
Love, K.M., Chafetz, H.S., 1990. Petrology of Quaternary travertine deposits, Arbuckle Mountains, Oklahoma. In: Herman, J.S., Hubbard, D.A. (Eds), Travertine-Marl: Stream Deposits in Virginia. Department of Mines, Minerals and Energy Division of Mineral Resources, Charlottesville, Virginia, Commonwealth of Virginia, pp. 65–78.
Malesani, P., Vannucchi, S., 1975. Precipitazione di calcite o di aragonite dalle acque termominerale in relazione alla genesi e all'evoluzione dei travertini. Atti della R. Academia d'Italia 58, 761–776.
Matsuoka, J., Kano, A., Oba, T., Watanabe, T., Sakai, S.S., Seto, K., 2001. Seasonal variation of stable isotopic composition recorded in a laminated tufa, SW Japan. Earth and Planetary Science Letters 192, 31–44.
Melezhik, V.A., Fallick, A.E., 2001. Palaeoproterozoic travertines of volcanic affiliation from a ^{13}C-rich rift lake environment. Chemical Geology 173, 293–312.
Meyer, H.J., 1984. Influence of impurity on the growth rate of calcite. Journal of Crystal Growth 66, 639–646.
Milliman, J.D., Hook, J.A., Golubic, S., 1985. Meaning and usage of micrite cement and matrix-reply to discussion. Journal of Sedimentary Petrology 55, 777–784.
Monty, C.L.V., 1976. The origin and development of cryptalgal fabrics. In: Walter, M.R. (Ed.), Stromatolites. Developments in Sedimentology. Elsevier, Amsterdam, pp. 193–250.
Moore, G.W., 1956. Aragonite speleothems as indicators of paleotemperature. American Journal of Science 254, 746–753.
Müller, G., Irion, G., Förstner, U., 1972. Formation and diagenesis of inorganic Ca-Mg carbonates in the lacustrine environment. Naturwissenschaften 59, 158–164.
Nicoll, K., Giegengack, R., Kleindienst, M., 1999. Petrogenesis of artifact-bearing fossil-spring deposits from Kharga Oasis, Egypt. Geoarchaeology 14, 849–863.
O'Brien, G.R., Kaufman, D.S., Sharp, W.D., Atudorei, V., Parnell, R.A., Crossey, L.J., 2006. Oxygen isotope composition of annually banded modern and mid-Holocene travertine and evidence of paleomonsoon floods, Grand Canyon, Arizona, USA. Quaternary Research 65, 366–379.
Ordoñez, D.S., García del Cura, M.A., 1980. Carbonatos biogénicos actuales en la cuenca del rio Dulce (provincia Guadalajara). Boletín de la Real Sociedad Española de Historia Natural. Sección Geológica 78, 303–315.
Owen, R.B., Renaut, R.W., Hover, V., Ashley, G.M., Muasya, A.M., 2004. Swamps, springs and diatoms: wetland regions of the semi-arid Bogoria-Baringo Rift, Kenya. Hydrobiologia 518, 59–78.
Owen, R.B., Renaut, R.W., Jones, B., 2008. Geothermal diatoms: a comparative study of floras in hot spring systems of Iceland, New Zealand, and Kenya. Hydrobiologia 610, 175–192.
Paprocki, H., Holzenthal, R.W., 2003. A new species of *Smicridea* McLachlan (Trichoptera: Hydropsychidae) from Venezuela and its role in travertine biogenesis. Journal of the North American Benthological Society 22, 401–409.
Pavlović, G., Prohić, E., Miko, S., Tibljaš, D., 2002. Geochemical and petrographic evidence of meteoric diagenesis in tufa deposits in northern Dalmatia (Zrmanja and Krupa Rivers, Croatia). Facies 46, 27–34.

Pedley, H.M., 1990. Classification and environmental models of cool freshwater tufas. Sedimentary Geology 68, 143–154.
Pedley, H.M., 1992. Freshwater (phytoherm) reefs: the role of biofilms and their bearing on marine reef cementation. Sedimentary Geology 79, 255–274.
Pedley, H.M., 1993. Sedimentology of the late Quaternary barrage tufas in the Wye and Lathkill valleys, north Derbyshire. Proceedings of the Yorkshire Geological Society 49, 197–206.
Pedley, H.M., 1994. Prokaryotic-microphyte biofilms and tufa: a sedimentological perspective. Kaupia (Darmstädter Beiträge zur Naturgeschichte) 4, 45–60.
Pedley, M., Andrews, J., Ordoñez, S., García del Cura, M.A., Martín, J.-A.G., Taylor, D., 1996. Does climate control the morphological fabric of freshwater carbonates? A comparative study of Holocene barrage tufas from Spain and Britain. Palaeogeography, Palaeoclimatology, Palaeoecology 121, 239–257.
Pedley, H.M., Martín, J.A.G., Ordóñez, D.S., García del Cura, M.A., 2003. Sedimentology of Quaternary perched springline and paludal tufas: criteria for recognition with examples from Guadalajara Province, Spain. Sedimentology 50, 23–44.
Pedone, V.A., Dickson, J.A.D., 2000. Replacement of aragonite by quasi-rhombohedral dolomite in a late Pleistocene tufa mound, Great Salt Lake, Utah, U.S.A. Journal of Sedimentary Research 70, 1152–1159.
Pentecost, A., 1990. The formation of travertine shrubs: Mammoth Hot Springs, Wyoming. Geological Magazine 127, 159–168.
Pentecost, A., 1994. Formation of laminate travertines at Bagno Vignone, Italy. Geomicrobiology Journal 12, 239–251.
Pentecost, A., 1995. The Quaternary travertine deposits of Europe and Asia Minor. Quaternary Science Reviews 14, 1005–1028.
Pentecost, A., 2005. Travertine. Springer, Berlin, 445 pp.
Pentecost, A., Jones, B., Renaut, R.W., 2003. What is a hot spring? Canadian Journal of Earth Sciences 40, 1443–1446.
Pentecost, A., Lord, T.C., 1988. Postglacial tufas and travertines from the Craven district of Yorkshire. Cave Science 15, 15–19.
Pentecost, A., Spiro, B., 1990. Stable carbon and oxygen isotope composition of calcites associated with modern freshwater cyanobacteria and algae. Geomicrobiology Journal 8, 17–26.
Pentecost, A., Viles, H., 1994. A review and reassessment of travertine classification. Géographie Physique et Quaternaire 48, 305–314.
Perry, C.T., 1994. Freshwater tufa stromatolites in the basal Purbeck Formation (Upper Jurassic), Isle of Portland, Dorset. Geological Journal 29, 119–135.
Peryt, T.M., 1983a. Classification of coated grains. In: Peryt, T.M. (Ed.), Coated Grains. Springer, Berlin, pp. 3–6.
Peryt, T.M., 1983b. Vadoids. In: Peryt, T.M. (Ed.), Coated Grains. Springer, Berlin, pp. 437–449.
Pia, J., 1933. Die rezenten Kalksteine. Mineralogische und Petrographische Mitteilungen, Ergänzungsband, 420 pp.
Popp, B.N., Wilkinson, B.H., 1983. Holocene lacustrine ooids from Pyramid Lake, Nevada. In: Peryt, T.M. (Ed.), Coated Grains. Springer, Berlin, pp. 142–153.
Privittera, M., Guidice, R.L., 1986. Sulla briovegetation dei Tufi Calcarei dell'Ennese (Sicilia). Cryptogamie Bryologie-Lichénologie 7, 129–140.
Rainey, D.K., Jones, B., 2005. Radiating calcite dendrites – precursors for coated grain formation in the Fairmont hot springs travertine, Canada. In: Özkul, M., Yagiz, S., Jones, B. (Eds), Proceedings of the 1st International Symposium on Travertine. Kozan Ofset Matbaacilik San. ve Tic. Ltd. Sti. Ankara, pp. 25–32.

Rainey, D.K., Jones, B., 2007. Rapid cold water formation and recrystallization of relict bryophyte tufa at the Fall Creek cold springs, Alberta, Canada. Canadian Journal of Earth Sciences 44, 889–909.

Rehman, J., Jones, B., Hagan, T.H., Coniglio, M., 1994. The influence of sponge borings on aragonite-to-calcite inversion in Late Pleistocene *Strombus gigas* from Grand Cayman, British West Indies. Journal of Sedimentary Research 64, 174–179.

Reis, O.M., 1926. Zusammenfassung über die im Ries südlich von Nördlingen auftretenden Süsswasserkalke und ihre Entstehung. Jahresberichte und Mitteilungen des Oberrheinischen Geologischen Vereins, Neue Folge 14, 176–190.

Renaut, R.W., Jones, B., 1997. Controls on aragonite and calcite precipitation in hot spring travertines at Chemurkeu, Lake Bogoria, Kenya. Canadian Journal of Earth Sciences 34, 801–814.

Renaut, R.W., Jones, B., 2000. Microbial precipitates around continental hot springs and geysers. In: Riding, R.E., Awramik, S.M. (Eds), Microbial Sediments. Springer, Berlin, pp. 187–195.

Renaut, R. W., Morley, C.K., Jones, B., 2002. Fossil hot-spring travertine in the Turkana Basin, northern Kenya: structure, facies, and genesis. In: Renaut, R.W., Ashley, G.M. (Eds), Sedimentation in Continental Rifts. SEPM Special Publication 73, pp. 123–141.

Richter, D.V., Besenecker, H., 1983. Subrecent high-Sr aragonite ooids from hot springs near Tekke Ilica (Turkey). In: Peryt, T.M. (Ed.), Coated Grains. Springer, Berlin, pp. 154–162.

Riding, R., 1991. Classification of microbial carbonates. In: Riding, R. (Ed.), Calcareous Algae and Stromatolites. Springer, Berlin, pp. 21–51.

Risacher, F., Eugster, H.P., 1979. Holocene pisoliths and encrustations associated with spring-fed surface pools. Sedimentology 26, 253–270.

Rosen, M.R., Arehart, G.B., Lico, M.S., 2004. Exceptionally fast growth rate of <100-yr-old tufa, Big Soda Lake, Nevada: implications for using tufa as a paleoclimate proxy. Geology 32, 409–412.

Sandberg, P., 1985. Aragonite cements and their occurrence in ancient limestones. In: Schneidermann, N., Harris, M. (Eds), Carbonate Cements. SEPM Special Publication 36, pp. 33–57.

Sant'Anna, L.G., Riccomini, C., Rodrigues-Francisco, B.H., Sial, A.N., Carvalho, M.D., Moura, C.A.V., 2004. The Paleocene travertine system of the Itaboraí basin, southeastern Brazil. Journal of South American Earth Sciences 18, 11–25.

Scholl, D.W., Taft, W.H., 1964. Algae, contributors to the formation of calcareous tufa, Mono Lake, California. Journal of Sedimentary Research 34, 309–319.

Schreiber, B.C., Smith, D., Schreiber, E., 1981. Spring peas from New York State: nucleation and growth of fresh water hollow ooliths and pisoliths. Journal of Sedimentary Petrology 51, 1341–1346.

Shearman, D.J., McGugan, A., Stein, C., Smith, A.J., 1989. Ikaite, $CaCO_3.6H_2O$, precursor of the thinolites in the Quaternary tufas and tufa mounds of the Lahontan and Mono Lake basins, western United States. Geological Society of America Bulletin 101, 913–917.

Siegel, F.R., 1965. Aspects of calcium carbonate deposition in Great Onyx Cave, Kentucky. Sedimentology 4, 285–299.

Smith, J.R., Giegengack, R., Schwarcz, H.P., McDonald, M.M.A., Kleindienst, M.R., Hawkins, A.L., Churcher, C.S., 2004. A reconstruction of Quaternary pluvial environments and human occupations using stratigraphy and geochronology of fossil-tufas, Kharga Oasis, Egypt. Geoarchaeology 19, 407–439.

Stamatakis, M.G., Renaut, R.W., Kostakis, K., Tsivilis, S., Stamatakis, G., Kakali, G., 2007. The hydromagnesite deposits of the Atlin area, British Columbia, Canada, and their

1. Introduction

Calcretes or caliches are one of the types of non-marine carbonates that have received much attention from a variety of scientists including geomorphologists, pedologists and sedimentologists. This interest in calcretes results from their widespread occurrence in continental settings, and because they are important repositories of information about the reconstruction of past ecosystems and environments, and about the tectonic, climatic and sedimentary regimes in which they formed. Exhaustive reviews on calcretes have been provided by Esteban and Klappa (1983), Wright and Tucker (1991), Paquet and Ruellan (1997), Watson and Nash (1997), Alonso-Zarza (2003) and Wright (2007). The most important aspects of calcretes are clearly developed in these works. The present review aims to provide a concise summary of the general aspects of calcretes, such as classification, morphology, micromorphology and geochemistry, with a focus on their palaeoenvironmental significance for reconstructing geomorphic and stratigraphic controls.

A practical definition of a calcrete is that proposed by Watts (1980) after modifying that of Goudie (1973): 'pedogenic calcretes are terrestrial materials composed dominantly, but not exclusively, of $CaCO_3$, which occur in states ranging from nodular and powdery to highly indurated and result mainly from the displacive and/or replacive introduction of vadose carbonate into greater or lesser quantities of soil, rock or sediment within a soil profile'. This definition only refers to pedogenic calcretes. Wright and Tucker (1991) later proposed a wider use of the term 'calcrete' that is not restricted to pedogenic occurrences; in some semi-arid to arid regions, extensive precipitation occurs in the shallow phreatic zone and produces large bodies of authigenic carbonates with many of the characteristics of pedogenic calcretes and palustrine carbonates (see Alonso-Zarza and Wright, 2010). Where calcium carbonate is introduced into a non-carbonate host, its authigenic origin is clear, but calcretes can also develop in carbonate bedrocks and sediments, including those formed around lake margins, seasonal wetlands and groundwater-discharge zones, creating a spectrum of complex relationships (Tandon and Andrews, 2001; Alonso-Zarza, 2003).

2. Classification

The classification of calcretes is complex since various criteria may be used. Purely descriptive classifications consider mineralogy and morphology. A fundamental distinction must be made between calcretes that are formed in soil profiles within the vadose zone (pedogenic calcretes) and ones formed around the water-table capillary fringe or below due to laterally moving waters, in some cases at considerable depth (Carlisle, 1980, 1983). The former owe their origin to the addition or redistribution of calcium carbonate associated with eluvial/illuvial processes, whereas the latter are due to precipitation from groundwater (Arakel and McConchie, 1982), occasionally under the influence of phreatophytic plants (Semeniuk and Meagher, 1981), hence the term 'groundwater calcrete' (or phreatic, channel or valley calcrete). A less common type is referred to as 'gully bed cementation' and takes place where carbonate-rich run-off infiltrates channel sediments, leading to the plugging of the sediment layer by carbonate cement and the production of laminar layers (Mack et al., 2000). It is pedogenic calcrete that can show well-developed profiles and has received the most attention from researchers (Figure 1A and B). Groundwater calcretes may be difficult to distinguish from those formed under pedogenic conditions (Pimentel et al., 1996; Mack et al., 2000; Tandon and Andrews, 2001). In some cases, it may also be difficult to distinguish groundwater calcretes or dolocretes from carbonates formed due to the effects of non-exclusively meteoric diagenesis (Williams and Krause, 1998).

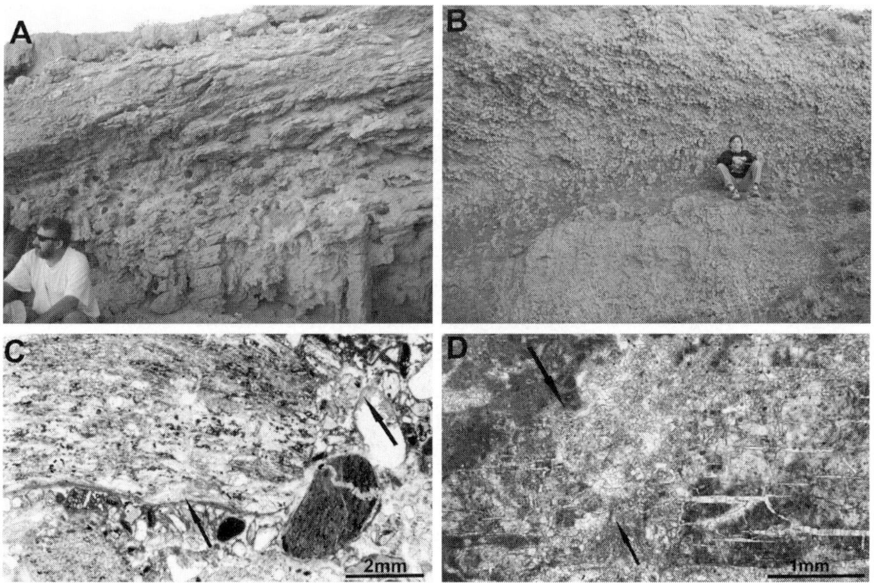

Figure 1 Calcrete macro- and micromorphology. (A) Thick pedogenic calcrete showing a lower nodular horizon and an upper laminar one forming a hardpan; they are separated by gravel clasts (Pleistocene of Gran Canaria, Spain). (B) Two pedogenic calcrete profiles (mostly nodular), separated by a bed of red clays that also contains calcrete nodules (the person is seated in this clay bed) (Palaeogene of the Teruel Basin, Spain). (C) Clasts of metamorphic rocks showing thin micritic coatings (arrowed) formed under the influence of microbial biofilms (Pleistocene of the Guadix Basin, Spain). (D) Coarse calcite crystals (arrows) replacing the feldspar that occupies most of the space (Palaeogene of the Sado Basin, Portugal).

Despite the wide use of the term 'calcrete' and its synonyms 'caliche' and 'cornstone' (Allen, 1960), and many more terms (Goudie, 1973), none are included in any soil classification, either as a soil name or as a horizon. Within a soil, the horizon of prominent carbonate accumulation has been named the K horizon (Gile et al., 1965); it has a diagnostic K-fabric. Pedogenic calcretes form within soil profiles where they constitute several discrete horizons of carbonate accumulation, forming a sub-profile within the main soil profile (Wright and Tucker, 1991). Aridisols, Vertisols, Mollisols and Alfisols (Soil Survey Staff, 1975) are the more typical soils containing calcretes (Wright and Tucker, 1991), developing in the B or C horizons as illuvial concentrations. In palaeosol classifications, calcrete-bearing palaeosols are considered Aridisols (Retallack, 1993), Calcisols (Mack et al., 1993) or palaeo-Aridisols (Nettleton et al., 2000).

With the dominant carbonate mineral and the amount of dolomite in mind, a simple classification was proposed by Netterberg (1980), who distinguished between calcretes, magnesian calcretes, dolomitic calcretes and dolocretes. The dominant mineralogy of calcretes is low-Mg calcite (Wright and Tucker, 1991), although complex reactions should be expected as a result of microbial and physicochemical processes (Watts, 1980). Dolomite is commonly recorded, being either primary or formed by replacement, but rarely constitutes the main mineral (Bustillo and Alonso-Zarza, 2007). In some groundwater dolocretes, dolomite may be the main mineral because it forms when the Mg/Ca ratio is high, which is more likely to happen in evolved groundwaters where Ca has been removed by precipitation of calcite. More complex carbonate mineralogies have been recorded from pedogenic carbonates associated with Mg-rich bedrock (e.g., Podwojewski, 1995) and volcanic rocks (Molina, 1988).

The morphology of calcretes and their different horizons has given rise to a large number of names (Netterberg, 1980; Goudie, 1983), which have been summarised by Wright and Tucker (1991). These include calcareous soil, calcified soil, powder calcrete, pedotubule calcrete, nodular calcrete, honeycomb, hardpan, laminar calcrete and boulder/cobble calcrete.

3. Calcrete Distribution

Soils in which significant amounts of calcium-carbonate accumulate are widespread at present. One estimate (Yaalon, 1988) suggests that such soils cover 13% of the Earth's land surface; in Australia, currently the driest continent, calcretes occupy 21% of the land surface (Chen et al., 2002), but this figure may include non-pedogenic forms. The distribution of ground-water calcretes on a global scale is not known. Calcium carbonate will accumulate in a soil where there is a moisture deficit, allowing any carbonate fixed in the soil during a dry season to survive leaching in seasons where rainfall exceeds evaporation. As organisms, especially plants and fungi, play a critical role in fixing carbonate, a critical aspect is the relationship of organic activity and the soil-moisture regime. There are certain relationships between the distribution of pedogenic calcretes and climatic regimes. For example, most such calcretes are found today in climatic regimes with a mean annual temperature of 16–20°C (Goudie, 1983), but they are also known from cold deserts, implying that rainfall is the critical factor. Most 'warm' calcretes form where the rainfall ranges between 100 and 500 mm (Goudie, 1983), but they can form in areas with higher rainfall, up to 1,000 mm (Mack and James, 1994). Other studies, such as that of Royer (1999), show that carbonate-bearing soils are mainly found where rainfall is <760 mm, but there are many exceptions to this pattern (Birkeland, 1999; Retallack, 2000). For example, Strong et al. (1992) document strongly biogenic calcrete features in soils from North Yorkshire in the UK where local highly permeable carbonate gravels have triggered soil-moisture deficits large enough to produce features typical of drier, warmer regions. Similar calcretes have been recorded from Pleistocene deposits from the UK from a phase of climatic amelioration within the Anglian (Elsterian, early Kansan) glacial stage (Candy, 2002). Thin calcretes are known today in Arctic areas (Lauriol and Clarke, 1999), in glacial tills from Antarctica (Foley et al., 2006), as well as from Pleistocene cold settings (Vogt and Corte, 1996).

Calcrete-bearing soils commonly support a sparse vegetation cover, including grasses, trees and shrubs. Goudie (1973) has compiled a wide list of species; many are xerophytic, but not all, and root development may vary from horizontal to oblique or vertical. Macroflora is the most obvious contributor to soil formation, but microflora and soil fauna are important, and commonly they start to prepare the soil for the colonisation of higher plants.

Calcrete profiles require long periods for their formation (see below), and climate changes may take place so that what is left as the final product in a well-developed profile might be the sum of many significant environmental changes (Wright, 2007). Indeed, once carbonate accumulates it may not remain as a record of the initial climatic phase, as shown by the Holocene calcrete-bearing soils of the Gangetic Plains (Srivastava and Parkash, 2002), where carbonate accumulations that had been formed around 6,500 BP under a dryer climate were removed over large areas during a more recent wetter phase.

The distribution of ancient pedogenic calcretes mirrors this modern day distribution in the sense that such examples are commonly found in ancient dry-land (especially 'red bed') successions. Ancient and recent calcretes from the Himalayas and India contain valuable palaeoclimatic information of several geological periods, including the Cretaceous (Ghosh, 1997), Palaeogene (Singh and Lee, 2007), Miocene (Ghosh et al., 2004) and Quaternary (Srivastava, 2001; Srivastava et al., 2007). The special significance of these calcrete-bearing palaeosols is that they provide palaeoclimatic data that may contribute to the understanding of the palaeomonsoonal regimes over time. Some ancient fluvial units such as the classic Old Red Sandstone in the southern UK contain hundreds of profiles representing tens of millions of years of non-marine deposition (Allen, 1986). They are also prominent features in many shallow-water carbonate successions, where they mark emersion surfaces (Wright, 1994).

4. Source of Calcium Carbonate and Mechanisms of Carbonate Precipitation

4.1. Provenance of calcium carbonate

Two main models have been proposed for the supply of carbonate to calcrete profiles (Goudie, 1983). In the '*per descensum*' model, the carbonate enters the profile in solution from the upper soil levels to a lower horizon; this is the common process in pedogenic calcretes. In contrast, carbonate obtained from groundwater defines the '*per ascensum*' model, which is dominant in phreatic calcretes. The Ca in pedogenic calcretes can be derived from many sources, including local rocks, rainwater, dust, the biota and sea spray in coastal areas (Goudie 1983; Cailleau et al., 2004). In most semi-arid and arid areas, the main source – excluding local rocks – must be dust and rainfall. Monger and Gallegos (2000) have, for example, evaluated the contributions from dust and rainfall in the Las Cruces region in New Mexico, noting that the main source of Ca is rainfall. However, the pathway of Ca into the soil may be complex, as some Ca may be fixed by the vegetation before later being released into the soil. Garvie (2004) has shown that a significant pathway could be cacti that concentrate Ca, which can be released later, when the plant decays (see below).

The sources of Ca can be assessed in calcretes using the $^{87}Sr/^{86}Sr$ ratios as proxies for the sources of Ca. Sr and Ca have similar chemical properties, and Sr therefore shows a strong affinity with Ca. This makes $^{87}Sr/^{86}Sr$ ratios a good tracer for the provenance of Ca (Dart et al., 2007). This ratio depends on the source of carbonate (e.g., old continental rock, volcanic rocks, marine carbonate, aeolian dust) (Quade et al., 1995). In one of the earliest applications in calcretes, Quade et al. (1995) were able to identify the sources of Sr for calcretes from coastal and inland areas in South Australia and Victoria; Sr isotopes revealed that, near the coast, the ocean was the principal source, but that volcanic sources were important locally, and that further inland the contribution of marine-derived Sr decreased, reflecting the addition of inland dust sources. In most cases, the contribution of the Ca from weathering of the host rock is minimum in comparison with the contribution of Ca from aeolian dust or atmospheric input (Chiquet et al., 1999; Capo and Chadwick,

1999; Naiman et al., 2000), even in the presence of adjacent marine carbonates (Hamidi et al., 2001). Only in some cases does the $^{87}Sr/^{86}Sr$ ratio of the carbonates yield values close to the silicate host rock, as in the calcrete of Tamil Nadu in South India (Durand et al., 2006); however, influx of Ca dust also contributed to the formation of that calcrete. If calcretes develop on silicate rocks, the chemical weathering of the minerals consumes CO_2, so pedogenic carbonates play an important role in CO_2 cycle as a possible sink for it (Lal and Kimble, 2000).

Groundwater calcretes and dolocretes are most likely to have complex sources, reflecting the geology in the drainage basin. Spötl and Wright (1992), measuring Sr ratios in Triassic dolocretes from the Paris Basin, France, concluded that the source was a complex basement terrain and not marine waters, as had previously been thought.

4.2. Mechanisms of accumulation

Pedogenic carbonate accumulation can generally be regarded as an eluvial/illuvial process (*per descensum* model of Goudie, 1983). Calcium carbonate in surface layers, mainly as dust, is taken into solution and infiltrates into the soil. Some may be transported in colloidal form (Baghernejad and Dalrymple, 1993). Calcium bicarbonate is also supplied directly via rainwater, and rarely from groundwater (Knuteson et al., 1989). However, it is clear that both biogenic and abiogenic processes are at work and result in soil carbonate with varied types of microfabrics (see below). The calcium carbonate added to the soil, by whatever mechanism, is emplaced in three ways. Firstly, if generated as distinct components in biofilms (Figure 1C), fungal mycelia, released from decayed cells, or as colloidal material, it constitutes a particulate component of the soil. Secondly, it can fill pore spaces in a passive manner as simple cement. Thirdly, it can be introduced as a displacive phase, commonly seen in non-carbonate hosts where the calcite, being unable to form adhesive bonds with non-carbonate grains (Chadwick and Nettleton, 1990), can displace the host grains (Figure 1D) to produce a range of macro- and micro-textures (reviewed by Wright and Tucker, 1991).

Abiogenic (inorganic) carbonate precipitation is driven by evaporation, degassing and evapotranspiration (Wright and Tucker, 1991). McFadden and Tinsley (1985) presented simulation models for inorganic carbonate precipitation, and by integrating climate data, soil properties and chemical thermodynamic relations it was possible to calculate the amount and depth of calcium carbonate depending on time. However, the models also have a biotic component because evapotranspiration controls the amount of water in the soil and respiration of soil organisms control the pCO_2 (Monger and Gallegos, 2000).

Apart from the influence on evapotranspiration, it appears that, in many calcretes, plants and other organisms drive most of the carbonate

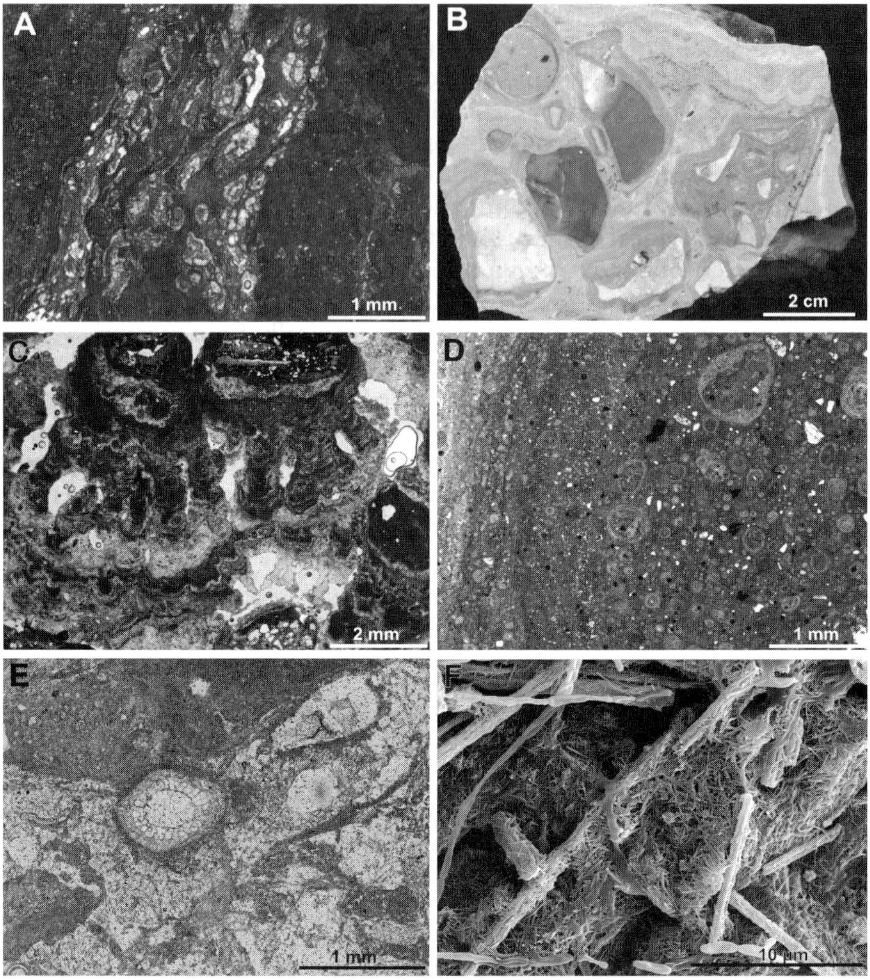

Figure 2 Biogenic features. (A) Alveolar-septal structure consisting of large interwoven micritic filaments; the porosity is filled by calcite spar (Miocene of the Teruel Basin). (B) Polished slab of a calcrete showing coarse gravel clasts and a topmost laminar horizon (Miocene of the Madrid Basin, Spain). (C) Asymmetrically coated (lower sides) gravel clasts (Teruel Basin). (D) Sand-sized micritic coated grains (Pleistocene of Canary Islands). (E) Root section showing calcified root cells (almost in the centre of the image); needle-fibre calcite crystals are seen in the lighter areas (recent calcrete in the neighbourhood of Brihuega, Spain). (F) Needle-fibre calcite crystals of different sizes; organic filaments (not collapsed) are also visible (same calcrete as E).

micro-organisms and root hairs may also form these filaments. Verrecchia et al. (1993) have demonstrated that, in some cases, biomineralised fungal filaments could have been composed of calcium oxalates that later were transformed to carbonate.

(4) Rhizoliths, calcified root cells and *Microcodium* are very common in calcretes (Figure 2E). They are discussed in detail in Klappa (1980) and Kosir (2004). The origin of *Microcodium* has been a matter of controversy. It is commonly interpreted as the product of calcification of root cells (Kosir, 2004), but some doubts have recently arisen on its rhizogenic origin (Singh et al., 2007; Kabanov et al., 2008).

(5) Needle-fibre calcite crystals are up to 10 μm wide and up to 200 μm in length, but can be very variable in size (Figure 2F). They vary from monocrystalline rods to polycrystalline chains, showing different morphologies (Verrecchia and Verrecchia, 1994). Their formation is due to either high levels of supersaturation or microbial activity, especially that of fungi or cyanobacteria (Callot et al., 1985; Phillips and Self, 1987) or specifically to the calcification of fungal sheaths (Bajnóczi and Kovács-Kis, 2006).

(6) Spherulites have been one of the most controversial features of laminar calcretes. They consist of low-Mg calcite fibro-radial spherulitic polycrystals, the diameter of which varies from 0.5 μm to more than 100 μm (Verrecchia et al., 1995). Contrary to some of the discussion on their origin (Verrecchia et al., 1995; Wright et al., 1996; Freytet et al., 1997), growth experiments and the fact that they occur at the very top of laminar calcretes support the idea that their formation is associated with cyanobacterial and bacterial mats that developed as thin films in ponds that later dried up, meaning they have to form at the sediment/atmosphere interface.

(7) Microborings are more or less cylindrical cavities of about 300 μm long and 1 μm in diameter. They have no preferred orientation and can be empty, coated by mucilage or lined by calcite crystals. They are similar in size and morphology to the porosity left by filaments or root hairs (Alonso-Zarza and Jones, 2007).

(8) Spherical to slightly polygonal bodies, 0.5–1 μm in diameter, with calcite walls of 0.1 μm thick have been attributed to spores (Jones, 1992; Alonso-Zarza and Jones, 2007) and bacteria (Alonso-Zarza and Arenas, 2004).

(9) Other biogenic features include those resulting from the activity of soil fauna, including such features as faecal pellets, and traces of bees, wasps, termites and ants. Some of these structures or ichofabrics are described in Genise et al. (2010).

Beta calcretes are only known really from the late Palaeozoic to the present, and it is surprising that more records are not available from the Devonian, during which there was significant plant cover. The occurrence of a specific biogenic feature is related to the presence of the organism that produced it, which was the case for the distribution of 'classic' *Microcodium* during the Cretaceous and Lower Tertiary (Klappa, 1978), whereas

Microcodium β is widespread in the geological record since the Carboniferous (Goldstein, 1988; Alonso-Zarza et al., 1998a). In addition, considerations of preservation potential, for example the effects of diagenesis on some of the delicate biogenic structures such as the spherulites or the needle-fibre calcite, cause these features to be more commonly recognised in recent (Quaternary) calcretes than in older rocks (Alonso-Zarza and Arenas, 2004).

6. Groundwater Calcretes

Groundwater calcretes are non-pedogenic carbonates whose formation is due to interstratal cementation, displacement and replacement of sediment bodies by carbonates within shallow aquifer systems (Netterberg, 1969; Mann and Horwitz, 1979). The mechanisms of carbonate precipitation are mostly evaporation, evapotranspiration, CO_2 degassing and the common ion effect (Wright and Tucker, 1991). Groundwater calcretes were initially referred to as 'valley calcretes' (Butt et al., 1977) to describe the massive carbonate bodies associated with drainage channels. However, this term may include both pedogenic and non-pedogenic types, so the terms 'groundwater calcretes'or 'phreatic calcretes' are generally preferred.

Extensive groundwater calcretes (widths of more than 50 km) are well documented (Chen et al., 2002). Their formation depends not only on the availability of water, but also on the permeability of the material that they replace. Their shape is controlled by the drainage topography (Mann and Horwitz, 1979). In general they vary in thickness from several centimetres (Tandon and Gibling, 1997) to several metres (Arakel, 1986). The identification of very thick groundwater calcretes (Pimentel et al., 1996) has not always been clear; in the Sado Basin, for example, originally considered thick groundwater calcrete beds were later interpreted as palustrine carbonates (Pimentel and Alonso-Zarza, 1999). However, recent work by Jutras et al. (2007) described massive phreatic calcretes (>10 m), from the Carboniferous of Canada, similar to those forming around salt lakes in Australia.

The morphology and characteristics of groundwater calcretes are varied, and sometimes difficult to differentiate from pedogenic calcretes (Figure 3A). The six main types of groundwater calcretes are described in the following paragraphs.

(1) The first type consists of thick massive beds of carbonate that have been deposited by the lateral flow of groundwater. The thicknesses are commonly >1.5 m (Mack et al., 2000), but some authors consider that the beds should be thicker than 3 m (Wright and Tucker, 1991) or even 10 m (Jutras et al., 2007). Regardless, this type is commonly referred to as groundwater calcrete; the microfabric is alpha-dominated, the lower

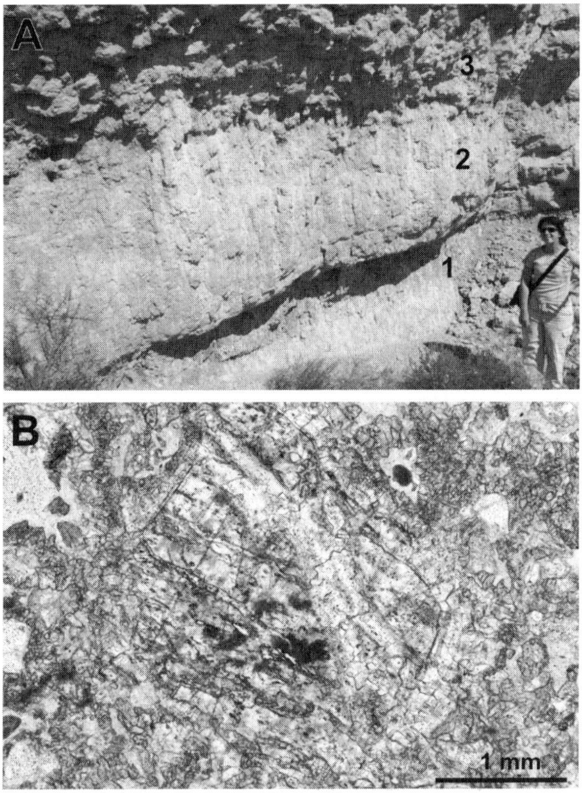

Figure 3 Groundwater calcretes. (A) Groundwater calcrete developed on previous pedogenic ones (1 and 2). Lighter vertical traces are due to roots. The deposition of a thin fluvial gravel bed (3) favoured the rise in groundwater and subsequent cementation of 1 and 2 (Miocene of the Madrid Basin). (B) Quartz grain etched and partially displaced/replaced by calcite (Sado Basin).

boundary is sharp and there is intense cementation and replacement of sediments by carbonate (Jutras et al., 2007), either calcite or dolomite (Spötl and Wright, 1992). This type of calcrete is classically described from around modern salt lakes in Central Australia (Arakel and McConchie, 1982; Arakel, 1986; Jacobson et al., 1988). In some cases, upward displacive growth of carbonate creates isolated or coalescent gentle mounds (Mann and Horwitz, 1979).

(2) The second type consists of soft carbonate nodules with diffuse boundaries that occur in layers which conform to the stratification of the sediment body, or which follow stratal planes of channels and mimic the convex geometry of the channel-fill deposits (Khadkikar et al., 1998).

(3) Cemented layers forming lenses up to 20 cm thick and 3 m long are the third type. They contain locally vertically elongated nodules (Tandon and Gibling, 1997). In some cases, they have an upper fringe of nodules and tubules precipitated at the water table and in the capillary fringe (Mack et al., 2000).
(4) Cemented proximal and medial alluvial fan facies as well as fluvial channels have been considered as a fourth type of groundwater calcretes (Tandon and Narayan, 1981; Nash and Smith, 1998). Their thickness and extent depend on the dimensions of the sedimentary body which undergoes cementation.
(5) The fifth type consists of thin calcified root mats; these have been included as groundwater carbonates because they may have been developed by phreatophytes in relation to very surficial, perched groundwater tables (Semeniuk and Meagher, 1981). In the Upper Cretaceous of India, for example, a shallow groundwater table favoured the formation of both root mats and gleyed horizons (Ghosh, 1997).
(6) Thin sheets (10–50 cm) in the subsurface of fluvio-aeolian sands (Purvis and Wright, 1989) form the sixth type. These sheets consist of aggregates of calcium carbonate developed just above the water table, so they may transect stratigraphic boundaries and unconformities (Semeniuk and Meagher, 1981).

Several attempts have been made to identify criteria to differentiate between groundwater and pedogenic calcretes (Wright, 1995; Pimentel et al., 1996). Groundwater calcretes commonly show sharp basal and top contacts. They are mostly massive bodies lacking any internal horizons. They commonly lack vertical root traces and peds (Mack et al., 2000), and are not overlain by horizons of illuviated clays (Mack and James, 1992). They also lack lacustrine biota or any indication of carbonate precipitation within a free water body. Groundwater calcretes are more common in more permeable coarse channel sediments (Nash and McLaren, 2003; Wright, 2007).

The micromorphology of groundwater calcretes is characterised by the absence of biogenic features. They are therefore encased in the so-called 'alpha' microfabrics (Wright and Tucker, 1991). They consist frequently of crystalline mosaics, with crystals varying in size from microns to millimetres, etched and floating grains (Figure 3B), nodules and a variety of desiccation features. The chemistry of phreatic water controls the mineralogy of the groundwater precipitates. In the proximity of the catchment areas, groundwater is commonly of low salinity and calcite is the main precipitate, but groundwater movement from the catchment area down to the playa-lake marginal discharge areas favours progressive concentration (Arakel, 1986). Changes in groundwater chemistry explain the formation of groundwater dolocretes and gypcretes towards the distal part of closed basins (Arakel, 1986; Armenteros et al., 1995; Schmid et al., 2006; Khalaf, 2007). Groundwater dolocretes show a wide range of crystal sizes and include spheroidal (Spötl and Wright, 1992) and zoned dolomite crystals, as well as dolomite with cloudy nuclei (Pimentel et al., 1996). Groundwater-dolocrete formation may also be favoured by the mixing of groundwaters and lake brines (Colson and Cojan, 1996), or with sea water (Williams and Krause, 1998). Barite is uncommon but found where groundwaters are more saline (Khalaf, 2007).

Groundwater calcretes generally occur in arid to semi-arid climates. Climate controls their formation for three reasons (Mann and Horwitz, 1979): (1) conditions of continual moisture favour carbonate dissolution, not concentration; (2) intermittent heavy rains tend to form better groundwater systems (due to more effective infiltration) than the equivalent rainfall spread over a longer timespan; and (3) high evaporation and evapotranspiration rates are essential for chemical precipitation of carbonate. In Western Australia, the active zone of groundwater formation occurs where the water table lies at depths of 2–5 m. In such arid environments, evaporation and evapotranspiration from the water table are insignificant below 5 m.

Rates of groundwater calcrete formation are difficult to establish, because the relationships with the under- and overlying sediments are not necessarily ordered stratigraphically, that is, groundwater calcrete is commonly younger than the overlying sediments. Whatever the case, as groundwater calcretes are supplied by calcium carbonate from the phreatic waters, and not limited by atmospheric input or weathering, it is likely that the rate of formation exceeds that of pedogenic forms (Wright, 2007). For example, studies on the Kalahari calcretes have effectively shown that 4-m-thick groundwater calcretes may form in less than 3,000 years (Nash and McLaren, 2003).

7. Pedogenic Profile Development

7.1. The idealised or classic profile

Gile et al. (1966) proposed that the morphology of calcic soils can be seen as a sequence of morphological stages that reflect the different degrees of development (relative time of development) of the soil. Within Stages I–III, the gravel content is important and is different in fine and coarse clastic deposits (Figure 4A). In gravel-rich calcic soils, Stage I is characterised by a thin, discontinuous coating on pebbles. In Stage II, the coatings are continuous and vary in thickness. Massive accumulations between clasts and fully cemented gravels are included in Stage III. In gravel-poor soils, Stage I shows few filaments or faint coatings on ped surfaces. Soft nodules, 5–40 mm in diameter, are indicative of Stage II, whereas coalescent nodules

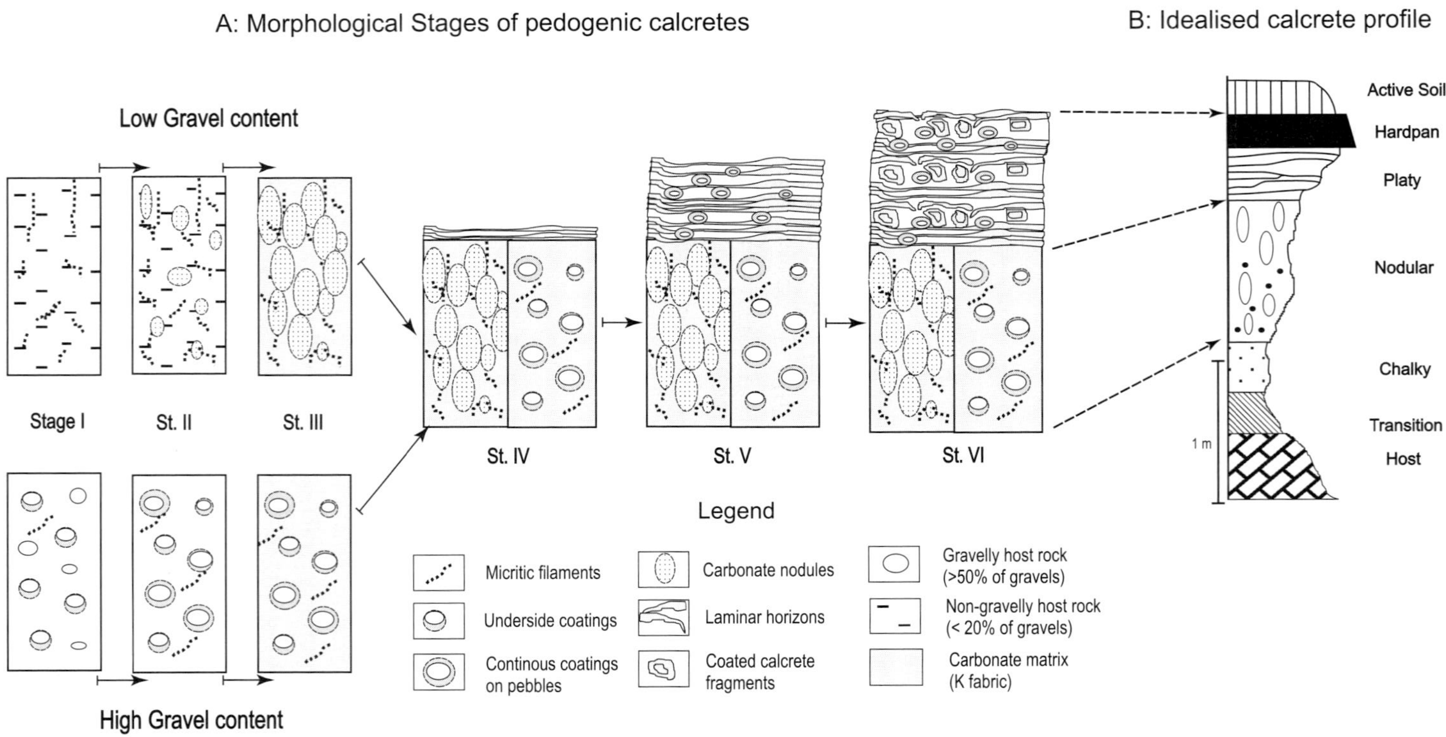

Figure 4 Stages of development of pedogenic calcretes (modified from Alonso-Zarza, 2003). (A) Based on Machette (1985). (B) Schematic idealised pedogenic profile (Esteban and Klappa, 1983), showing the relationship between the different horizons and the more advanced stages of development.

characterise Stage III. Machette (1985) established six stages, the first three similar to those previously established by Gile et al. (1966). Stage IV is characterised by carbonate-rich laminae less than 1 cm thick. Thicker laminae and pisoliths are indicators of Stage V, while Stage VI is characterised by multiple phases of brecciation, pisolith formation and recementation.

Based on numerous observations, Esteban and Klappa (1983) described an idealised calcrete profile (Figure 4B), in which the stages of formation of the different horizons are similar to those of the Gile et al. (1966) and Machette (1985) models. The profile consists of the following horizons (from base to top and including the host).

(1) *Host material.* This may be of any composition, texture and degree of compaction. Permeability and calcium-carbonate content may affect the degree of calcrete development (Wright, 1990c). The host material lacks any calcrete features and so is distinguished from the overlying calcrete horizons.
(2) *The transitional horizon.* This is the zone of *in situ* weathering of the host through mechanical, physicochemical and biological processes to form a regolith of weathered detritus. Relic primary structures of the host are commonly preserved. Its lower boundary is difficult to define and lies between the host material and the well-defined uppermost calcrete horizons. It indicates incipient soil development through changes produced by the action of organisms and by the movement of water through the host rock.
(3) *Chalky horizon.* This is a soft horizon consisting of a micrite and/or microspar matrix that contains etched detrital grains and peloids. It tends to be homogeneous, texturally and structurally, although some nodules are spatially related to roots. It is commonly located between the transitional and the nodular horizons, but it may occupy any other position within the profile or be absent. Precipitation of carbonate without significant induration leads to the formation of this horizon.
(4) *Nodular horizon.* This horizon is formed by powdery to indurated nodules of calcium carbonate embedded in a less carbonate-rich matrix. The nodules vary in morphology between vertical, horizontal, irregular or even branching. In cases where the nodules are vertically elongated (Figure 5A), the horizon is sometimes called the 'prismatic horizon'. Nodular horizons tend to show diffuse lower and upper boundaries (Figure 1B). Microscopically, the nodules are composed of micrite rich in etched grains, relics from the host material. Coated grains in which the nucleus is an etched grain are also common. Carbonate precipitation takes place in discontinuous areas in close association with roots and related micro-organisms. The biological components of the soil become calcified, forming rhizoliths, calcified filaments and

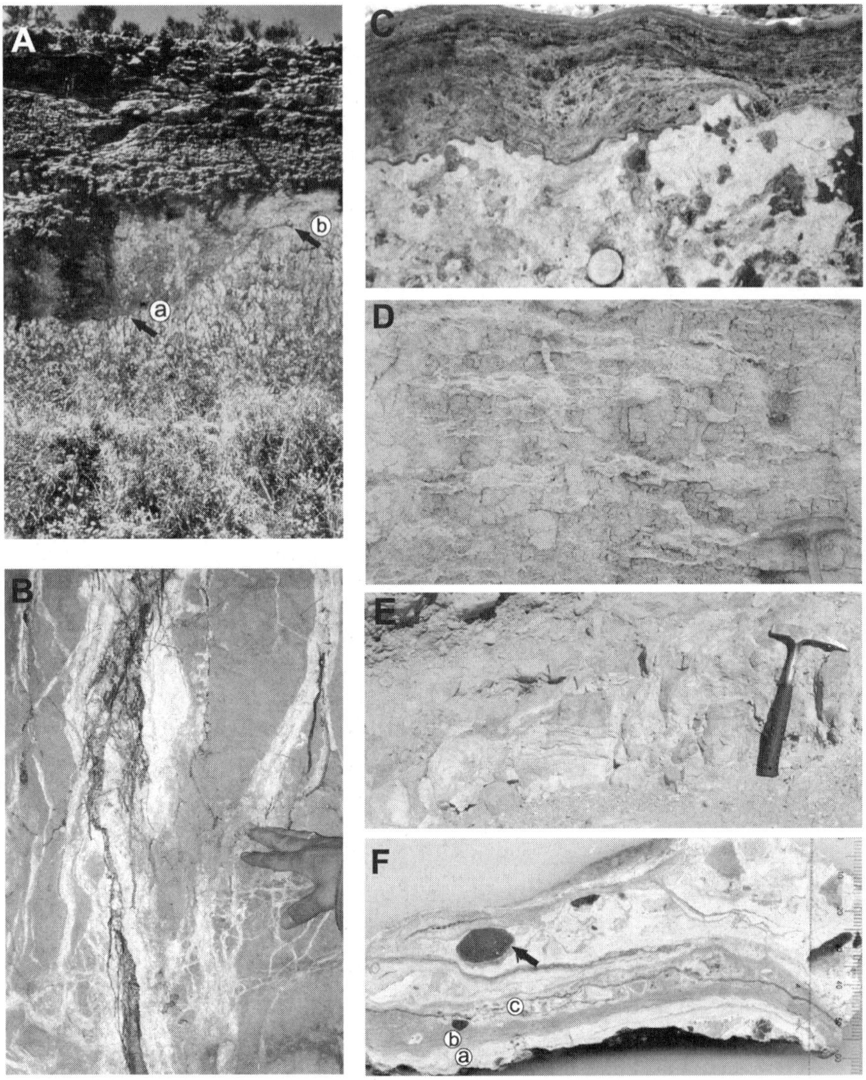

Figure 5 Laminar calcretes. (A) Thick 'multi-episodic' calcrete profile. The nodular horizon is partially eroded and the trough is filled by coarse sand (a). The laminar horizon is situated at the top. Although the nodular and laminar horizons are in lateral contact (b), important erosion is recorded in this calcrete (Pleistocene, Murcia, Spain). (B) Vertical calcrete development by penetration of tree roots (recent, Brihuega). (C) Laminar calcrete at the top of a limestone bed. The contact is very irregular (Quaternary, Florida). (D) Incipient laminar calcrete interbedded with clays (Miocene, Madrid Basin). (E) Calcrete developed at the Jurassic/Palaeogene unconformity; coarse coated clasts are visible (Teruel Basin). (F) 'Multi-episodic' laminar calcrete. Laminated micrite (a and b) alternate with laminae containing gravel and sand-sized clasts (c). Some of the clasts show coatings (arrowed) at their lower sides (Teruel Basin).

nodules. Vertical water movements and vertical root systems favour the formation of vertically oriented carbonate nodules. Initially they are much dispersed, but with time they coalesce to form the nodular horizon. These processes lead to the formation of the morphological Stages I–III of Machette (1985).

(5) *Platy horizon*. This commonly occurs overlying the nodular horizon (Figure 5A). When a hardpan is present at the top of the profile, the platy horizon underlies it. However, if a hardpan is lacking, the platy horizon is the topmost horizon of the calcrete profile. This horizon has also been called the 'laminar horizon'. Some calcretes composed almost exclusively of the laminar horizon are named 'laminar calcretes'. These are discussed in more detail below. The platy or laminar horizon has a wavy to thinly bedded habit, planar fracture porosity, and an abundance of alveolar-septal structures, rhizoliths and needle-fibre calcite. Internally, the different laminae show varied microfabrics that include micritic layers, micritic layers with tubiform pores, laminae very rich in alveolar-septal structures and laminae including micrite coated grains.

Once the nodules coalesce to form a dense horizon, root systems cannot easily penetrate the nodular horizon. In addition, water is mainly confined to the uppermost part of the profile in the still-unconsolidated zone above the nodular horizon. The morphology of the root systems therefore changes. Roots trying to maximise the amount of water they absorb tend to extend laterally, promoting the development of sub-horizontal networks. The laminar horizon starts to form in the still-unconsolidated zone. This stage includes Stages IV and V of Machette (1985).

(6) *Hardpan*. In very well developed profiles, this is commonly the topmost horizon. It is well indurated and the porosity is very low. Macroscopically it may be structureless, massive, laminated (Figure 1A) or nodular. This horizon is commonly formed by micrite containing corroded grains that are rarely coated. Laminated micritic layers may also be present. Thick hardpans are commonly fractured and brecciated, allowing the identification of the *brecciated horizon* (Stages V and VI of Machette,1985). Differentiation between the platy horizon and the hardpan may be difficult, but hardpans have a more massive appearance in the field.

(7) *Pisolithic horizons*. These consist of sand- or gravel-sized clasts coated by laminated micrite and are very common in calcretes developed on coarse-grained host rocks (Figure 2B). In many cases, the horizons follow the geometry of the coarse deposit on which they developed (Alonso-Zarza et al., 1998b). The micrite laminae may coat all the clasts or only their undersides. The laminae are composed of dense micrite, micrite with alveolar-septal structures and/or microspar. The coated clasts are embedded in a dense matrix of irregular masses of microspar including some detrital grains and micrite with alveolar-septal

structures. Pisolithic horizons are common at the top of the calcrete profile when intense brecciation favours the formation of calcrete-sourced clasts, but also occur interbedded within the profile.

7.2. The rhizogenic model: laminar calcretes

Studies carried on in various calcretes from southern Europe (Wright et al., 1988, 1995) have shown that isolated (in millimetres) to thick (>2 m) laminar calcrete beds can form through the calcification of root mats. The root mats show fenestral textures representing fine root tubules with or without alveolar-septal structures, but in other cases, they are formed entirely of *Microcodium* remains.

Laminar calcretes may occur in different situations (Figure 5A–E): (1) at the top of thick calcrete profiles, but below the top soil (Figure 5A), (2) cutting any rock or sediments, being oblique to the stratification or even with a mainly vertical development (Figure 5B), (3) at the top of any bedrock (Figure 5C) or (4) interbedded within sedimentary deposits (Figure 5D). The formation of these laminar calcretes has been widely discussed (Verrecchia et al., 1995; Wright et al., 1996; Freytet et al., 1997; Alonso-Zarza, 1999), and there is a substantial consensus that they should be interpreted as rootcretes (Jones, 1992) or rhizogenic calcretes (Wright et al., 1995), since it is commonly accepted that the main agents responsible for their formation are horizontal (Mack and James, 1992; Wright et al., 1988) and vertical (Alonso-Zarza and Jones, 2007) root systems. However, when the calcretes occur at the top of the soils or bedrock, other organisms can operate. Cyanobacteria (Vogt, 1984; Verrecchia et al., 1995), bacteria, fungi (Verrecchia and Verrecchia, 1994) and lichens (Klappa, 1979) may account for the formation of laminar carbonate horizons.

Laminar calcretes can be misinterpreted as stromatolites (Wright, 1989). The main features of calcretes that allow their differentiation from stromatolites are: (1) the laminae are very irregular and show micro-unconformities attributable to phases of dissolution (Wright, 1989); (2) the laminae include etched grains, ooids and clays; (3) rhizoliths, alveolar-septal structures and spherulites are very common in laminar calcrete horizons (in some cases they may form the whole horizon) (Wright et al., 1988); and (4) alternations between micrite laminae and others rich in detrital sediments, ooids or coated micritic grains are common (Fedoroff et al., 1994; Sanz and Wright, 1994). Calcified filaments are common in both laminar calcretes and stromatolites, but in the latter they are commonly oriented perpendicular to the lamination, whereas in calcretes they show no preferred orientation.

When laminar calcretes occur at the top of a calcrete/soil profile or at the top of any type of bed rock (Figure 5E), their formation may be controlled by the presence of shallow-water tables and thin pond-water films on top of the soil surface. Very commonly, these calcretes contain spherulites and features related to the calcification of algae and/or cyanobacteria, which are phototrophic communities (Verrecchia et al., 1995). Both biochemical and physicochemical processes can contribute to the lithification of the laminae, for example through the precipitation of micrite by increased carbonate concentration of the ponded or capillary zone water, and also to the formation of non-pedogenic structures, such as coarse spar calcite cements. As laminar calcretes form at the soil/atmosphere interface, their presence indicates phases of direct subaerial exposure (Blumel, 1982) and so may reveal a phase of exhumation of the soil or bed rock.

7.3. Alteration model

This model relates to calcrete development on a calcareous host and does not involve the net addition of carbonate to the soil profile, but involves largely the redistribution of existing carbonate, although some new material is likely to have been added. As discussed by Rabenhorst et al. (1991), the calcrete development is via an alteration front. No clearly defined stages occur comparable to those seen in Aridisols as discussed above. In studies of calcretes developed on weakly consolidated Quaternary limestones in carbonate terrains such as the Bahamas and Yucatan, a key feature of profile development is the formation of calcrete stringers and mottled textures (Wright, 1994). In the case of the latter, the lack of displacive growth of carbonate leads to mottling effects in the porous host, rather than to nodule development. Stringers are calcified root zones and can be very extensive, being traceable in anastomosing sets for tens of metres or more within the profile in excavations of cliff sections.

7.4. Calcrete dynamics: sedimentation/erosion/calcretisation relationships

Classically calcretes have been interpreted as a result of continuous and aggrading pedogenesis in which the ratio between sedimentation rate and calcrete development results in the different stages of calcrete development (Leeder, 1975; Wright and Marriott, 1996). Even if the rainfall and the atmospheric input of carbonate could remain constant during long periods of time, the position of the land surface relative to the zone of carbonate accumulation in the soils is unlikely to have remained unchanged (Wright, 2007). Studies of Quaternary (Alonso-Zarza et al., 1998b) and even Palaeozoic (Marriott and Wright, 1993) calcretes have revealed the interplay of pedogenesis, episodic sedimentation and erosion. The interplay of these processes can be imprinted at a profile-horizon scale (decimetres to metres), and even at very small scale, such as the individual laminae (millimetres to centimetres) of the laminar calcretes (Figures 5F and 6) (Alonso-Zarza and Silva, 2002).

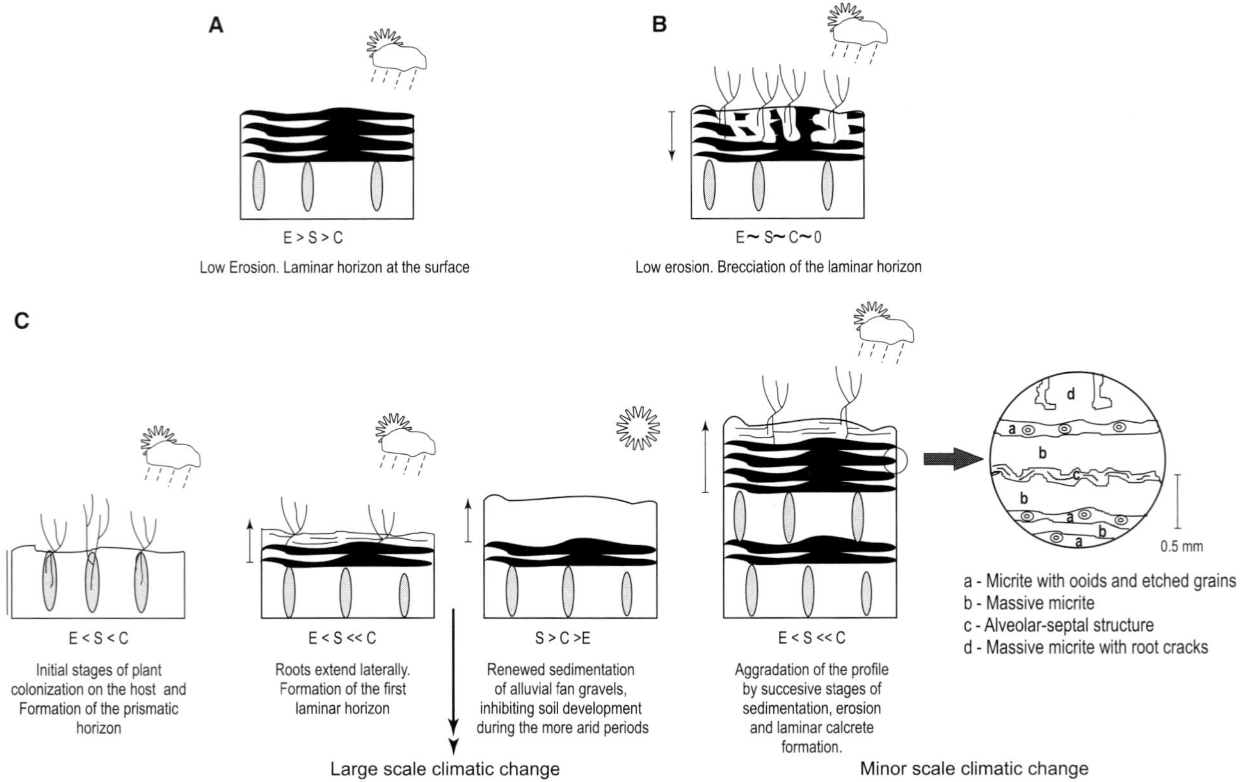

Figure 6 Final stages of the development of calcrete profiles depending on the overall rates of E (erosion), S (sedimentation) and C (calcrete formation). (A) Erosion is low, but higher than sedimentation, and calcrete formation is very slow. The development will finish with the exposure of the laminar horizon. (B) If erosion, sedimentation and calcrete formation are all close to zero, the result is the brecciation of the laminar horizon. (C) The development of thick calcrete profiles includes complex processes, with phases of erosion and sedimentation during the calcrete formation, so thick laminar calcrete profiles will form. Small-scale climatic changes cause the formation of different microstructures in the same laminar horizon, whereas larger scale changes will induce changes in the main horizon (from nodular to laminar). Different E, S and C ratios are shown to indicate how they interplay to form thick calcrete profiles. Based on Alonso-Zarza et al. (1998b) and Alonso-Zarza and Silva (2002).

At the profile-horizon scale, calcretes may be relatively complex, as similar horizons may occur at different positions within the calcrete profile, indicating composite profiles. In contrast, significant horizons may be lacking due to truncation by erosion during profile development (Alonso-Zarza et al., 1998b). Erosion causes reworking and the lowering of the carbonate profile by leaching and translocation to lower levels (Elbersen, 1982). In some cases, small-scale channel incision has truncated calcrete horizons, and the channel deposits have later been capped by a laminar calcrete that extends from the truncated to the non-truncated parts of the profile (Figure 5A). This indicates the complexity of calcrete profiles at the local scale of some profiles, as such erosive events that cause incompleteness may be not recognised.

The formation of thick calcrete profiles is often the result of several phases of erosion, soil formation and sedimentation (Alonso-Zarza and Silva, 2002) which, in many cases, are repeated over time. This is especially noticeable in the development of calcretes with a thick laminar calcrete at top (Stages V and VI). The degree of development, as well as the characteristics and thickness, depend on the time the root systems can be supported in the upper soil horizon. Alonso-Zarza et al. (1998b) considered the following three different situations (Figure 6).

(1) Cases where the erosion rate is low but exceeds the sedimentation rate. The upper part of the calcrete profile (B horizon) is removed and the laminar horizon is exposed directly to the atmosphere (Figure 6A). Lichens (Klappa, 1979), spherulites may grow in these superficial conditions, even in the presence of thin water bodies. Karstic microforms may also be present.
(2) Erosion and sedimentation at the top of the profile are reduced. Accumulation of calcium carbonate is continuously increasing, replacing even the uppermost soil. Once the top of the soil is totally replaced, there is no space any more for new carbonate accumulation, so calcrete development is inhibited. Pedogenic and later diagenetic processes lead to the lithification of the soil profile and to the formation of the hardpan. Later weathering of the profile causes the brecciation of the uppermost part (Figure 6B). This is Stage VI of Machette (1985).
(3) Deposition is low and episodic, but exceeds the erosion rate. This contributes new surface sediments for soil organisms and the subsequent development or maintenance of root systems. This favours the formation of very thick laminar horizons (Figure 6C); these do not represent individual events, but rather the addition of multiple phases of sedimentation and soil-formation processes, giving place to the formation of thick, cumulate profiles, such as those described from the western USA (Machette, 1985) or the Pleistocene of the Canary Islands (Alonso-Zarza and Silva, 2002).

At a lower scale (millimetres to centimetres), these laminar calcretes are formed by centimetre-scale alternations in which the different laminae may consist of micritic layers, micritic layers with fine tubiform pores, ooids, detrital grains and clays, and/or of micrite with an alveolar-septal structure (Figure 6C). These alternations reflect the small-scale periods of sedimentation, erosion and soil formation (Fedoroff et al., 1994) in the upper part of a relatively stable surface, and may indicate climate-related vegetation changes (Alonso-Zarza and Silva, 2002). The occurrence of these laminae interbedded with detrital sediments characterises environments in which sedimentation was slow and episodic. Therefore, after detrital sediment input, the surfaces became stable and root mats developed. Renewed sedimentation accounted for the development of new laminae on the new surfaces. The formation of these calcretes is essentially through vertical aggradation of successive laminae of ooids, sediment and massive micrite towards the top of the profiles (Figure 6C).

Less is known about the dynamic behaviour of calcrete profiles on carbonate substrates. Both Davies (1991) and Sattler et al. (2005) recorded a marked lateral variability in the types of calcrete development at exposure surfaces in Carboniferous and Cretaceous successions, respectively. These differences likely reflect local micro-topographic differences and preservational effects. Budd et al. (2002) noted that, in their studies of the development of macroscopic features at a range of carbonate exposure surfaces of different durations, the types of features developed were not directly related to the duration of exposure. These studies indicated that a simple model of calcrete development as a linear progression, rather like the illuvial models of Gile et al. (1966) and Machette (1985), may not apply to calcretes developed in carbonate-platform settings.

8. Calcretes in Quaternary Landscape Systems

Many of the 'classic' studies on the use of soils – and specially calcretes – for local and regional correlations are based on Quaternary case studies. The work of Machette (1985), in calcic soils of the southwestern USA, demonstrated that the total-profile index of secondary carbonate can be used to correlate and differentiate Quaternary sediments, to estimate the age of pedogenic calcretes, and even to estimate the recurrence intervals and the frequency of fault movements across fault zones.

Absolute dating of Quaternary calcretes allows more precision in the establishment of the time needed to form the different morphological stages. Radiocarbon dating of Quaternary calcretes has been carried out by several authors (Rowe and Maher, 2000; Deutz et al., 2002, amongst many others), pointing out the problems of mixing and later alteration by diagenesis, but providing good results. In addition, the development of the

U–Th series method since the last decade allows the determination of the age of calcretes with certain accuracy. The age is calculated from $^{230}Th/^{234}U$ and $^{234}U/^{238}U$ ratios of the sample (Ivanovich and Harmon, 1992). Uranium is very soluble in oxidised groundwaters, and co-precipitate with calcium during calcrete formation, whereas ^{230}Th, is immobile in near-surface environments and does not co-precipitate in the calcrete. The subsequent increase of ^{230}Th provides a tool to measure the time since carbonate was precipitated because it is produced later on. The upper limit of the technique is 350 ka (Rowe and Maher, 2000). This method has been applied to date calcretes in alluvial terraces of the Sorbas Basin in Spain (Kelly et al., 2000; Candy et al., 2004). In addition to U–Th, the U–Pb system can also be used to date much older calcretes (Wang et al., 1998; Rasbury et al., 2006), with different grades of uncertainty from more than 20 Ma to less than 3 Ma. The maximum age of calcretes that can thus be dated reliably may reach the Carboniferous (Rasbury et al., 2000).

In the Cinca and Gállego rivers, southern Pyrenees, calcretes that developed on stabilised surfaces on terraces, show a clear tendency of enrichment of carbonate from the lower to the upper terraces. The chronological data obtained by palaeomagnetism and luminiscense indicate that Stage I forms in 10 ka, while Stages IV and V need 600–800 ka (Sancho et al., 2004).

The calcrete developed at the top of the infill of the Guadix Basin has provided numerous data for discussion. Its age (42.6 $\pm$ 5.6 ka) is considered to be the age of the cessation of the active sedimentation within the basin. Since then, the incision of the Arroyo Gor began, and the calcrete age gave an estimation of the incision rate which was high (4 mm year^{-1}) (Azañón et al., 2006). Detailed isotopic analyses in this laminar calcrete also suggest different processes of formation of the laminae, which are the result of differences in climatic conditions. The massive micritic laminae show the heaviest carbon- and oxygen-isotope values and represent the most arid conditions, whereas laminae dominated by alveolar-septal structures are lighter in carbon, which is interpreted as an indication of a denser vegetation cover under a less arid climate.

Similar climatic alternations seem to be reflected in many Quaternary calcretes, as in those described from the Canary Island (Alonso-Zarza and Silva, 2002). However, more studies are needed to better constrain the climatic conditions and the time represented by each lamina. In doing so, Quaternary calcretes will potentially provide good data sources for the interpretation of the effects of climatic changes that can be applied to older calcretes. For example, the presence of poorly developed calcretes containing megarhizoliths in the aeolianites of Gran Canary in the Canary Islands has been considered as a proof that less arid conditions prevailed during certain Quaternary periods, allowing the development of a sparse vegetation cover (Alonso-Zarza et al., 2008). Pedogenic Quaternary calcretes also appear as bodies of restricted lateral continuity on colluvial deposits. In this case, calcretisation is a response of the periods of lesser tectonic activity and greater stability of the slope (Jiménez-Espinosa and Jiménez-Millán, 2003).

Groundwater calcretes are common in mudflat environments of playa lakes that act as outlets for discharge of regional groundwater (Arakel, 1991) in the vicinity of evaporitic basins (Jutras et al., 2007), as well as in different types of alluvial fans (Mack et al., 2000). Groundwater calcretes influence the evolution of the fan and overall drainage network (Nash and Smith, 2003). This is because both the decrease in water/sediment supply and the calcrete lithification may isolate the alluvial fans from the main drainage network (Stokes et al., 2007), favouring stable surfaces and pedogenic development (Nash and Smith, 1998; Wright, 2007).

Mixing between groundwater and pedogenic calcretes is common in alluvial fan and fluvial deposits. Groundwater cementation commonly occurs from the boundary with the finer deposits towards the top of the gravel channels, whereas pedogenic calcretes occur at top of the fluvial channels and finer deposits to the base of the sedimentary bodies. Pedogenesis occurs after the sedimentation and preceding the following sedimentary bed. On the contrary, groundwater cementation may proceed in different periods. In some cases, the increase of tectonic activity may introduce increased amounts of groundwater through the more permeable beds, favouring groundwater calcrete formation (Stokes et al., 2007). Mixed pedogenic and groundwater calcretes are also common in the deposits of interfluve channels in Late Quaternary interfluves from the Ganga Plains (Sinha et al., 2006). Their distribution in the alluvial architecture is a response to aggradation and degradation of the plains caused by monsoon-driven rhythms. The common imprint of pedogenic effects on groundwater features and vice versa offers new data for a better understanding of the overall alluvial system; groundwater calcretes must therefore be included within the pedo-stratigraphic models for alluvial fans (Stokes et al., 2007).

9. Calcretes in a Stratigraphic Framework

The spatial variability of pedogenic calcrete development within a sedimentary basin will ultimately reflect variations in the residence time in the soil of the zone of carbonate accumulation, as well as the particle size of the host sediment (Wright, 2007). These factors are controlled by the sedimentation, subsidence and calcretisation rates, which are all strongly related to tectonism and climate. Erosive processes must also be considered. When erosion is insignificant, the variation in palaeosol types is a response to the type and rate of sedimentation versus the rate of pedogenesis

(Marriott and Wright, 1993). If sedimentation is rapid and unsteady, weakly developed and vertically stacked profiles separated by minimally weathered sediment (compound palaeosols) form. Vertically successive profiles may partially overlap (composite palaeosols) if the rate of pedogenesis exceeds the rate of sedimentation. If sedimentation is steady, thick cumulative soils can form (Marriott and Wright, 1993; Kraus, 1999); thick laminar calcretes are a good example of this situation (Alonso-Zarza et al., 1998b). The presence or absence of a particular genetic stage of calcrete may be used as a rough estimate of ancient flood-basin accretion rates (Leeder, 1975; Wright and Marriott, 1996).

Variations in sedimentation, subsidence and erosion rates within a specific basin are controlled by both autochthonous and allochthonous processes. In alluvial basins, the sedimentation rate decreases across the floodplain with distance from the channel; this is an autochthonous process. The term 'pedofacies' refers to 'laterally contiguous bodies of sedimentary rocks that differ in their contained laterally contiguous palaeosols as a result of their distance (during formation) from areas of relatively high sediment accumulation' (Bown and Kraus, 1987). The pedofacies relationship explains why areas closest to the alluvial channels have thick, poorly developed palaeosol profiles, whereas more distal floodplain areas commonly exhibit different types and better developed soils.

Pedofacies relationships have been recognised in many ancient alluvial successions (Bown and Kraus, 1987; Smith, 1990; Alonso-Zarza et al., 1992b) and seem to be most applicable to overbank deposits (Kraus, 1997). However, they have not been seen in all floodplain palaeosol successions (Wright, 1992). Three main causes may explain the lack of pedofacies relationships in floodplain settings: (1) very low sedimentation rates may favour soils reaching a steady state, thus erasing pedofacies variations (Kraus, 1999); (2) floodplain aggradation may not only be the result of true suspension-load settling but also of deposition of laterally extensive crevasse-splay lobes (Behrensmeyer et al., 1995) and even by sheet floods containing pedogenic mud aggregates (Muller et al., 2004); and (3) in poorly drained soils, the intensity of soil development may mask the lateral variations in maturity (Kraus, 1997), and soil properties are more directly controlled by hydrology. In sum, pedofacies mostly result from changes in sedimentation rate at the same stratigraphical level, although they can also involve differences in depth to the water table at different locations on the floodplain.

Changes in floodplain aggradation through stratigraphic sections (with time) can also be estimated. Wright and Marriott (1996) developed a quantitative and sophisticated model to estimate the rates of floodplain aggradation using calcrete-bearing palaeosols. These authors considered that the residence time of the sediments in the zone of active pedogenesis is controlled by the frequency of the depositional events and by the thickness of sediment deposited during each event. Both can be represented in a plot

Clearly, climate controls the degree of development of palaeosols, in part because of its influence on sedimentation rates, and also because the climatic effect may overprint the effects of other controls (Tanner, 2010). Stable-isotope analyses are now considered a crucial tool for obtaining palaeoclimatic information (Tanner, 2010). In palaeosols from the late Eocene of India, for example, the C and O isotopes reveal that calcrete formation occurred during drier intervals, whereas the host mudstones were deposited during warm and wet intervals (Singh and Lee, 2007).

Tanner and Lucas (2006) demonstrated that the upward stratigraphic change from kaolinitic palaeosols with gley features (in Carnian palaeosols) to Stage IV calcretes (in Norian palaeosols) in the Chinle Group in the southwestern USA relates to gradual aridification during the Upper Triassic. However, in many situations it is difficult to determine if the variation in the type and development of calcretes is due to intra- or extrabasinal forces. For example, the environmental changes reflected in the calcrete-bearing palaeosols in the Triassic of NW Argentina respond to the geomorphological position (catenary relationship) and the structural setting of the palaeosol (Tabor et al., 2006).

Calcrete-bearing soils and palaeosols are common in arid and semi-arid alluvial fans. Their formation takes place on stable fan surfaces, on which levels of erosion or sedimentation are reduced. This stability may occur over the whole fan due to a climatic change, or because the locus of deposition/erosion has moved by entrenchment, avulsion or dissection (Wright, 1992), which will create different distributions of the soils along the fan (Wright and Alonso-Zarza, 1990). Non-entrenched fans show weakly developed soils in proximal areas and more developed ones in the distal parts (according to McCraw, 1968). In entrenched fans, either partially or totally, the isolated (entrenched) areas will show more prolonged pedogenesis, while the areas where sedimentation takes place will show less-developed soils (McCraw, 1968; Talbot and Williams, 1979). In cases where climatic changes have caused the retardation or cessation of sedimentation, soils will develop on the whole fan surface (Talbot and Williams, 1979). Similar patterns can be observed if the causes of non-sedimentation are induced by base-level changes (Gile et al., 1981).

In many cases, alluvial basins are closed drainage systems in which the central areas are occupied by lakes, either fresh-water or saline. It seems that the existence of a long-lasting evaporite basin is a major control on the formation of thick groundwater calcretes (Jutras et al., 2007). Closed carbonate or evaporitic lacustrine basins commonly show low relief, flat topographies; small changes in the amount of surficial water input, or in climate, therefore favour movements of groundwaters that result in the mixing of pedogenic, groundwater and palustrine or playa water. Consequently, it can be difficult to differentiate amongst pedogenic and groundwater calcretes on one hand, and palustrine carbonates on the other (Alonso-Zarza, 2003). Figure 7 illustrates how a gradual rise of the water

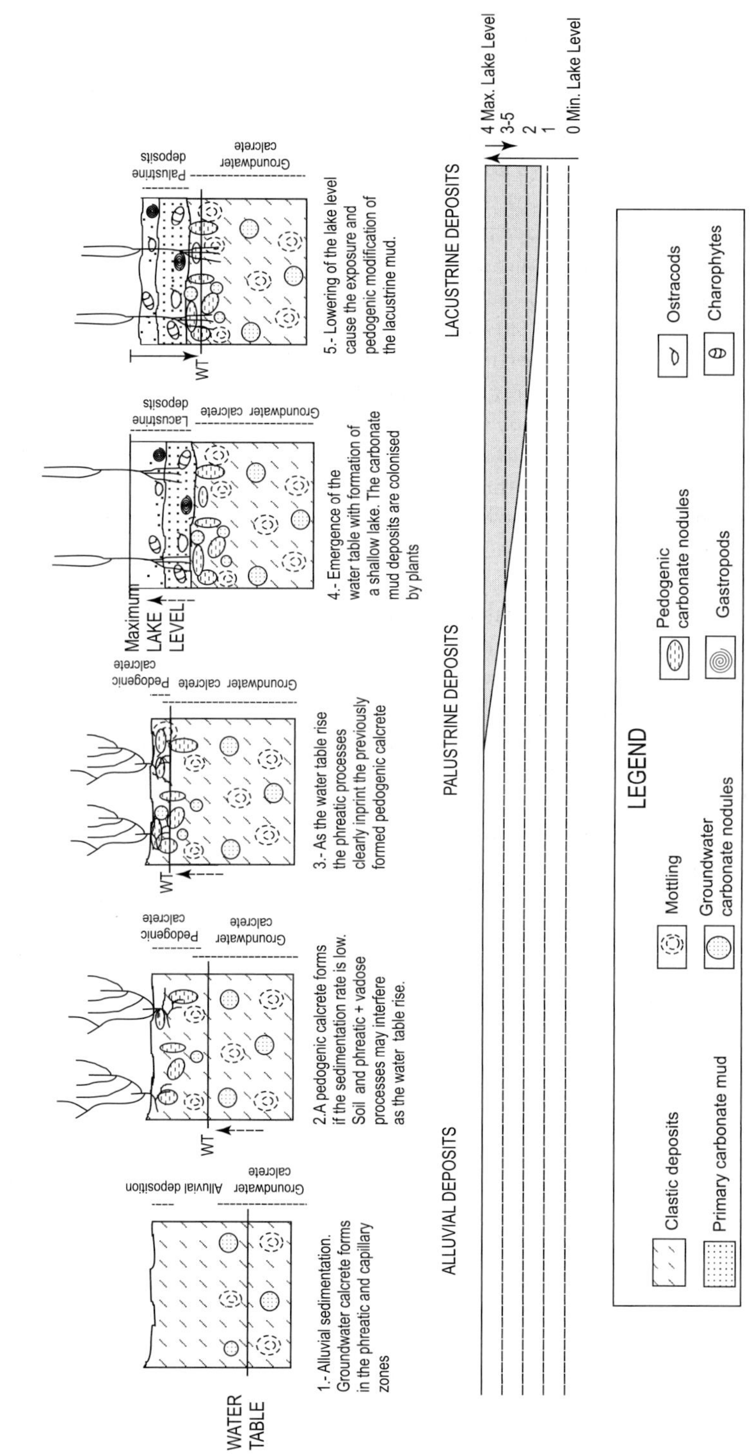

Figure 7 Sketch of the lateral and vertical transitions of alluvial, palustrine and lacustrine deposits in a distal alluvial and/or floodplain environment. The different situations 1–5 indicate the position of the water table (Modified from Alonso-Zarza, 2003).

table results first in the modification of previously formed pedogenic carbonates by groundwater, and later in the emergence of a free water body on the surface. A further lowering of the water table causes the previously deposited lacustrine sediments to be situated in either the vadose or the phreatic zone, where they can be modified by either pedogenic or groundwater process. The transition from pedogenic to groundwater calcretes, ending in palustrine carbonates, is seen in the Tertiary from the Duero Basin (Armenteros and Huerta, 2006) and has been interpreted as the result of a progressive decrease in sediment supply, very probably controlled by climate.

Calcretes, and indeed all other types of palaeosols, are also commonly associated with unconformities. Therefore, they can often be used as sequence boundaries (McCarthy and Plint, 1998) in non-marine deposits. The maturity of the palaeosols may give an idea of the type and range of the sedimentary discontinuity. In the Mississippian of the Appalachian Basin, Etthenson et al. (1988) described well-developed and thick caliche profiles on disconformities related to important tectonic activity or regional regressions; less mature profiles occur within sedimentary discontinuities related to local regressions. Laminar calcretes are frequently located at important unconformities, such as in Minorca, where Carboniferous deposits are penetrated and overlain by a laminar calcrete developed prior to the deposition of the Permian red-bed deposits (Gómez-Gras and Alonso-Zarza, 2003). In the Teruel Basin in NE Spain, a laminar calcrete, probably early Palaeogene in age, developed on Jurassic marine deposits. This calcrete contains *Microcodium* and probably marks the Mesozoic/Cenozoic boundary (Alonso-Zarza and Arenas, 2004). Calcretes as indicators of unconformities are also recognised and used in subsurface stratigraphy (Hanneman et al., 1994; Hanneman and Wideman, 2006, 2010), providing one more element of basin-architecture analysis in buried deposits.

10. Conclusions

Calcretes are near-surface accumulations of calcium carbonate of pedogenic origin or formed by laterally flowing, shallow groundwaters, mainly in alluvial aquifers. Earlier views that pedogenic calcretes were formed by precipitation from groundwaters have long since been dismissed. Now we better appreciate that they truly are pedogenic precipitates, with carbonate sourced largely from the atmosphere, and with precipitation triggered in many cases by the activities of roots and fungi. In some cases, the biochemical pathways leading to calcite formation are complex. Progressive models for pedogenic profile development originally envisaged a simple linear evolution, but Quaternary examples have revealed complex stratigraphies controlled by climate, amongst other factors, that induce intervals of erosion, deposition, vegetation changes and carbonate accumulation. As a result, a more dynamic model has developed that views calcrete profiles as sensitive to environmental changes (Figure 6). Under certain circumstances, calcrete can be removed from a profile, or moved to other levels within the profile as a consequence of changes in precipitation. As pedogenic calcretes are abundant in some ancient alluvial–fluvial successions, their occurrence and degree of development have been used to develop complex models of landscape evolution and to assess the relative importance of large-scale controls, such as tectonism and eustasy.

Geochemical studies, especially of the stable C and O isotopes, have suggested many important applications for pedogenic calcretes. Analysis of these isotopes can provide information on palaeotemperatures, palaeoatmospheric composition and palaeovegetation types. Due to the relative abundance of pedogenic calcretes in the stratigraphic record, we have glimpses of the history of these factors in deep time.

ACKNOWLEDGEMENTS

This work was funded by project CGL-2005-05953-C02-02/2008-05584-C02-02 from the Spanish *Ministerio de Educación y Ciencia* and CCG07-UCM/AMB2299 and 910404 from CM-UCM. B.P. Singh is thanked for his review of the paper. The editors, L.H. Tanner and T. van Loon, notably contributed to improve the manuscript.

REFERENCES

Adams, A.E., 1980. Calcrete profiles in the Eyam Limestone (Carboniferous) of Derbyshire: petrology and regional significance. Sedimentology 27, 651–660.

Allen, J.R.L., 1960. Cornstone. Geological Magazine 97, 43–48.

Allen, J.R.L., 1986. Pedogenic calcretes in the Old Red Sandstone Facies (Late Silurian–Early Carboniferous) of the Anglo-Welsh area, Southern England. In: Wright, V. P. (Ed.), Paleosols: Their Recognition and Interpretation. Blackwell Scientific Publications, Oxford, pp. 58–86.

Allen, J.R.L., Williams, B.P.J., 1979. Interfluvial drainage on Siluro-Devonian alluvial plains in Wales and the Welsh borders. Journal of the Geological Society 136, 361–366.

Alonso-Zarza, A.J., Jones, B.J., 2007. Root calcrete formation on Quaternary karstic surfaces of Grand Cayman. Geologica Acta 5, 77–88.

Alonso-Zarza, A.M., 1999. Initial stages of laminar calcrete formation by roots: examples from the Neogene of central Spain. Sedimentary Geology 126, 177–191.

Alonso-Zarza, A.M., 2003. Palaeoenvironmental significance of palustrine carbonates and calcretes in the geological record. Earth-Science Reviews 60, 261–298.

Alonso-Zarza, A.M., Arenas, C., 2004. Cenozoic calcretes from the Teruel Graben, Spain: microstructure, stable isotope geochemistry and environmental significance. Sedimentary Geology 167, 91–108.

Alonso-Zarza, A.M., Calvo, J.P., García del Cura, M.A., 1992a. Palustrine sedimentation and associated features – grainification and pseudo-microkarst – in the Middle Miocene (Intermediate Unit) of the Madrid Basin, Spain. Sedimentary Geology 76, 43–61.

Alonso-Zarza, A.M., Genise, J.F., Cabrera, M.C., Mangas, J., Martín-Pérez, A., Valdeolmillos, A., Dorado-Valiño, M., 2008. Megarhizoliths in Pleistocene aeolian deposits from Gran Canaria (Spain): ichnological and palaeoenvironmental significance. Palaeogeography, Palaeoclimatology, Palaeoecology 265, 39–51.

Alonso-Zarza, A.M., Sanz, M.E., Calvo, J.P., Estévez, P., 1998a. Calcified root cells in Miocene pedogenic carbonates of the Madrid Basin: evidence for the origin of *Microcodium* b. Sedimentary Geology 116, 81–97.

Alonso-Zarza, A.M., Silva, P., Goy, J.L., Zazo, C., 1998b. Fan-surface dynamics and biogenic calcrete development: interactions during ultimate phases of fan evolution in the semiarid SE Spain (Murcia). Geomorphology 24, 147–167.

Alonso-Zarza, A.M., Silva, P.G., 2002. Quaternary laminar calcretes with bee nests: evidences of small scale climatic fluctuations, Eastern Canary Islands, Spain. Palaeogeography, Palaeoclimatology, Palaeoecology 178, 119–135.

Alonso-Zarza, A.M., Sopeña, A., Sánchez-Moya, Y., 1999. Contrasting paleosol development in two different tectonic settings: the Upper Buntsandstein of the Western Iberian Ranges, Central Spain. Terra Nova 11, 23–29.

Alonso-Zarza, A.M., Wright, V.M., 2010. Palustrine carbonates. In: Alonso-Zarza, A.M., Tanner, L.H. (Eds), Carbonates in Continental Settings: Facies, Environments and Processes. Developments in Sedimentology (Elsevier, Amsterdam) 61, pp. 103–132.

Alonso-Zarza, A.M., Wright, V.P., Calvo, J.P., García del Cura, M.A., 1992b. Soil-landscape and climatic relationships in the Middle Miocene of the Madrid Basin. Sedimentology 39, 17–35.

Arakel, A.V., 1986. Evolution of calcrete in palaeodrainages of the Lake Narpperby area, Central Australia. Palaeogeography, Palaeoclimatology, Palaeoecology 54, 283–303.

Arakel, A.V., 1991. Evolution of Quaternary duricrusts in Karinga Creek drainage system, central Australian groundwater discharge zone. Australian Journal of Earth Sciences 38, 333–347.

Arakel, A.V., McConchie, D., 1982. Classification and genesis of calcrete and gypsite lithofacies in palaeodrainage basins of inland Australia and their relationship to carnotite mineralization. Journal of Sedimentary Petrology 52, 1147–1170.

Armenteros, I., Bustillo, M.A., Blanco, J.A., 1995. Pedogenic and groundwater processes in a closed Miocene basin (northern Spain). Sedimentary Geology 99, 17–36.

Armenteros, I., Huerta, P., 2006. The role of clastic sediment influx in the formation of calcrete and palustrine facies: a response to paleogeographic and climatic conditions in the southeastern Tertiary Duero basin (northern Spain). In: Alonso-Zarza, A.M., Tanner, L.H. (Eds), Paleoenvironmental Record and Applications of Calcretes and Palustrine Carbonates. Geological Society of America, Special Paper 416, pp. 119–132.

Atkinson, C.D., 1986. Tectonic control on alluvial sedimentation as revealed by an ancient catena in the Capella Formation (Eocene) of Northern Spain. In: Wright, V. P. (Ed.), Paleosols: Their Recognition and Interpretation. Blackwell Scientific Publications, Oxford, pp. 139–179.

Azañón., J.M., Tuccimei, P., Azor, A., Sánchez-Almazo, I.M., Alonso-Zarza, A.M., Soligo, M., Pérez-Peña, V., 2006. Calcrete features and age estimates from U/Th dating: implications for the analysis of Quaternary erosion rates in the northern limb of the Sierra Nevada Range (Betic Cordillera, southeast Spain). In: Alonso-Zarza, A.M., Tanner, L.H. (Eds), Paleoenvironmental Record and Applications of Calcretes and Palustrine Carbonates. Geological Society of America, Special Paper 416, pp. 223–239.

Baghernejad, M., Dalrymple, J.B., 1993. Colloidal suspensions of calcium carbonate in soils and their likely significance in the formation of calcic horizons. Geoderma 58, 17–41.

Bajnóczi, B., Kovács-Kis, V., 2006. Origin of pedogenic needle-fiber calcite revealed by micromorphology and stable isotope composition – a case study of a Quaternary paleosol from Hungary. Chemie der Erde 66, 203–212.

Behrensmeyer, A.K., Willis, B.J., Quade, J., 1995. Floodplains and paleosols of Pakistan Neogene and Wyoming Paleogene deposits: a comparative study. Palaeogeography, Palaeoclimatology, Palaeoecology 115, 37–60.
Birkeland, P.W., 1999. Soils and Geomorphology (3rd Edition). Oxford University Press, New York, NY, 430 pp.
Blumel, W.D., 1982. Calcrete in Namibia and SE-Spain. Relations to substratum, soil formation and geomorphic factors. Catena Supplement 1, 67–82.
Bown, T.M., Kraus, M.J., 1987. Integration of channel and floodplain suites. I. Development sequence and lateral relations of alluvial paleosols. Journal of Sedimentary Petrology 57, 587–601.
Braissant, O., Guillaume, C., Aragno, M., Verrecchia, E.P., 2004. Biologically induced mineralization in the tree *Milicia excelsa* (Moraceae): its causes and consequences to the environment. Geobiology 2, 59–66.
Braithwaite, C.J.R., 1983. Calcrete and other soils in Quaternary limestones: structures, processes and applications. Journal of the Geological Society of London 140, 351–363.
Braithwaite, C.J.R., 1989. Displacive calcite and grain breakage in sandstones. Journal of Sedimentary Petrology 59, 258–266.
Brewer, R., 1964. Fabric and Mineral Analysis of Soils. Wiley, New York, NY, 470 pp.
Budd, D.A., Gaswith, S.B., Oliver, W.L., 2002. Quantification of macroscopic diagenetic subaerial exposure features in carbonate rocks. Journal of Sedimentary Research 72, 917–928.
Bustillo, M.A., Alonso-Zarza, A.M., 2007. Overlapping of pedogenesis and meteoric diagenesis in distal alluvial and shallow lacustrine deposits in the Madrid Miocene Basin, Spain. Sedimentary Geology 198, 255–271.
Butt, C.R.M., Horwitz, R.C., Mann, A.W., 1977. Uranium occurrences in calcrete and associated sediments in Western Australia. CSIRO Mineral Research Laboratories, Australia, Report FP 16.
Cailleau, G., Braissant, O., Verrecchia, E.P., 2004. Biomineralization in plants as a long term carbon sink. Naturwissenschaften 91, 191–194.
Callot, G., Guyon, A., Mousain, D., 1985. Interrelations entre aiguilles de calcite et hyphes mycéliens. Agronomie 5, 209–216.
Calvet, F., Juliá, R., 1983. Pisoids in the caliche profiles of Tarragona (NE Spain). In: Peryt, T. M. (Ed.), Coated Grains. Springer, Berlin, pp. 73–79.
Candy, I., 2002. Formation of a rhizogenic calcrete during a glacial stage (Oxygen Isotope Stage 12): its palaeoenvironmental stratigraphic significance. Proceedings of the Geologists' Association 113, 259–270.
Candy, I., Black, S., Sellwood, B.W., 2004. Quantifying time scales of pedogenic calcrete formation using U-series disequilibria. Sedimentary Geology 170, 177–187.
Canti, M., 1998. Origin of calcium carbonate granules found in buried soils and Quaternary deposits. Boreas 27, 275–288.
Capo, R.C., Chadwick, O.A., 1999. Sources of strontium and calcium in desert soil and calcrete. Earth and Planetary Science Letters 170, 61–72.
Carlisle, D., 1980. Possible variations in the calcrete–gypcrete uranium model. Open File Report, US Department of Energy, GJBX-53 (80), 38 pp.
Carlisle, D., 1983. Concentration of uranium and vanadium in calcretes and gypcretes. In: Wilson, R.C.L. (Ed.), Residual Deposits. Geological Society of London Special Publication 11, pp. 185–195.
Chadwick, O.A., Nettleton, W.D., 1990. Micromorphological evidence of adhesive and cohesive forces in soil cementation. Developments in Soil Science 19, 207–212.
Chen, X.Y., McKenzie, N.J., Roach, I.C., 2002. Distribution in Australia: calcrete landscapes. In: Chen, X. Y., Lintern, M. J., and Roach, I. C. (Eds), Calcrete: Characteristics, Distribution and Use in Mineral Exploration. Cooperative Research

Stokes, M., Nash, D.J., Harvey, A.M., 2007. Calcrete 'fossilisation' of alluvial fans in SE Spain: the roles of groundwater, pedogenic processes and fan dynamics in calcrete development. Geomorphology 85, 63–84.

Strong, G.E., Giles, J.R.A., Wright, V.P., 1992. A Holocene calcrete from North Yorkshire, England: implications for interpreting palaeoclimates using calcretes. Sedimentology 39, 333–347.

Tabor, N.J., Montañez, I.P., Kelso, K.A., Currie, B., Shipman, T., Colombi, C., 2006. A Late Triassic soil catena: landscape and climate controls on paleosol morphology and chemistry across the Carnian-age Ischigualasto-Villa Unión basin, northwestern Argentina. In: Alonso-Zarza, A.M., Tanner, L.H. (Eds), Paleoenvironmental Record and Applications of Calcretes and Palustrine Carbonates. Geological Society of America, Special Paper 416, pp. 17–41.

Talbot, M.R., Williams, M.A., 1979. Cyclic alluvial fan sedimentation on the flanks of fixed dunes, Janjari, Central Niger. Catena 6, 43–62.

Tandon, S.K., Andrews, J.E., 2001. Lithofacies associations and stable isotopes of palustrine and calcrete carbonates: examples from an Indian Maastrichtian regolith. Sedimentology 48, 339–355.

Tandon, S.K., Gibling, M.R., 1997. Calcretes at sequence boundaries in Upper Carboniferous cyclothems of the Sydney Basin, Atlantic Canada. Sedimentary Geology 112, 43–67.

Tandon, S.K., Narayan, D., 1981. Calcrete conglomerate, case-hardened conglomerate and cornstone: a comparative account of pedogenic and non-pedogenic carbonates from the continental Siwalik Group, Punjab, India. Sedimentology 28, 353–367.

Tanner, L.H., 2010. Terrestrial carbonates as indicators of paleoclimate. In: Alonso-Zarza, A.M., Tanner, L.H. (Eds), Carbonates in Continental Settings: Geochemistry, Diagenesis and Applications. Developments in Sedimentology (Elsevier, Amsterdam) 62, pp. 179–214.

Tanner, L.H., Lucas, S.G., 2006. Calcareous paleosols of the Upper Triassic Chinle Group, Four Corner region, southwestern United States. In: Alonso-Zarza, A.M., Tanner, L.H. (Eds), Paleoenvironmental Record and Applications of Calcretes and Palustrine Carbonates. Geological Society of America, Special Paper 416, pp. 53–74.

Verrecchia, E.P., Braissant, O., Cailleay, G., 2006. The oxalate–carbonate pathway in soil carbon storage: the role of fungi and oxalotrophic bacteria. In: Gadd, G. M. (Ed.), Fungi in Biogeochemical Cycles. Cambridge University Press, Cambridge, pp. 298–310.

Verrecchia, E.P., Dumont, J.-L., Verrecchia, K.E., 1993. Role of calcium oxalate biomineralization by fungi in the formation of calcretes: a case study from Nazareth, Israel. Journal of Sedimentary Research 63, 1000–1006.

Verrecchia, E.P., Freytet, P., Verrecchia, K.E., Dumont, J.L., 1995. Spherulites in calcrete laminar crusts: biogenic $CaCO_3$ precipitation as a major contributor to crust formation. Journal of Sedimentary Research A65, 690–700.

Verrecchia, E.P., Verrecchia, K.E., 1994. Needle-fiber calcite: a critical review and a proposed classification. Journal of Sedimentary Research A64, 650–664.

Vogt, T., 1984. Croûtes calcaires: types et genèse. Dissertation Thèse, Université Louis Pasteur, Strasbourg, 239 pp.

Vogt, T., Corte, A.E., 1996. Secondary precipitates in Pleistocene and present cryogenic environments (Mendoza Precordillera, Argentina, Transbailalia, Siberia, and Seymour Island, Antarctica). Sedimentology 43, 53–64.

Wang, Z.S., Rasbury, E.T., Hanson, G.N., Meyers, W.J., 1998. Using the U–Pb system of calcretes to date the time of sedimentation of clastic sedimentary rocks. Geochimica et Cosmochimica Acta 62, 2823–2835.

Watson, A., Nash, D.J., 1997. Desert crusts and varnishes. In: Thomas, D. S. G. (Ed.), Arid Zone Geomorphology: Process, Form and Change in Drylands. John Wiley and Sons, Chichester, pp. 69–107.

Watts, N.L., 1980. Quaternary pedogenic calcretes from the Kalahari (southern Africa): mineralogy, genesis and diagenesis. Sedimentology 27, 661–686.
Williams, C.A., Krause, F.F., 1998. Pedogenic-phreatic carbonates on a Middle Devonian (Givetian) terrigenous alluvial-deltaic plain, Gilwood Member (Watt Mountain Formation), northcentral Alberta, Canada. Sedimentology 45, 1105–1124.
Wright, V.P., 1986. The role of fungal biomineralization in the formation of early Carboniferous soil fabrics. Sedimentology 33, 831–838.
Wright, V.P., 1989. Terrestrial stromatolites: a review. Sedimentary Geology 65, 1–13.
Wright, V.P., 1990a. A micromorphological classification of fossil and recent calcic and petrocalcic microstructures. In: Douglas, L. A. (Ed.), Soil Micromorphology: A Basic and Applied Science. Developments in Soil Science, Vol. 19. Elsevier, Amsterdam, pp. 401–407.
Wright, V.P., 1990b. Syngenetic formation of grainstones and pisolites from fenestral carbonates in peritidal settings: discussion. Journal of Sedimentary Petrology 60, 309–310.
Wright, V.P., 1990c. Estimating rates of calcrete formation and sediment accretion in ancient alluvial deposits. Geological Magazine 127, 273–276.
Wright, V.P., 1992. Paleopedology: stratigraphic relationships and empirical models. In: Martini, I. P., Chesworth, W. (Eds), Weathering, Soils and Paleosols. Elsevier, Amsterdam, pp. 475–499.
Wright, V.P., 1994. Paleosols in shallow marine carbonate sequences. Earth-Science Reviews 35, 367–395.
Wright, V.P., 1995. Losses and gains in weathering profiles and duripans. In: Parker, A., Selwood, B. W. (Eds), Quantitative Diagenesis: Recent Developments and Applications to Reservoir Geology. Kluwer Academic Publishers, Dordrecht, pp. 95–123.
Wright, V.P., 2007. Calcretes. In: Nash, D., McLaren, S. (Eds), Geochemical Sediments and Landscapes. Wiley–Blackwell, Oxford, UK, pp. 10–45.
Wright, V.P., Alonso-Zarza, A.M., 1990. Pedostratigraphic models for alluvial fan deposits: a tool for interpreting ancient sequences. Journal of the Geological Society, London 147, 8–10.
Wright, V.P., Beck, V.H., Sanz-Montero, M.E., 1996. Spherulites in calcrete laminar crusts: biogenic $CaCO_3$ precipitation as a major contributor to crust formation. Discussion. Journal of Sedimentary Research 66, 1040–1041.
Wright, V.P., Marriott, S.B., 1993. The sequence stratigraphy of fluvial depositional systems: the role of floodplain sediment storage. Sedimentary Geology 86, 203–210.
Wright, V.P., Marriott, S.B., 1996. A quantitative approach to soil occurrence in alluvial deposits and its application to the Old Red Sandstone of Britain. Journal of the Geological Society, London 153, 907–913.
Wright, V.P., Peeters, C., 1989. Origins of some early Carboniferous calcrete fabrics revealed by cathodoluminescence: implications for interpreting the sites of calcrete formation. Sedimentary Geology 65, 345–353.
Wright, V.P., Platt, N.H., Marriot, S.B., Beck, V.H., 1995. A classification of rhizogenic (root-formed) calcretes, with examples from the Upper Jurassic–Lower Carboniferous of Spain and Upper Cretaceous of southern France. Sedimentary Geology 100, 143–158.
Wright, V.P., Platt, N.H., Wimbledon, W., 1988. Biogenic laminar calcretes: evidence of calcified root mat horizons in palaeosols. Sedimentology 35, 603–620.
Wright, V.P., Tucker, M.E., 1991. Calcretes: an introduction. In: Wright, V. P., Tucker, M. E. (Eds), Calcretes. IAS Reprint Series, Vol. 2. Blackwell Scientific Publications, Oxford, pp. 1–22.
Yaalon, D.H., 1988. Calcic horizons and calcrete in Aridic soils and paleosols: progress in last twenty two years. Soil Science Society of America Agronomy Abstracts. Cited in Wright and Tucker (1991).

as dating of karst deposits. Since the last decade of the 20th century, research has focused on the extraction and interpretation of climate and environmental proxy data from karst archives, in particular from stalagmites. The nature of the karst environment, and karst processes provide all the ingredients necessary for the capture and preservation of environmental signals (Fairchild et al., 2007).

Karst environments develop in many locations around the globe, where limestones and dolomites (and less commonly, gypsum) crop out, and where there is sufficient flowing water for dissolution weathering to dominate. This interaction between soluble rock and weathering by dissolution produces a unique suite of surface and subterranean landforms and a terrain normally dominated by subsurface drainage. Whilst karst landscapes are characterized by landforms derived from dissolution, some spectacular depositional forms such as tufa barrages and cascades (Ford and Pedley, 1996) can develop within river systems, draining karst terrains.

Because dissolution and precipitation of carbonates in karst regions mobilize carbon from the surface and subsequently store it in calcium-carbonate minerals within the subsurface zone, karst processes have an important role in the carbon cycle. The flux of carbon is largely controlled by the water volume and the concentration of CO_2, the latter being derived mostly from the soil zone (Reardon et al., 1979). It is, thus, dependent on rainfall, temperature and vegetation changes. In summary, the carbon flux in karst environment depends on global changes. Global warming leads to increasing CO_2 emissions that will enhance karst dissolution, due to higher root and microbial respiration rates. The capacity of karst systems to sequester and store carbon for a long period in the form of calcium carbonate, is still far from being quantified, and geochemical modelling is needed to understand if and how effectively karst processes may reduce the transport of inorganic carbon into the oceans (Szramek and Walter, 2004). Karst processes are of interest because they yield spectacular landforms, they archive climate changes, and they participate in the carbon cycling.

The aim of this chapter is to illustrate the link between surface and subsurface karst processes which result in the formation of unique archives that record a variety of physical and chemical proxy data that allow the reconstruction of past climate and environmental changes, including variability in the carbon cycle.

1.1. Surface and shallow subsurface: dissolution processes

The first step in the formation of karst archives is the dissolution of a carbonate or evaporite rock by meteoric water combined with carbon dioxide. The dissolution of carbonates at the surface and in the shallow subsurface (the first few metres) provides most of the building material for mineral deposits formed in caves cut in karstified host rocks. About 70% of dissolution is accomplished within the top 10 m and diminishes with increasing depth from surface CO_2 supply (Williams, 2008). The eluviation of carbonates within a soil is called decalcification and may lead to the complete removal of carbonate from the soil profile, and to accumulation of insoluble material. Soil, however, is not always present. In arid or high mountain regions there might be little or no soil, and CO_2 is supplied directly from the atmosphere.

Below the soil, a zone known as epikarst can extend from 3 up to 30 m or more. The epikarst has been defined as a near surface zone of weathered carbonate rocks with high secondary porosity, from 10 to 30% (Williams, 2008). The epikarst is characterized by a high water-storage capacity and a variable flow (fissure and pore flow), which distinguishes it from the underlying transmission zone, where water storage is minimal (Williams, 2008). Collectively, soil (when present), epikarst and the transmission zone constitute the vadose zone, which is where the mobilization, transportation and reprecipitation of ions mostly occur.

Karst dissolution at the surface and in the epikarst is the passage from carbonate minerals (or sulphates such as gypsum) to their single ions or molecule components which diffuse in rainwater containing dissolved carbon dioxide. The rate of dissolution of CO_2 in water ranges from a few minutes to a few hours and the aqueous solution of CO_2 forms weak carbonic acid:

$$CO_2(g) + H_2O(l) \leftrightarrow H_2CO_3(aq) \qquad (1)$$

The carbonic acid dissociates in two steps:

$$H_2CO_3(aq) \leftrightarrow HCO_3^-(aq) + H^+(aq) \qquad (2)$$

$$HCO_3^-(aq) \leftrightarrow H^+(aq) + CO_3^{2-}(aq) \qquad (3)$$

HCO_3^- is the bicarbonate ion and CO_3^{2-} the carbonate ion. At 25°C, the dissociation constant K for reaction (2) is 4.2×10^{-7}, while for equation (3) it is 5.0×10^{-11}. It is, therefore, clear that the greatest contributor of H^+ to the solution is reaction (2) and that bicarbonate is the most common carbonate species in karst waters.

The dissociation of the carbonic acid in bicarbonate (and carbonate) ions produces hydrogen ions, resulting in increasing concentration (or better, activity) of the H^+ ion in the solution. The measure of the concentration of H^+ in solution is the pH, which is defined as the negative logarithm to the base ten of the numerical value of the H^+ ion concentration

$$pH = -\log[H^+] \qquad (4)$$

Natural rainwater which had been exposed only to atmospheric CO_2 commonly has a pH of 5.6. The solubility of CO_2 in water is proportional to its concentration in the medium traversed by the water but is inversely proportional to temperature, which is one of the reasons why, in some cold-climate and high-altitude settings, karst dissolution occurs in the absence of soil.

When soils are present, karst waters absorb more CO_2 produced by plants and microbial respiration and become, thus, more aggressive. Green plants respire into the soil about 40% of the CO_2 that they extract from the atmosphere and their root activity increases with rising temperature. Thus, warmer temperatures favour the formation of more acidic solution (higher CO_2 dissolved in the soil water) and the subsequent dissolution of karst rocks. As an example, the pH of soil waters in temperate karst environments is <5 (Saether and De Caritat, 1997).

The amount of soil CO_2 depends also on depth and drainage of the soil, the type of plant community it sustains, and seasonality. In well-drained, tropical karst soils, under warm and wet climate conditions and little seasonal temperature contrast, soil CO_2 is produced in abundance throughout the year, with concentrations of about 10,000 ppm (Dever, 1985). By contrast, at mid-latitudes, the strong seasonality results in a low soil CO_2 concentration in the cold season (about 1,000 ppm), but in a very high CO_2 concentration in the warm season, which can exceed 10,000 ppm (Dever, 1985; Baldini et al., 2008). Therefore, the increase of soil CO_2 concentration with rising temperature has a greater effect on karst dissolution than the temperature effect on the solubility of CO_2. This is one of the reasons why tropical and temperate karst landscapes can be characterized by spectacular landforms.

In the presence of carbon dioxide, calcite dissolution is much greater than for pure water (10 mg/l) and is described by the following equation:

$$CaCO_3(s) + CO_2(g) + H_2O(l) \leftrightarrow Ca^{2+}(aq) + 2HCO_3^-(aq) \qquad (5)$$

At temperatures of about 25°C, and depending on the degree of saturation with CO_2, the solubility of calcite commonly ranges from 10 to 50 mg/l, but the higher figure may be well exceeded in those karst environments were soil CO_2 is particularly abundant.

Once the dissolution of calcite progresses, it changes the pH of karst waters. Equation (5) can be rewritten, as follows:

$$CaCO_3(s) + HCO_3^-(aq) + H^+(aq) \leftrightarrow Ca^{2+}(aq) + 2HCO_3^-(aq) \qquad (6)$$

From equation (6) it becomes clear that calcite dissolution 'consumes' H^+, and the pH may consequently increase to values exceeding 7.

On bare karst, surface runoff enriched in dissolved atmospheric carbon dioxide dissolves the bare rock starting from the highest points and proceeding to solution flutes, larger solution grooves and large underground drainage systems.

A karst landscape develops over million of years of dissolution. Kaufmann (2002) calculated denudation ranges from less than 10 up to 100 mm/1,000 years, depending on the discharge and the partial pressure of the dissolved CO_2. Dissolution leaves behind the insoluble minerals such as silicates of aluminium (clays), silica and iron oxides.

1.2. Precipitation processes and the formation of karst environmental archives

The process of deposition in karst environments is mostly a subsurface phenomenon, with the exception of some riverine and spring systems, and is typical of the vadose zone. In practice, there is a transmission of dissolved material from the surface to the subsurface, where the dissolved ions recombine to form minerals. In addition to solution transport, karst waters transport particles of various sizes, from pebbles to clay, which may form thick deposits in underground passages.

Most of the karst environmental archives consist of inorganic and bio-mediated layered calcite deposits. The precipitation of calcite is controlled by the reaction:

$$Ca^{2+}(aq) + 2HCO_3^-(aq) \leftrightarrow CaCO_3(s) + CO_2(g) + H_2O(l) \qquad (7)$$

The saturation state of waters with respect to calcite is defined by a saturation index (Si):

$$Si_{cc} = \log\left\{\frac{(Ca^{2+})(CO_3^-)}{K_s}\right\} \qquad (8)$$

K_s is the solubility product constant and the term $(Ca^{2+})(CO_3^{2-})$ is the ion activity product. For calcite dissolved in water at 25°C, K_s has been variously reported in tables as being 4.5×10^{-9} (Krauskopf, 1967) or 3.6×10^{-9} (Williams, 2001).

It is worth mentioning that some karst waters, both at the surface and underground, are at or below the saturation value for calcite (Kempe, 1982). At saturation, $Si_{cc} = 0$ and the solution is at equilibrium, thus, there is no net precipitation or dissolution. Positive values of Si_{cc} indicate that the solution is supersaturated, and precipitation of calcite from the solution should occur. To initiate calcite deposition in caves, as well as in karst rivers, and at the emergence of springs, nucleation must occur, which implies overcoming of an energy barrier, most commonly through raising the supersaturation. Nucleation is also favoured by the presence of nucleation sites, which can be provided by pre-existing calcium-carbonate deposits, or by clastic sediments.

Supersaturation is attained in karst surface and underground waters through degassing, in which CO_2 is passed from aqueous solution to the surface air or the cave atmosphere (Figure 1). Degassing shifts reaction (7) to the right, and the amount of calcium and bicarbonate ions that can remain in the solution is reduced, leading to calcite precipitation. Where karst

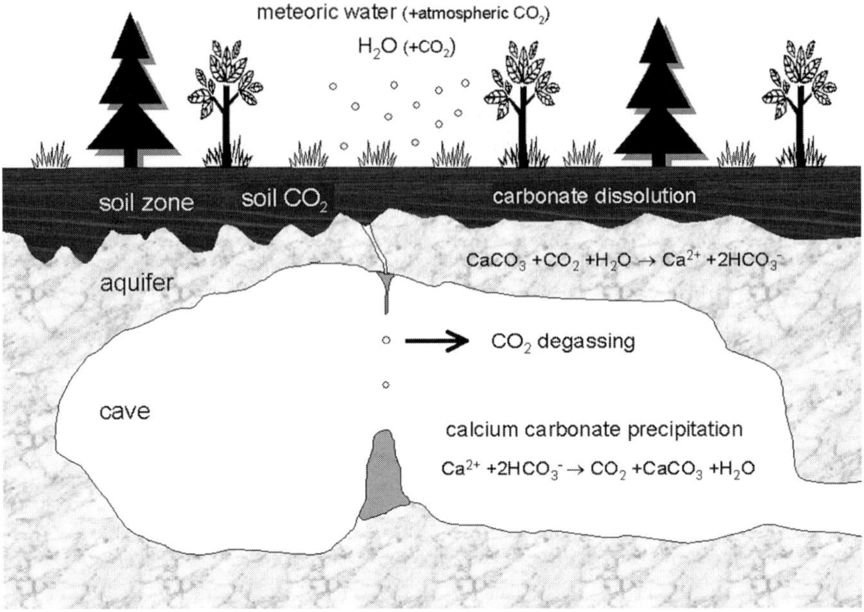

Figure 1 Schematic representation of the karst processes of dissolution and reprecipitation. In the atmosphere, droplets combine with atmospheric CO_2 to form a diluted acidic solution. This solution then combines with soil CO_2 (gas) and produces a more efficient acid. The acid reacts with calcium carbonate in the karst rock, which dissolves. When the new solution emerges at the tip of a stalactite, CO_2 is released to the cave atmosphere (bold arrow). Degassing decreases the amount of CO_2 in solution, which becomes supersaturated with respect to $CaCO_3$. The droplet impacts with the stalagmite growing from the cave floor and precipitates calcium carbonate at the tip of the speleothem. The process of addition of calcium carbonate 'units' at the surface allows speleothem growth.

groundwaters come to the surface at springs, or discharge along a river, CO_2 degassing typically results in calcite supersaturation, allowing purely inorganic calcite precipitation (Lu et al., 2000). Rapid CO_2 degassing is aided by turbulence, which promotes the passage of carbon dioxide from the aqueous solution to the gas phase in air (Chen et al., 2004). Dreybrodt (1999) indicated that the concentration of the calcium ion in solution in relation to the partial pressure of CO_2 (pCO_2) also controls calcite growth in purely inorganic conditions. Laboratory experiments and models show that the rate of growth of inorganic calcite increases by increasing the concentration of calcium ions (Van der Weijden et al., 1997; Baker et al., 1998). This means that calcite precipitation is favoured in karst regions where the limestone rock is subject to enhanced dissolution, that is in wet and warm settings. In some temperate to relatively cold cave environments, karst waters are close to saturation, or even undersaturated, and show a constant calcium-ion concentration throughout the year (Frisia et al., 2000). In this case, changes in the pCO_2 within the cave air, which is possibly driven by surface processes (see Fairchild et al., 2007), promotes seasonal variations in calcite growth rates, which is higher when the cave pCO_2 diminishes, thus promoting the passage of carbon dioxide from the aqueous phase to the gaseous phase.

In surface karst environments, calcium-carbonate deposition may be actively promoted by CO_2 removal related to organism respiration and/or photosynthesis. Microbial biofilms and a variety of organisms, including animals (Drysdale, 1999) seem to heavily influence the formation of layered calcareous tufa and travertines in karst streams, rivers and springs. In the dark cave environment, apart from near-entrance areas, the role of biologic activity in calcium-carbonate precipitation is still poorly known, although there is increasing evidence of a possible role of bacteria in promoting speleothem deposition (Cacchio et al., 2004; Cañaveras et al., 2006; Blyth and Frisia, 2008). Microbiological mediation is more evident in layered whitish to yellowish soft calcareous tufa deposits that form rims, dams and cascades near, but not necessarily into, the photic zone adjacent to cave entrances (Borsato et al., 2007).

2. Karst Deposits and their Macro-Morphology

Karst deposits are produced by physical, chemical and bio-mediated depositional processes. The physical processes are most typical in tectonically young regions, whereas large-scale chemical precipitation is common when the karst is mature. Clastic karst sediments include all clast sizes, from boulders to clay. Breccias form from cave roof collapse, and conglomerates from deposition by underground rivers. Clay is the common product of limestone and dolomite weathering and may form thick deposits on cave floors. Research on karst clastic sediments is limited with respect to the study of chemical and bio-mediated precipitates because of the difficulty of obtaining accurate ages and their complex stratigraphy. By contrast, chemical and bio-mediated precipitates have been the subject of high-quality studies, involving recovering of various aspects of climatic and environmental signals.

Chemical and biochemical karst deposits of palaeoclimatic and palaeoenvironmental significance commonly consist of calcite, the thermodynamically stable form of $CaCO_3$ at surface pressure and temperature, and aragonite, the high-pressure polymorph of the same compound, which is metastably precipitated from high Mg/Ca waters (Tucker, 2008). These two minerals may occur as powdery, plastic and highly porous sediments or as hard, compact, crystalline deposits. These physical characteristics may be used to classify speleothems as summarized in Table 1. The 'softer' deposits, moonmilk and calcareous tufa, share in common the possibility of an active biomediation component in their formation (Figure 2). By contrast, in the 'hard' deposits (sparitic speleothems) calcium-carbonate precipitation seems to be mostly driven by inorganic processes.

Karst deposits are archives of past global (and local) climate and environmental changes because they can be dated through a series of techniques, from visible to UV-fluorescent annual lamina counting, to radiocarbon and ^{230}Th-^{234}U-^{238}U disequilibrium radiometric dating. The U-Pb series methodology is rapidly improving and specimens older than 1 million years can now be dated. Woodhead et al. (2006), for example obtained U-Pb ages of about 4 million years for stalagmites from the Nullarbor (Australia).

Another emerging methodology to obtain extremely high-resolution ages is through detection of annual chemical cycles, which can be used to refine the age model between two radiometrically dated intervals (Fairchild et al., 2007). Flow- and cave-related factors, which change seasonally in cave water supersaturation, influence cave carbonate chemistry, and, thus, the high-resolution properties of speleothems. Chemical cycles, therefore, should be used to refine age models when seasonal variability in cave water and atmosphere have been assessed through extensive monitoring (Frisia et al., 2003; Borsato et al., 2007; Mattey et al., 2008). Cave deposits may also contain magnetic minerals and thus acquire a natural remnant magnetization, thus preserving records of palaeomagnetic variations throughout geological time (Latham et al., 1982; Perkins, 1996). There is the possibility, therefore, to reconstruct secular variations of the Earth's magnetic field from karst deposits and to use palaeomagnetic variations as an additional dating tool. The climate and environmental data extracted from karst-deposit archives, thus, can be framed in a precise time span and be annually resolved, similar to ice-core and tree-ring records. The immense potential of karst deposits as palaeoclimate archives with respect to other archives, however, is in the length of time that they record, which may well be unparalleled by other climate archives.

2.1. Macro-morphological characteristics of speleothems

Speleothems occur in a variety of morphologies, from thick and wide coatings on cave floors and walls to paper-thin calcite rafts formed at the water/air interface in cave pools. The most common forms used for palaeoclimatic research are stalagmites, stalactites and flowstones (Figure 2).

Table 1 Summary of speleothem types based on over 10 years of monitoring in caves at low-to-high elevation in Italy (modified and completed after Borsato, 1996; Frisia et al., 2000; Borsato et al., 2000, 2007).

Speleothem type	Physical properties					Parent water				Mechanism of formation
	Density (dry phase) g/cm^3	Porosity (%)	Water content weight (%)	Hardness	Lamination	Flow	Temperature (°C)	Ca^{2+} content (mg/l)	Calcite saturation Si_{cc}	
Moonmilk	<0.5	70–95	60–95	Plastic	Visible and faint	Laminar. Also fed by drips	3–5 constant*	35–50	0.0–0.2	Mostly bio-mediated
Calcareous tufa	0.8–1.6	30–70	20–50	Soft	Visible stromatolite-like	High, locally turbulent	7–12 fluctuating*	40–100	0.4–1.0	Bio-mediated and inorganic
Sparitic (crystalline)	2.4–2.7	<4	<1	Hard	Visible (not always). When present commonly due to fabric change	Fed by drips, mostly laminar	>5 constant*	30–120	0.1–0.8	Mostly inorganic

Notes: Each type is characterized according to its specific physical properties, typical parameters of the parent fluid, and mechanisms of formation. The summary is not intended to include all existing speleothem types and all environments of speleothem formation. More information based on long-term monitoring in other climate settings, and in particular in tropical regions, is needed to increase our knowledge on speleothems as related to their environment of formation. The asterisk * indicates that constant and fluctuating temperatures refer to daily and seasonal scales.

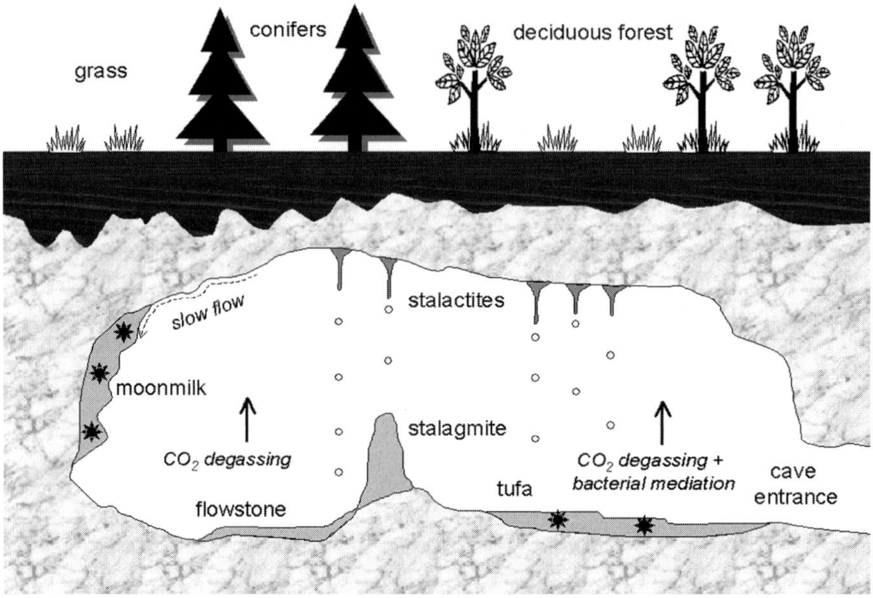

Figure 2 The most common speleothems discussed in the present chapter and their environment of formation. Stalactites form in any passage and 'hang' from the roof. Stalagmites are fed by a single drip and grow from the floor upward. Massive, rigid, inorganic flowstones develop in the inner passages of the cave and are fed by several drips, or by a spring. The soft, calcareous tufa forms near the cave entrance, is fed by several drips (or a spring), and its formation requires the intervention of microbial activity (indicated by stars). Moonmilk forms under conditions of a constant water flow where microbial respiration (indicated by stars) increases the supersaturation state of waters at, or below, saturation for calcite.

2.1.1. Stalagmites

Stalagmites commonly form under water dripping from a stalactite and grow from the floor toward the ceiling. Stalagmites show a variety of morphologies, from cylindrical to conical forms. Petal-like extensions from a cylindrical core have been observed where the fall height (distance from the tip of the feeding stalactite to the top of the stalagmite) is several metres or where feeble air currents are present. A great fall height may also give rise to forms resembling stacked dishes or cylindrical stalagmites with a sunken central splash cup (Fairchild et al., 2007). Kaufmann (2003) demonstrated that the growth rate and radius of stalagmites are controlled by the climate. In particular, the stalagmite diameter is largely controlled by the drip rate (the time interval between successive drops) which is related to rainfall: large stalagmite diameters form during wet periods.

In stalagmites, crystals grow with their growth axis perpendicular to the substrate. Repeated fabric patterns, or cyclic mineralogical changes have been observed in stalagmites formed in climates characterized by seasonal contrasts, and are interpreted as reflecting seasonal variations in the composition of the drip waters. Laminated stalagmites are, therefore, very accurate archives of the past climate and environmental changes.

2.1.2. Stalactites

Stalactites are cone- or cylinder-shaped speleothems that extend downwards from the cave ceiling. They commonly form from dripping water. Conical stalactites may be several tens of centimetres wide and taper downward, and commonly consist of concentric layers of crystals elongated perpendicular to the central vertical growth axis (the axis which extends from the ceiling toward the floor of the cave). The original hollow centre of the stalactite may be clogged by sparite crystals. Once the central conduit is closed by calcite growth, water flows along the outer surface of the speleothem, which may show undulations.

Cylinder-shaped stalactites or soda straws commonly show a hollow centre and relatively thin walls (average of about 0.2 mm). Water is drawn slowly to the tip of the soda straw, and exits from these hollow stalactites in the form of single droplets. The diameter of soda straws is thus determined by the surface tension of the water drop at their tip. Soda straws, therefore, accrete downwards, rather than outwards. Their outer surface is relatively smooth, whereas the inner wall is characterized by growth steps which are formed by the emergence of crystal terminations. Soda straws commonly show a regularly spaced banding which suggests an annual origin (Huang et al., 2001; Fairchild et al., 2007).

2.1.3. Flowstones

Flowstones or stalagmitic flowstones (Hill and Forti, 1997) commonly consist of stacked layers of crystals elongated normal to the substrate. Flowstones may be several metres thick, and tens or hundreds of metres long in the downstream direction, and are formed through degassing of a thin water current which is slightly supersaturated with respect to calcium carbonate. The growth rate of stalagmitic flowstones varies from a few micrometres (μm) up to about 100 μm/year. They can grow over tens of thousands of years when water flows over their surface, and cease to grow during dry and (or) very cold periods (Fairchild et al., 2007). Stalagmitic flowstones are, therefore, ideal for addressing the chronology of warm events during glacials, and for the timing of glacial to interglacial transitions (Spötl et al., 2002; Drysdale et al., 2007).

tufa is bio-mediated precipitation (Borsato et al., 2007), whereas flowstones should develop by purely (or predominant) inorganic precipitation. If biomediation plays an important role in the removal of CO_2 from the aqueous solution, its effect must be accounted for in the investigation of calcareous tufa for palaeoclimate and palaeoenvironmental studies.

The physical aspect of calcareous tufa laminae encodes the response of the deposit to seasonal or decadal changes in one or the other of the factors controlling their growth. Laminae can be the result of changes in discharge rate, which is related to the recharge of the vadose zone (epikarst and transmission zone) and in the ability of bacteria to trap detrital particles or induce carbonate precipitation (Borsato, 1996; Kano et al., 2003; Andrews and Brasier, 2005). The annual changes in thickness and nature of calcareous tufa laminae appear to be modulated by degassing, which, in turn, is a function of both current and bacterial activity (Borsato et al., 2007). The critical parameter controlling tufa growth, however, is high discharge. Turbulent flow and high discharge favour degassing. A decrease in discharge and a change to laminar flow may, therefore, result in the demise of calcareous-tufa formation.

Calcareous tufa shares its dependence on the flow rate with its surface counterpart. A progressive decline of both surface and hypogean calcareous-tufa formation in Europe and North Africa starting about 6,000 years ago has been related to the diminution in mean annual rainfall possibly paced by the decrease in July insolation in the northern hemisphere (Soligo et al., 2002; Žak et al., 2002; Andrews, 2006; Borsato et al., 2007). Frančišković-Bilinski et al. (2004), however, inferred that the cessation of much tufa deposition in Croatia might be a consequence of the increase in atmospheric CO_2 levels, rather than of hydrological changes. There is potential for hypogean and surface calcareous tufas to be archives of both hydrology and atmospheric CO_2 variability, which possibility deserves to be fully explored.

A prerequisite for the full exploitation of calcareous tufas as palaeoclimate and palaeoenvironmental archives is the possibility to obtain reliable and accurate ages. With respect to the sparitic speleothems, tufas are heterogeneous deposits with both an organic and a mineral component. Radiocarbon dating on the organic component highlighted discrepancies between stratigraphy and radiometric ages, possibly related to the opening of the system (Drysdale and Head, 1994; Horvatinčić et al., 2000; Borsato et al., 2007). Recent advancements on dating based on the U-series method could circumvent the dating problem (Mallick and Frank, 2002) but, so far, few attempts have been carried out to use the U/Th method to obtain the age of tufas because of the complexity related to open-system U-leaching and the presence of detrital Th.

The presence of allochthonous material also has important implications for the interpretation of stable-isotope variations in tufa, which may reflect admixture of autochthonous calcite precipitated from the formation waters with allochthonous material from the host rock. Microbially induced precipitation could also result in kinetic effects, in particular, on carbon-isotope fractionation. The carbon isotopic composition of tufa developed in temperate regions from both surface and cave environments commonly shows $\delta^{13}C$ values ranging from about −8.0‰ (PDB) to about −10.0‰ (Horvatinčić et al., 2003; Žak et al., 2002; Borsato et al., 2007), which are typical of calcium carbonate formed from waters where the dissolved CO_2 originated in the soil and where the predominant vegetation consisted of C3 (most trees and temperate grass) plants (Cerling et al, 1989; Andrews, 2006). Borsato et al. (2007) interpreted shifts in the $\delta^{13}C$ values of calcareous tufa from a 'baseline' of approximately −9.5‰ (PDB) to approximately −5‰ to the contribution of Jurassic limestone detrital particles trapped in the laminae in periods of high discharge (high rainfall episodes).

In surface deposits, where flow-related degassing may result in calcite precipitated in isotopic disequilibrium, it is the biological component of tufa that can be used for palaeoenvironmental analysis. For example, Durand (2002) analysed the carbon- and oxygen-isotope composition of annual growth increments in shells of the fresh-water mollusc *Alathyria pertexta* occurring in tufa of the Barkley Karst in Queensland, Australia. By calibrating the isotope signal of modern shells with present-day temperature and water dissolved inorganic carbon (DIC), Durand (2002) was able to demonstrate that the $\delta^{18}O$ of the single growth layers of the shell preserved a temperature signal close to the instrumental measurements, and that the $\delta^{13}C$ reflected the water DIC. Durand (2002) concluded that the investigation of well-dated fresh-water molluscs from karst regions has the potential to provide a deep understanding on climate-related hydrological changes.

Hypogean and surface calcareous tufa deposits have the potential to be amongst the most accurate archives of hydrology. The role of degassing and biomediation has to be recognized and used for the correct interpretation of the isotopic signal before we can attempt a quantification of climate and environmental changes from these deposits (Dreybrodt, 1999). Given the fact that, in many karst environments, tufa deposits can provide a history of hydrological changes for hundreds of thousands of years, we believe that they should deserve the same attention that has been given to sparitic, mostly inorganic speleothems as a reliable source of information about the past climate.

4. Calcite Moonmilk

Moonmilk is a very peculiar type of karst deposit, the formation of which appears to be dependent on the presence of micro-organisms even more than calcareous tufa. Moonmilk is a cave deposit and it has not been

Figure 5 SEM photograph of a specimen taken from a calcite moonmilk deposit in the Grotta Cesare Battisti at 2,000 m above sea level (Alps of NE Italy). The specimen consists of filaments (F) which are elongated threads tens of nanometres wide and several micrometres long, and fibres (Fb) which are thicker, pliable threads showing some evidence of crystallinity (some stacked rhombohedra are visible). Fibres and filaments embedded rigid rods, visible in the foreground, which consist of stacked rhombohedra platelets (P).

An example of the significance of moonmilk as a palaeoenvironmental archive is provided by a massive deposit, about 1 m thick and forming canopies on the walls of the cave and flowstones on the floor, which developed at about 2,000 m above sea level in the Grotta Cesare Battisti cave (N. Italy). The cave is located above the present-day treeline (Borsato et al., 2007). The moonmilk from this cave is coarsely laminated, with thin reddish layers (a few millimetres thick) capping thick (about 5 cm thick or more) white layers. The reddish portion of the laminae is rich in clay, and suggests periods of slow calcite growth. The moonmilk deposit commenced to grow in the Grotta Cesare Battisti during the Early Holocene, and from ~8,100 cal years to ~6,500 cal years BP (Before Present), where Present is, by convention, the year 1950, its growth rate was ~300 μm/year. Given water availability, the limiting factor for moonmilk formation in the Cesare Battisti high-elevation cave has been observed to be soil CO_2 production, which is a function of temperature and vegetation cover. The high growth rate should, therefore, be indicative of a warm and humid climate, and of a treeline that rose well above the cave. This inference is supported by the $\delta^{13}C$ value of the moonmilk calcite, which, for the Early to Mid-Holocene, is about −7‰ (Figure 6) and which reflects a higher proportion of light soil carbon evolved under an open conifer wood (Borsato et al., 2000). The $\delta^{13}C$ value of moonmilk calcite is, thus, inversely correlated with temperature at the surface because a higher

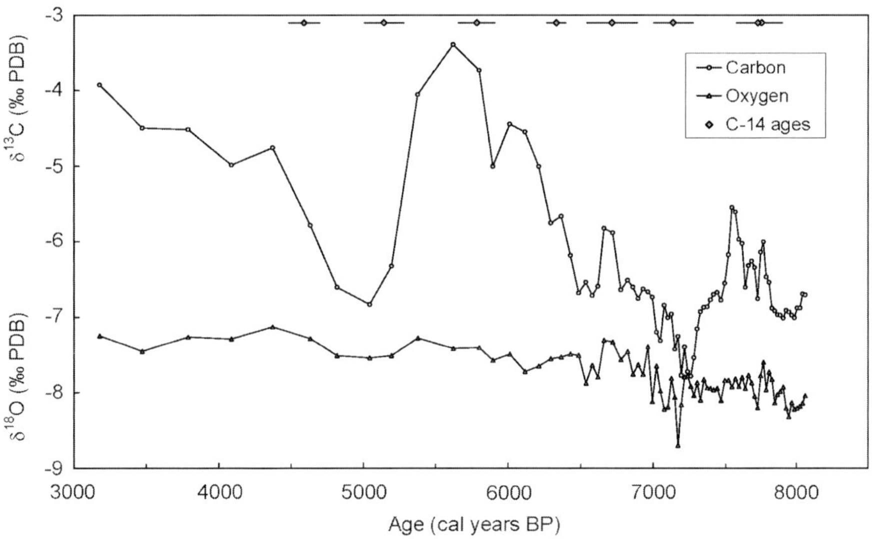

Figure 6 The C and O isotope composition of calcite moonmilk flowstone from the Grotta Cesare Battisti. Radiocarbon ages with the error bars are shown in the top part of the figure. The growth rate decreased after approximately 6,000 cal year BP until the speleothem ceased to grow at about 3,000 cal year BP. The resolution of the isotope data is, consequently, much lower for this portion of the speleothem than during the previous phase (about 8,000–6,000 cal year BP) (from Borsato et al., 2007).

temperature resulted in a higher soil CO_2 production, and in more negative $\delta^{13}C$ calcite values.

From ~6,500 to ~3,000 cal years BP on, the growth rate of the moonmilk decreased to ~35 μm/year, probably as a result of a long-term climate cooling in the Alps. Lower surface temperatures can be inferred also from a shift to more positive values of the $\delta^{13}C$ value at ~6,500 cal years BP, which is probably due to a higher proportion of isotopically heavier atmospheric and rock carbon in the carbon pool. Cooler climatic conditions would have resulted in the progressive descent of about 200 m of the treeline (Tinner and Theurillat, 2003) and in less efficient soil CO_2 production. At about 3,000 cal year BP, the formation of massive moonmilk stopped. Present-day deposits consist of individual patches of moonmilk on cave walls, rather than of massive flowstones. It can be concluded that during a warm and humid phase in the Holocene, moonmilk efficiently sequestered carbon dioxide in the underground karst environment. The cooling trend which culminated in the Little Ice Age probably hindered massive moonmilk formation. The warming that followed the end of the Little Ice Age, and is continuing into the 21st century, is occurring within an environmental context that is greatly perturbed by human-related activities

5.2.1. Columnar fabric

Columnar is the most common fabric type observed in sparitic speleothems (Kendall and Broughton, 1978). Macroscopically, it is composed of crystals elongated perpendicular to the growing surface (Figure 8A–F) with serrated crystal boundaries, or of crystals separated by voids parallel to the elongation, which may host fluid inclusions. Genty and Quinif (1996) proposed also the use of the term 'palisadic' for the macroscopic aspect of columnar calcite as

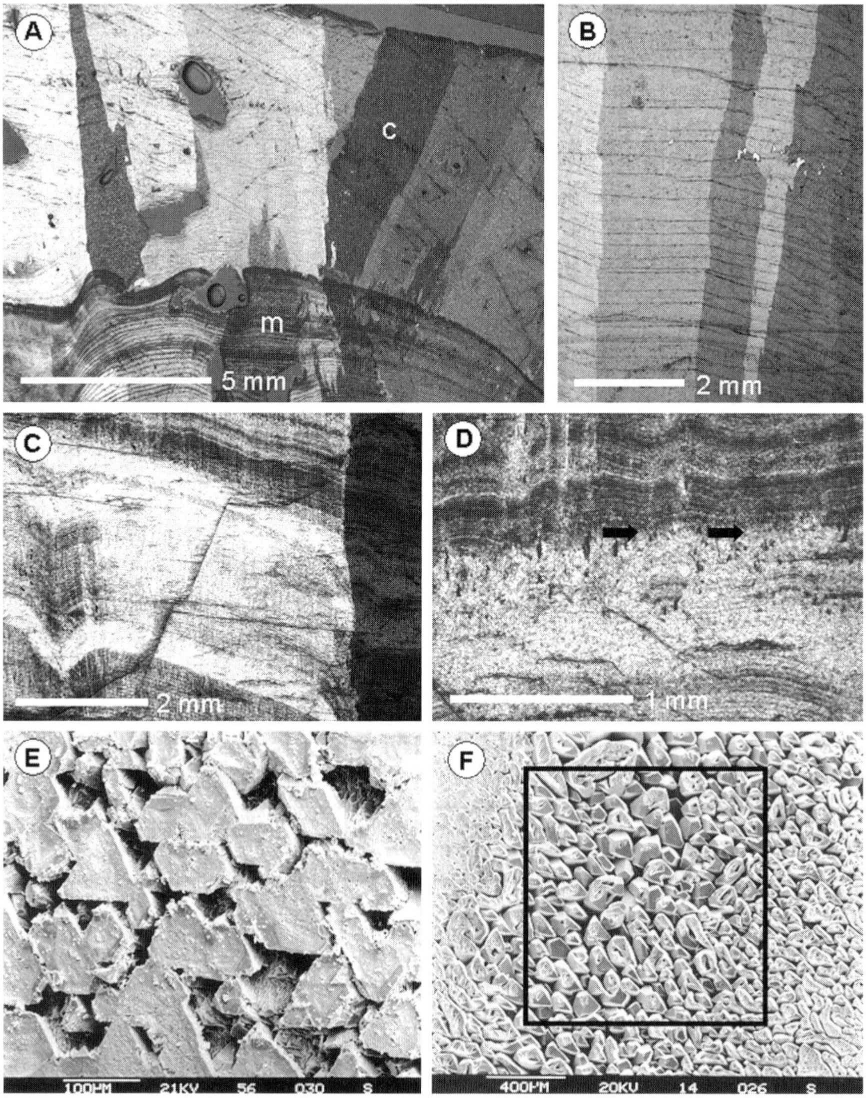

seen in the hand specimen. Microscopically, columnar calcite can be differentiated into five sub-types (Figure 7A) on the basis of the length/width (L/W) ratio and the microstructural characteristics of the crystals. Short columnar and columnar 'proper' sub-types consist of composite crystals with an L/W ratio ≤6:1 and about 6:1, respectively, which show relatively straight crystal boundaries and uniform extinction under crossed polars (Figure 8A–D). Each composite crystal is commonly ≥1 mm wide and ≥2 mm long, and consists of the regular stacking of rhombohedral crystallites typically 50–100 μm wide and ≥100 μm long (Figure 8E–F). Short columnar crystals are characterized by few microstructures, and show flat (F) rhombohedra faces at their emergence at the top of the growing speleothem (Figure 8E, F). Flat faces develop when the dominant growth mechanism is that of spiral growth. They are commonly the slowly growing faces and, eventually, dominate the crystal morphology. Fast growing faces, which show many crystal defects, such as steps and kinks, will tend to disappear during growth, being replaced by the flat faces. The elongated columnar fabric sub-type has an L/W ratio ≥6:1 and consists of rigid individuals. The open columnar fabric sub-type yields porous aggregates, and has a milky appearance in the polished slab. Compact columnar sub-types and an open columnar fabric may occur in different portions of the same speleothem, and commonly indicate a change in drip (flow) rate.

The microcrystalline fabric sub-type (Figures 8A and 9A–C) was distinguished by Frisia et al. (2000) from the columnar fabric type on the

Figure 8 Columnar fabric. A: Thin section showing short columnar fabric (c) with visible annual laminae which developed on microcrystalline sub-type (m) (stalagmite ER76 from the Grotta di Ernesto, NE Italy). The proportion of the dark layers within each lamina is higher in the microcrystalline sub-type. Short columnar individuals have straight boundaries and are translucent. The microcrystalline sub-type shows 'extinction domains' consisting of irregularly stacked crystallites with high intercrystalline porosity which confers milky appearance and irregular boundaries. B: Typical elongated columnar calcite with straight boundaries (from Crag Cave, SW Ireland). C: Annually laminated columnar calcite (stalagmite from Abisso Spiller, NE Italy). Laminae are defined by the proportion of dark and light layers (scale bar: 2 mm). D: Detail of Figure 8C, showing faint laminae in the light, translucent portion. Rhombohedral tips of crystallites are visible at the boundary with the darker portion at the top of the translucent, faintly laminated layer. E: SEM photograph of crystals outcropping at the surface of a sparitic flowstone, showing pseudo-hexagonal, most probably prismatic [0 0 0 1], terminations which are slightly concave, possibly as a result of preferred growth at the edges (Grotta di Costalta, NE Italy). F: SEM photograph of columnar calcite showing rhombohedra terminations outcropping at the tip of a stalagmite. At the point where the drop splashes onto the surface of the speleothem (impact point), well-developed rhombohedra faces developed (highlighted within the box). By contrast, towards the flank of the stalagmite (top left and bottom right), crystal terminations are flattened, more similar to those of the flowstone in a. The development of the faces is limited by flow and film thickness (Crag Cave, SW Ireland).

basis of its high concentration of crystal defects, which indicate either a high growth rate or the effect of growth inhibitors on crystallite stacking. The microcrystalline fabric has a milky, opaque, porous aspect in hand specimens. The composite crystals consist of stacked crystallites, typically $<4\,\mu m$ wide and up to $50\,\mu m$ long, which may not have grown in optical continuity with the substrate, and which show a patchy extinction when viewed under the crossed polars of a petrographic microscope (Figure 9B–C). Columnar fabric sub-types are commonly indicative of calcite formation in quasi-equilibrium conditions, including isotope quasi-equilibrium (Frisia et al., 2000). They appear to form under a wide range of temperatures, from about 5°C to >25°C and at relatively low Si_{cc} (from about 0.1 to 0.35: Frisia et al., 2000). The elongated columnar sub-type has been observed to develop when the Mg/Ca ratio (molar) in the feeding water exceeds 0.3 (Frisia et al., 2000). The drip rate appears to control the transition between columnar and microcrystalline sub-types. Short columnar and columnar fabrics have been observed to form in caves where drips rates are constant and range from 0.1 to 0.3 ml/min (Frisia et al., 2000).

The microcrystalline sub-type forms in the same caves, but under variable drips rates ranging from 30 ml to <0.1 ml/min, and in the presence of optically visible impurities (Figure 7A). Solid-state ^{31}P and ^{1}H magic angle spinning nuclear magnetic resonance (MAS NMR) and synchrotron analyses micro-XRF data provide indications about the nature of the

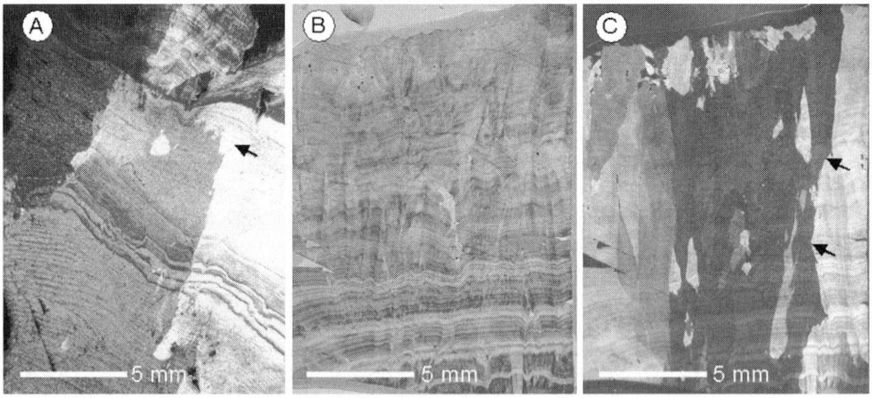

Figure 9 Microcrystalline. A: Typical 'extinction domains' in microcrystalline calcite: composite crystals show irregular, interfingered boundaries and 'patchy' extinction at crossed polars, rather than the straight boundaries and uniform extinction which characterizes columnar calcite. B: Microcrystalline calcite in annually laminated stalagmite from the Alps. C: The same specimen viewed at crossed polars. Note 'extinction domains'. In each 'extinction domain', microcrystals from overlying layers have a roughly similar optical orientation in the middle of the domain, but perturbation in stacking order occurs towards the border of each domain (arrows), resulting in interfingered, irregular crystal boundaries.

impurities. They may consist of individual P ions, crystalline P phases (Mason et al., 2007) and trace metals such as Zn, Cu and Pb (Borsato et al., 2007). Fourier transform infra-red (FT-IR) spectroscopy highlights the presence of organic molecules, most probably in the form of fulvic acids (Ramseyer et al., 1997). The development of a microcrystalline fabric must, therefore, be related to a combination of drip rate, nature and concentration of impurities in the drip water.

5.2.2. Fans

'Fans' is the macroscopic term which include fabrics showing crystals with an L/W ratio greatly exceeding 6:1; they are distinguished from columnar fabrics because the crystal radiate from a centre outwards (Figure 7B). This fabric is common in aragonite speleothems, but can also characterize calcite with a high concentration of Mg. At a microscope scale, the fabric consists of the sub-types 'rays' and 'acicular'.

5.2.2.1. Rays. Rays, as defined by Folk (1974), are straight-edged individuals, from a few millimetres to several centimetres long and 10 μm to 1 mm wide, elongated along the *c*-axis, with uniform extinction, and square terminations (Figure 10A–B). When pseudo-hexagonal cross-sections, due to twinning on {1 1 0}, and square terminations are observed by electron microscopy, the original mineralogy of the rays is aragonite (Frisia et al., 2002).

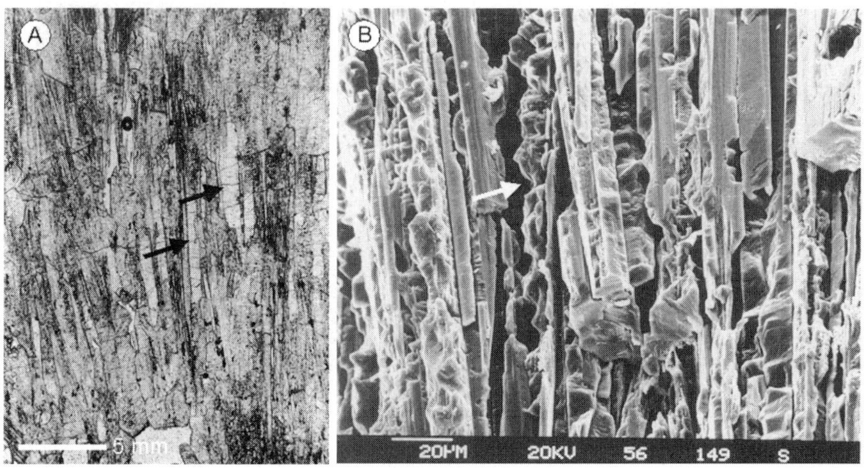

Figure 10 Rays consisting of aragonite have undergone partial diagenesis, with dissolution and reprecipitation of calcite (stalagmite CL27 from the Grotte de Clamouse, SW France). A: Aragonite rays with square terminations show the development of a brick-like fabric (arrows) where calcite precipitated in the voids left after aragonite dissolution. B: SEM photograph illustrating the dissolution of aragonite rays and calcite precipitated in voids formed through dissolution.

5.2.2.2. Acicular. In acicular fabrics, single individuals have an L/W ratio ≫6:1, are elongated along the *c*-axis, and commonly show needle-like terminations (Figure 11A–B). An acicular fabric is uncommon in speleothems consisting of calcite, where it is associated with micrite and clay. Possibly, the fine-grained portion of the speleothem influences nucleation and growth of these extremely elongated crystals. In aragonite speleothems, an acicular fabric is interpreted as due to precipitation from a highly supersaturated solution under extremely low drip rates (Frisia et al., 2002). Acicular (needle-like) crystals have been observed at the tips of

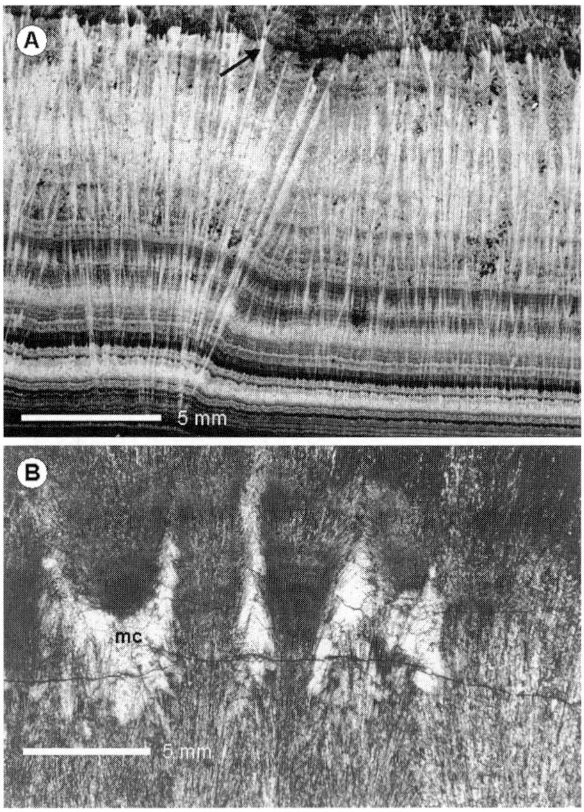

Figure 11 Acicular sub-type. A: Acicular fabric as observed in a flowstone of about 230,000 years old (Grotta di Collalto, NE Italy). Acicular crystals consist of calcite and cross-cut micrite and clay-rich layers. The tip of the acicular crystals became nucleation sites for other elongated individuals that radiated fan-like from the centre. There is no unequivocal evidence that this fabric resulted from diagenetic replacement. B: Late Holocene aragonite needle crystals in a stalagmite from the Grotte de Clamouse (S. France). Needles originally formed radiating structures separated by voids, which were subsequently filled by mosaic calcite (mc).

stalagmites and stalactites radiating from a central nucleus. The individual aragonite needle-like crystals typically show micro-twinning as the most typical microstructure. Aragonite needles commonly show $\delta^{13}C$ and $\delta^{18}O$ enrichment compared with aragonite rays, which is indicative of kinetic influence on the formation of these crystals. Holocene and pre-Holocene acicular aragonite and rays are commonly replaced by the stable polymorph calcite, which alters the original chemical (including isotopic) composition of the speleothems.

5.2.3. Fibrous fabric

Fibrous fabric sub-types, fibres and whiskers, consist of extremely elongated individual crystals which are different from elongated columnar and acicular crystals because of their microstructural characteristics. Fibres consist of crystals from a few to ~50 nm wide which bear striking similarity to organic fibres, including their flexibility. Commonly, fibre crystals do not show microstructural defects. Fibres are uncommon in sparitic speleothems and are typical of moonmilk and calcareous tufa.

Whiskers share with fibres their extremely elongated morphology (L/W ratio ≫6:1) and the absence of crystal defects. Whiskers, however, form rigid crystals and lack the quality of pliability.

5.2.4. Dendritic fabric

The dendritic fabric is characterized by composite calcite crystals consisting of crystallites arranged in branches similarly to a single dendrite crystal. The branching composite crystals, the tree-like appearance of which is best seen under crossed polars, consist of stacked rhombohedral crystallites 4–10 μm in diameter, forming scaffold-like structures where each 'branch' can be 20–100 μm wide and a few millimetres long. Branching probably occurs because of preferential nucleation at kink sites and at the emergence of crystal defects such as twins (Frisia et al., 2000). The dendritic fabric shows many crystal defects, the most typical being lamellae. Two sub-types have been observed: open and compact (Figure 7C). The open dendritic sub-type is probably the most common and shows primary pores between the scaffold-like arrangement of the crystals (Figure 12A–C). In the compact sub-type, no pores are visible between the branches. The dendritic fabric has been observed to develop at Si_{cc} similar to those of the columnar calcite types (or slightly higher), and at variable drip rates (Frisia et al., 2000). In most cases, the drip rate may stop for a few months, commonly coinciding with the dry season.

The dendritic fabric is also typical of sparitic speleothems near the cave entrance, subject to air currents. In this case, the sub-type is almost always open dendritic and the fabric formed under the influence of kinetic

calcrete. There is, therefore, the possibility that the speleothem micrite fabric results from depositional processes which involve the contribution of bacterial mediation or that of fine detritus flushed in the cave upon reactivation of the feeding system. Calcite mosaic fabric in sparitic speleothems (Figure 14) commonly consists of euhedral (rhomb-shaped) and sub-euhedral crystals. In speleothems, mosaic fabrics have massive, translucent appearance. Crystals show various dimensions, commonly $>10\,\mu m$. Mosaic fabric has been observed as cement filling primary voids, and as the final product of dissolution/reprecipitation processes, when an unstable phase, such as aragonite or fine-grained calcite, had been dissolved and replaced by a mosaic of coarse calcite crystals (see Figure 14). In this case, anhedral shapes, which are typically observed in marine, phreatic, void-filling cements have also been observed.

Speleothems consisting of euhedral calcite crystals have been described from phreatic micro-environments such as pools and gours (Hill and Forti, 1997). There is still little information available on the development of mosaic fabrics. On the basis of both experimental data and the observed development of euhedral, rhomb-shaped crystals in pools, the interplay between kinetics and thermodynamic equilibrium is one possible explanation for the development of euhedral, rhomb-shaped morphology. Dickinson et al. (2002) reported that, when crystallization of calcite was controlled only by the Ca^{2+} concentration in solution, the crystals morphology was that of perfect rhombohedra. By contrast, when degassing controlled growth, crystals developed significant defects, stepped growth and multi-nucleation sites. A mosaic fabric could, therefore, indicate that degassing did not play an important role in calcium-carbonate deposition.

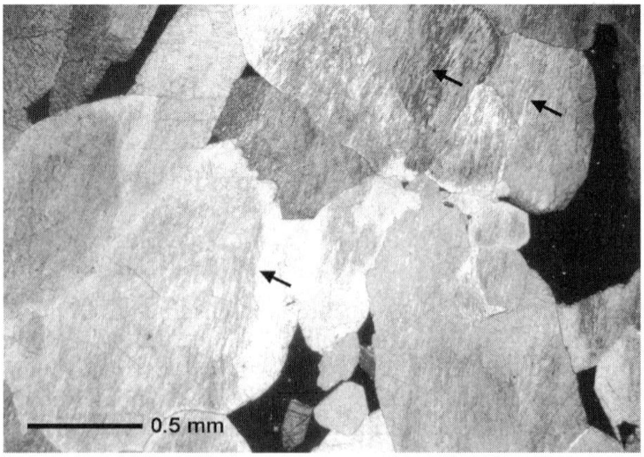

Figure 14 Mosaic fabric consisting of sub-euhedral to anhedral calcite crystals with relict aragonite needles (arrows) in a stalagmite from the Grotte de Clamouse (S. France).

5.3. Significance of stable isotopes in speleothems

Changes in climate and environment at the surface are transmitted to the cave through the physical and chemical properties of infiltration water (Figure 2). The ratio of stable C and O isotopes incorporated in speleothem carbonates mostly reflects soil dynamics (carbon) and atmospheric (oxygen) phenomena, as schematically illustrated in Figure 15. Trace elements are additional environmentally significant proxy data incorporated in speleothems and their variations in concentration are commonly a function of soil pH changes (Borsato et al., 2007). Aerosols formed by natural or anthropogenic emissions, such as sulphates, are also transported from the surface and incorporated in speleothems. Growth rate, fabrics and fluid inclusions complement the pool of environmental and climate proxy data from cave deposits. In particular, speleothem fabrics provide information on the deviation from equilibrium of the depositional system, which plays a role in equilibrium isotope incorporation (Frisia et al., 2000). Several complex processes influence the incorporation of C and O isotopes in speleothems and, consequently, the variability of $\delta^{13}C$ and $\delta^{18}O$ values of speleothem layers deposited at different times. These processes have been thoroughly discussed in the excellent review papers by McDermott (2004), McDermott et al. (2006) and Fairchild et al. (2007).

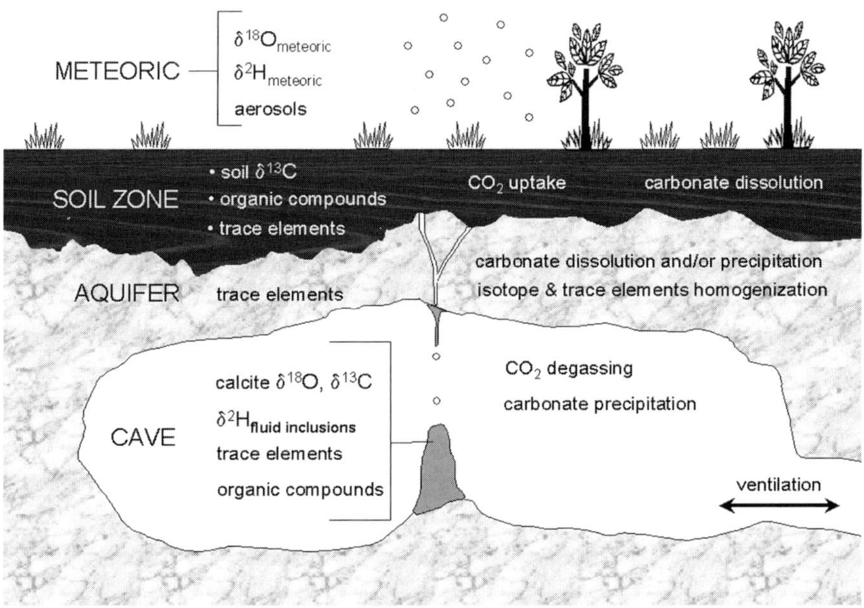

Figure 15 The most common relationships which link proxy data encoded in speleothems with climate and environmental parameters.

In synthesis, the $\delta^{13}C$ values of speleothem carbonates depend on the $\delta^{13}C$ value of the DIC of the parent water, provided that kinetic processes do not, or only slightly, modify the equilibrium calcite/water fractionation. Large shifts in the $\delta^{13}C$ values of DIC can occur when climate-driven change plant assemblages from C3 (trees and temperate grass) to C4 (succulents, cacti, dry-climate grass). When soil-respired CO_2 comes from C3 plants, the $\delta^{13}C$ values of DIC typically range from −26 to −20‰. Cave carbonates formed in equilibrium with CO_2-respired from C3 plants are characterized by $\delta^{13}C$ values ranging from approximately −14‰ to approximately −6‰ (McDermott et al., 2006). Soil CO_2 respired from C4 plants has $\delta^{13}C$ values from ~16 to −10‰ and the $\delta^{13}C$ of secondary carbonates formed in equilibrium with a range of respired CO_2 between approximately −6‰ and +2‰ (McDermott et al., 2006).

Prior calcite precipitation (PCP) in the unsaturated zone above the cave (Baker et al., 1997), duration of water/soil–gas interaction (Bar-Matthews et al., 1997) and temperature-dependent changes in the proportion of atmospheric carbon to the dissolved CO_2 (Genty et al., 2003) are other phenomena which result in $\delta^{13}C$ values of speleothem carbonates that are higher than predicted by equilibrium fractionation. The $\delta^{18}O$ value of secondary carbonates formed in caves ideally reflects the $\delta^{18}O$ values of drip waters in caves (which reflect the $\delta^{18}O$ of the precipitation at the surface, if deposition occurred at, or close to, isotopic equilibrium (McDermott et al., 2006). There are other factors that influence the $\delta^{18}O$, such as cave temperature and non-equilibrium effects.

Understanding the role of kinetics is extremely critical for the correct interpretation of stable-isotope series extracted from cave carbonates. The focus here is consequently on *in situ* kinetic processes.

Kinetic effects commonly take place within the cave and may alter the calcite/water isotopic fractionation at the fluid/mineral interface. The development of transfer functions that allow interpreting stable-isotope data in terms of temperature or rainfall amount thus becomes extremely complex. On the basis of the assumption that a positive correlation of $\delta^{13}C$ and $\delta^{18}O$ reflect kinetics rather than climate processes, and on the basis of an extensive review of published stable-isotope series extracted from speleothems, Mickler et al. (2006) suggested that the majority of the studied specimens formed in isotopic disequilibrium. Assessment of isotopic equilibrium is needed if one intends to use equilibrium equations to derive quantitative data relative to climate parameters from the stable-isotope values of speleothem calcite. Isotopic equilibrium is commonly tested by carrying out a 'Hendy test' (Hendy, 1971) on several synchronous crystallization layers consisting of the same fabric types. If C and O isotopes do not co-vary along the same layer with progressive distance from the vertical growth axis, kinetic effects are negligible. This means that degassing of the parent water was relatively slow, as non-equilibrium carbon dioxide degassing (rapid) of the fluid at the top of the speleothem would result in preferential removal of the light isotopes and subsequent enrichment in ^{13}C and ^{18}O of the parent water. Laboratory experiments of calcite precipitation conducted in cave-like conditions, by flowing a supersaturated solution along a 30°-sloped channel and by stimulating CO_2 degassing, have clearly shown kinetic fractionation along the slope, resulting in progressive stable oxygen and carbon-isotope ratio enrichment in calcite forming downslope in the tube with respect to the flowing parent water.

Wiedner et al. (2008) inferred that progressive degassing and decrease of flow rate were responsible for the formation of crystallites enriched in ^{13}C and ^{18}O in the down-flow direction. Their experiments also suggest the possibility that a certain degree of kinetic fractionation is always possible. Partin et al. (2007) proposed that kinetic fractionation in calcite formed in isotopic quasi-equilibrium conditions as indicated by several Hendy tests that account for about 0.5‰ of the $\delta^{18}O$ variability observed across single growth layers in stalagmites from Borneo. Mickler et al. (2006) suggested that positive co-variance through time (i.e. along the vertical growth axis of a stalagmite, which identifies the temporal dimension) of C and O isotope values may be related to kinetic effects that are not predicted by the Hendy test. Evapotranspiration and PCP, which are directly connected to climatic parameters, however, may result in strong positive co-variation of $\delta^{13}C$ and $\delta^{18}O$ along the vertical axis (time) dimension. The positive co-variation of $\delta^{13}C$ and $\delta^{18}O$ therefore appears to be another valuable climate proxy which can be extracted from speleothems. In this case, the analysis of fabrics becomes extremely important. For example, if fabric types do not considerably change, drip rate, or amount of impurities in the parent water, or intensity of degassing as related to seasonal changes in air flow are likely to have remained relatively constant through time (Frisia et al., 2000). We have already mentioned that stalagmites and flowstones consisting of columnar fabric types should have formed in conditions of quasi-isotopic equilibrium. The positive co-variance of C and O isotope signals in speleothems formed of columnar calcite therefore may be indicative of climate-related phenomena, such as a decrease in drip rate due to a decrease in rainfall (Frisia et al., 2006).

The flow rate at the top of a speleothem driven by drip rate is probably one of the most important phenomena which may, or may not, yield kinetic effects at the crystal/liquid interface (Garside, 1977). For example, in the case of compact columnar and microcrystalline fabrics, drip and flow rates are different. The typical columnar fabric develops under relatively constant flow, whereas the microcrystalline sub-type forms when the drip rate is variable. A variable flow rate influences crystal growth rates (Garside, 1977) and also the (more or less prolonged) residence time of the parent water at the speleothem tip. The observed preferential uptake of lighter isotopes in microcrystalline calcite has yet to be explained, but it may be the result of deposition under predominantly low degassing conditions.

The most extreme case of flow-rate variability has been observed in association with the dendritic fabric, which is characterized by ^{13}C enrichment with respect to other fabrics formed in similar conditions. In this case, calcite is mostly deposited under conditions of prolonged or rapid degassing during periods of a low drip rate, or during time intervals when the pCO_2 dissolved in the drip water is much higher than the pCO_2 in the cave air, for example in late summer or in the case of strong ventilation.

The role of impurities needs more investigation. It has been documented from laboratory experiments that the incorporation of some trace metals present in different concentrations in the parent waters causes morphological changes in the calcite crystals that are related to preferential incorporation of the metals at specific lattice sites on the growing calcite surface (Braybrook et al., 2002). Little is known, however, about the effects that calcite-morphology changes related to the incorporation of trace metals or, possibly, organic molecules may have on isotope fractionation.

5.4. Mineralogical changes

Dramatic seasonal changes in drip rate, in certain circumstances, may result in mineralogical changes (Fairchild et al., 2007). The most commonly observed minerals are calcite and aragonite, the latter precipitating in the dry season (Railsback et al., 1994). The formation of this unstable low-temperature, high-pressure $CaCO_3$ polymorph in the karst subsurface environment is still a matter of debate (see the discussion in Frisia et al., 2002).

At the Grotte de Clamouse (S. France), the water feeding actively growing aragonite ray crystals is characterized by a saturation index (Si) of aragonite of about 0.00 and a Si calcite of ~0.1 (Frisia et al., 2002). In this situation, calcite, rather than aragonite, should precipitate. Prismatic aragonite, however, precipitates instead of the stable calcite when the Mg/Ca molar ratio of the drips exceeds 1:1 (Frisia et al., 2002). González and Lohmann (1987) reported that aragonite forms in caves from solutions when the Mg/Ca molar ratio of the water is higher than 1:5. Frisia et al. (2002), however, reported that aragonite did not precipitate in pools where the water had an Mg/Ca molar ratio of ~2.0. This suggests that there must be another variable that plays a role in triggering the precipitation of aragonite. In the Grotte de Clamouse, the drip rate is a major factor, which for columnar calcite ranges from 0.03 to 0.05 ml/min, and which is for aragonite rays about 0.0003 ml/min. Cabrol (1978) had already observed that a very low discharge was a necessary condition for the occurrence of aragonite speleothems in caves in southern France.

Another key factor which favours aragonite over calcite at surface conditions is the pH. Aragonite has been precipitated from solutions with an Mg/Ca molar ratio of 4:1 and a pH between 8.2 and 10.8

(Kim et al., 2006). The results demonstrated that the $CaCO_3$ polymorph which at high pH preferentially incorporated CO_3^{2-} ions is aragonite.

Experiments targeted at controlling the precipitation of $CaCO_3$ polymorphs in solutions with a relatively low degree of supersaturation indicated that the presence of Mg ions shifts the Gibbs free energy of nucleation from surface-dominated to bulk lattice-dominated, and favour aragonite (Hou and Feng, 2005). Suitable organic macromolecules may also act as templates for aragonite precipitation by lowering the Gibbs free energy of nucleation (Hou and Feng, 2005). A higher pH and the presence of Mg seem therefore to be key factors in the formation of aragonite layers in sparitic speleothems.

Cave aragonite displays a variety of forms, which probably accounts for the extremely complex trace-element zoning within single stalagmite layers (Finch et al., 2003) and for the enrichment in both ^{13}C and ^{18}O which characterizes needle crystals (Frisia et al., 2002). Careful petrographic investigation aimed at the recognition of crystal morphologies in stalagmite layers becomes therefore a prerequisite for the correct interpretation of geochemical signals in layered aragonite stalagmites.

5.5. Organic components

Investigation of the potential of molecular organic geochemistry in speleothems as climate and environmental proxies has received little attention compared to stable isotopes and trace elements (Fairchild et al., 2007). In shallow caves, that is those where passages are located <50 m below the surface, and when filtering effects in the aquifer are negligible, drip waters are expected to carry organic molecules onto the speleothems (Baker et al., 2000; Perrette et al., 2005). By using FT-IR spectroscopy, Ramseyer et al. (1997) detected organic molecules, most probably acidic functional groups from the soil, in speleothems. They inferred that organic molecules could have been either adsorbed at the growing speleothem surface or be trapped in aqueous fluid inclusions. On the basis of UV fluorescence wavelength analyses of stalagmites and drip water, Baker et al. (2000) suggested that acidic functional groups occur in the fluid inclusions. We have carried out synchrotron radiation micro FT-IR spectroscopy, which has a resolution of $\sim 2\,\mu m$, on columnar fabric and noted several organic-related peaks. In particular, a peak at wave number about $3{,}400\,cm^{-1}$ is typical of O-H stretching bonds, and a peak at about $2{,}930\,cm^{-1}$ that Ramseyer et al. (1997) related to strongly bound organic components, most probably fulvic-acid structures. Synchrotron-generated maps suggest that columnar calcite incorporates functional groups from soil organic matter within the solid phase, rather than in fluid inclusion (Figure 16).

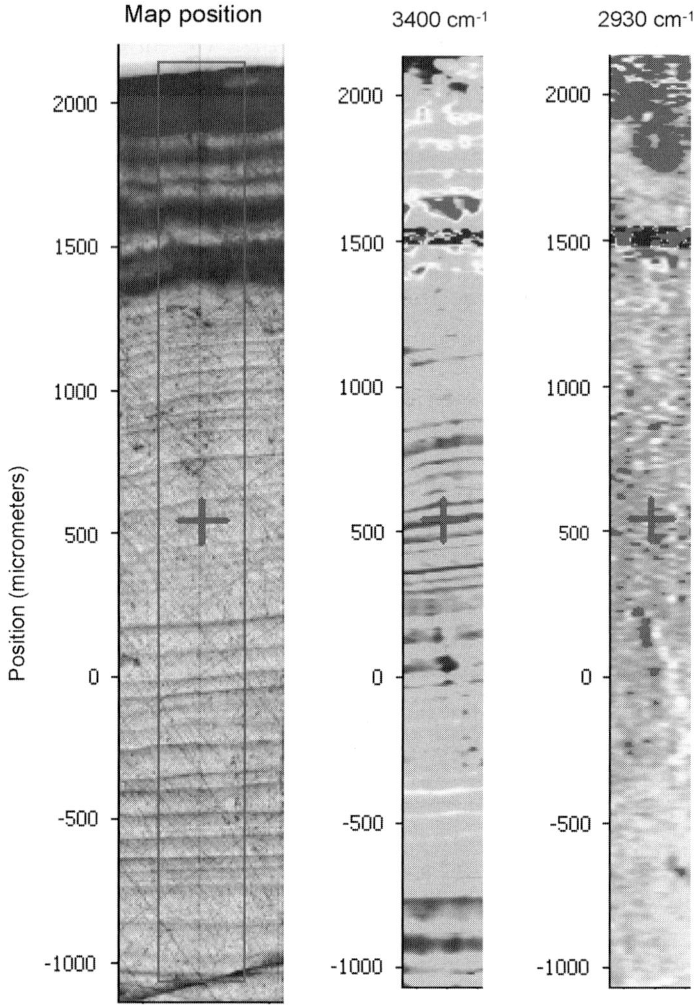

Figure 16 Synchrotron radiation FT-IR generated maps relative to the distribution within annually laminated columnar calcite of the organic components associated with the 3,400 cm^{-1} and 2,930 cm^{-1} wave numbers. The area covered by the maps is shown on the thin section. Black and dark gray: highest concentration. Light gray to white: lowest concentration. The maps highlight that the distribution of organic molecules is mostly within the translucent portion of the laminae, rather than in the impurity-rich darker portion, and is pervasive. Individual spectra show that the organic molecules are interspersed within the calcite lattice, not in fluid inclusions. The maps were generated at the European Synchrotron Radiation Facility FT-IR beamline (ID 21) in Grenoble.

The soil provenance of most organic ligands trapped within sparitic speleothems suggests that their organic content should enable reconstruction of climate-related soil processes. The study of Xie et al. (2003) is a good example of the potential of organic geochemistry applied to speleothem-based palaeoclimate reconstructions. Xie et al. (2003) analysed lipids from a subtropical Chinese stalagmite and observed that the portion of the specimen which grew during the cold Younger Dryas was characterized by an increase in low-molecular weight (LMW) *n*-alkanols, which are related to soil microbes. They interpreted the LMW increase as reflecting a decrease in vegetation above the cave resulting from the cold conditions. Warmer and humid conditions are marked by an increase in high-molecular weight (HMW) *n*-alkanols, which are related to an increase in vegetation cover. Comparison of lipid data with stable-isotope ratios in the same stalagmite provided support to the inferred ecosystem changes. The study indicates that lipid analysis is possibly one of the most powerful palaeovegetation proxies, and needs further exploration. New methodologies for the optimization of lipid extraction from stalagmites allowing access to chemically bound lipids (Blyth et al., 2006) and spectroscopic techniques that detect organic molecules in the solid sample will allow the pool of proxy data that we can obtain from karst deposits to be expanded. Fingerprinting of organic molecules in sparitic speleothems has just started, but will likely emerge as one of the most promising fields of study in palaeoclimate research.

Pollen grains are also part of the pool of proxies of organic origin which may be extracted from speleothems. The oxidising cave environment is, however, not the best for pollen preservation. Pollen taphonomy in the cave environment therefore has been poorly investigated. There is conspicuous absence of experimental background to understand pollen transport and dispersion for underground karst (Turner and Hannon, 1988). Studies of sediments and speleothems from several caves in Spain, however, suggest that the pollen assemblages preserved in the hypogean karst reflect regional vegetation even better than those preserved at the surface (Navarro et al., 2001). Speleothems showed a poor pollen sum and concentration with respect to cave sediments, but they appear to be potentially very useful to determine regional vegetation changes when located near cave entrances.

6. Karst Archives of Gravitational, Seismic and Volcanic Processes

Gravitational collapse of cave roofs and floors leads to the formation of collapse breccias, which may completely fill some passages. Collapse breccias have been defined as 'crackle' and chaotic. The crackle breccia occurs in the upper part of the collapse, and is characterized by clasts of various dimensions, but most commonly in the range of cobbles and boulders, separated by fractures. The clasts are derived from the fractured parent rock. Crackle breccias are clast-supported. Through time, crackle breccias can be consolidated by deposition of sparite cement in the fractures. Chaotic breccias are polymictic and range from clast-supported to matrix-supported. The clasts consist of parent-rock material and speleothems from the collapsed cave roofs or floors. The matrix is made up of small grains from the parent rock, clay from residual dissolution of the karst rock, and other residual grains. Chemically precipitated cements may separate successive collapse episodes.

The study of collapse breccias allows the reconstruction of cave evolution in relation to climate and environmental changes. The origin of collapse breccias in caves cut within Messinian reefal facies on the island of Mallorca (Spain) has been interpreted as being controlled by high-frequency sea-level fluctuations (Robledo Ardila et al., 2004). Cave passages were created during marine lowstands in the still poorly consolidated aragonitic reef limestone, whereas deposition during highstand increased the load and induced gravitational collapse in the underlying passages.

Broken and deformed stalagmites potentially provide information on palaeoseismicity at regional-to-local scale (Postpischl et al., 1991). Recognition of a common azimuthal orientation of damaged and collapsed stalagmites on cave floors, and common anomalous deviations of stalagmite vertical growth axes in growing individuals have been proposed as a criterion to detect seismic events (Postpischl et al., 1991). Borsato et al. (2007) observed, in stalagmites from Cogola di Giazzera (N. Italy), a growth interruption at about 1060 ± 70 AD, which coincides with a historical earthquake that was even mentioned by Dante in his 'Commedia'. The growth interruption is abrupt. There is no evidence of a progressive shift of the feeding drip, as would be expected in the case of gravitational displacement of the roof. The $\delta^{13}C$ and $\delta^{13}O$ values of the portion below the hiatus do not show positive co-variance, as would be expected in the case of growth interruption due to a shift towards dry-climate conditions, or to PCP that might have clogged the feeding system. Furthermore, most active stalagmites of the cave show similar characteristics. Borsato et al. (2007) therefore suggest that cave speleothems archive seismic activity. Apparently, seismic activity is best recorded by rupture and deformation of long, thin stalagmites. Deformation of other types of speleothems, such as columns or flowstones, should be related to other phenomena. Specific speleothems that formed along pool edges (shelfstones) have been found to be ideal to reconstruct the post-glacial isostatic rebound of the Sierra Nevada following the last deglaciation (Granger and Stock, 2004).

Shelfstones are geological tiltmeters in that they mark the horizontal surface at the time of deposition, having formed at the air/water interface. When these deposits have a tilted position with respect to the present-day horizontal surface, they probably record movements of the Earth's crust. Because the tilt can be measured and the deposit accurately dated by the uranium-series method, it is possible to reconstruct the timing of tilting of mountainous areas. Granger and Stock (2004) concluded that shelfstones offer an excellent method for the measurement of isostatic rebounds due to alpine glaciation '*likely to be suitable wherever karst is found*'.

Kempe (2004), however, proposed the formation of cave ice during cold periods as an alternative hypothesis for the occurrence of damaged speleothems in the Postojna caves. Cave ice pushes and breaks most speleothem types, including massive columns. Kempe (2004) suggested that the occurrence of extensive speleothem rupture episodes can be related to glacial climate conditions.

Volcanic activity is preserved in the sulphur chemistry of sparitic speleothems (Frisia et al., 2005a). The sulphur emitted from large volcanic eruptions becomes oxidised into sulphate and reaches the soil as wet and dry deposition (Figure 15). The propagation of the atmospheric sulphate load from the surface to the cave then depends on the transmission by the karst aquifer. Commonly, caves cut in fissured limestone are ideal candidates for the rapid transmission of the atmospheric signal to speleothems. A volcanically induced increase in local acidity levels may also mobilize trace metals, which become incorporated in stalagmites soon after a major eruption. The application of synchrotron radiation X-ray absorption near edge structure spectroscopy (XANES), coupled with synchrotron-radiation based X-ray fluorescence (XRF), indicated that large volcanic eruptions can be recognized in stalagmites as thin and continuous sulphate layers (Frisia et al., 2005a,b). Frisia et al. (2008) detected three distinct sulphate peaks in a portion of annually laminated stalagmite from NE Italy in the interval from 1800 to 2000 AD, coinciding with the eruptions of the Tambora (1815), Krakatau (1884) and Hekla (1947).

The use of speleothems and cave deposits to reconstruct seismic, volcanic and gravitational activity has great potential as a future direction of karst research. In particular, there are excellent perspectives for global correlation of past volcanic events by using the sulphate concentration encoded in karst deposits.

7. Conclusions

The karst environment is one of the richest sources of information about surface processes and their evolution through time. The karst

Braybrook, A.L., Heywood, B.R., Jackson, R.A., Pitt, K., 2002. Parallel computational and experimental studies of the morphological modification of calcium carbonate by cobalt. Journal of Crystal Growth 243, 336–344.

Cabrol, P., 1978. Contribution à l'étude du concrétionnement carbonaté des Grottes du Sud de la France: morphologie, genèse et diagenése. Memoires Centre d'Etudes et de Recherches Géologiques et Hydrogéologiques, Montpellier, 275 pp.

Cacchio, P., Contento, R., Ercole, C., Cappucchio, G., Preite Martinez, M., Lepidi, A., 2004. Involvement of microorganisms in the formation of carbonate speleothems in Cervo Cave (L'Aquila – Italy). Geomicrobiology Journal 21, 497–509.

Cañaveras, J.C., Cuezva, S., Sanchez-Moral, S., Lario, J., Gonzalez, J.M., Saiz-Jimenez, C., 2006. On the origin of fiber calcite crystals in moonmilk deposits. Naturwissenschaften 93, 27–32.

Cerling, T.E., Quade, J.M., Wang, Y., Bowman, J.R., 1989. Carbon isotopes in soils and palaeosols as ecology and palaeoecology indicators. Nature 341, 138–139.

Chen, J.A., Zhang, D.D., Wang, S.J., Xiao, T.F., Huang, R.E., 2004. Factors controlling tufa deposition in natural waters and waterfall sites. Sedimentary Geology 166, 353–366.

Chernov, A.A., 1961. The spiral growth of crystals. Soviet Physics Uspsekhi 4, 116–148.

Dever, M.L., 1985. Approches chimiques et isotopiques des interactions fluids-matrice en zone non saturée carbonatée. Thèse Docteur d'Etat, Université de Paris-Sud Centre d'Orsay, 196 pp.

Dickinson, S.R., Henderson, G.E., McGrath, K.M., 2002. Controlling the kinetic versus thermodynamic crystallisation of calcium carbonate. Journal of Crystal Growth 244, 369–378.

Dreybrodt, W., 1999. Chemical kinetics, speleothem growth and climate. Boreas 28, 347–356.

Drysdale, R.N., 1999. The sedimentological significance of hydropsychid caddisfly larvae (Order: Trichoptera) in a travertine-depositing stream: Louie Creek, Northwest Queensland, Australia. Journal of Sedimentary Research A 69, 145–150.

Drysdale, R.N., 2001. Factors controlling the hydrochemistry of Louie Creek, a travertine-depositing stream in the seasonally wet tropics of Northern Australia. Marine and Freshwater Research 52, 793–804.

Drysdale, R.N., Head, M.J., 1994. Geomorphology, stratigraphy and ^{14}C-chronology of ancient tufas in Louie Creek, northwest Queensland Australia. Géographie Physique et Quaternaire 48, 285–295.

Drysdale, R.N., Zanchetta, G., Hellstrom, J.C., Fallick, A.E., McDonald, J., Cartwrigth, I., 2007. Stalagmite evidence for the precise timing of North Atlantic cold events during the early last glacial. Geology 35, 77–80.

Durand, K., 2002. Preliminary investigation into the use of freshwater molluscs (*Alathyria pertexta* and *Velusonio wilsoni*) from the Barkly Karst, North West Queensland, as palaeoenvironmental indicators. Unpublished Honours Thesis, University of Newcastle, Australia, 106 pp.

Fairchild, I.J., Frisia, S., Borsato, A., Tooth, A., 2007. Speleothems. In: Nash, D.J., McLaren, S.J. (Eds), Geochemical Sediments and Landscapes. Blackwell Publishing, Oxford, pp. 200–245.

Finch, A.A., Shaw, P.A., Holmgren, K., Lee-Thorp, J., 2003. Corroborated rainfall record from aragonitic stalagmites. Earth and Planetary Science Letters 215, 265–273.

Folk, R.L., 1974. The natural history of crystalline calcium carbonate: effect of magnesium content and salinity. Journal of Sedimentary Petrology 44, 40–53.

Ford, D.C., Williams, P., 1989. Karst Geomorphology and Hydrology. Chapman and Hall, London, 601 pp.

Ford, T.D., Pedley, H.M., 1996. A review of tufa and travertine deposits of the world. Earth-Science Reviews 41, 117–175.
Frančišković-Bilinski, S., Barišić, D., Vertačnik, A., Bilinski, H., Prohic, E., 2004. Characterization of tufa from the Dinaric karst of Croatia: mineralogy, geochemistry and discussion of climate conditions. Facies 50, 183–193.
Frisia, S., Borsato, A., Fairchild, I.J., McDermott, F., 2000. Fabrics of speleothems from temperate-humid settings: growth mechanisms and environmental parameters. Journal of Sedimentary Research 70, 1183–1196.
Frisia, S., Borsato, A., Fairchild, I.J., McDermott, F., Selmo, E.M., 2002. Aragonite-calcite relationships in speleothems (Grotte de Clamouse, France): environment, fabrics, and carbonate geochemistry. Journal Sedimentary Research 72, 687–699.
Frisia, S., Borsato, A., Fairchild, I.J., Susini, J., 2005a. Variations in atmospheric sulphate recorded in stalagmites by synchrotron micro-XRF and XANES analyses. Earth and Planetary Science Letters 235, 729–740.
Frisia, S., Borsato, A., Mangini, A., Spötl, C., Madonia, G., Sauro, U., 2006. Holocene record of climate changes and land use in Sicily reconstructed from a stalagmite. Quaternary Research 66, 388–400.
Frisia, S., Borsato, A., Preto, N., McDermott, F., 2003. Late Holocene annual growth in three Alpine stalagmites records the influence of solar activity and the North Atlantic Oscillation on winter climate. Earth and Planetary Science Letters 216, 411–424.
Frisia, S., Borsato, A., Spötl, C., Miorandi, R., Villa, I., Cucchi, F., 2005b. Climate variability in the South-Eastern Alps of Italy over the last 17.000 years reconstructed from stalagmite records. Boreas 34, 445–455.
Frisia, S., Borsato, A., Susini, J., 2008. Synchrotron radiation applications to past volcanism archived in speleothems: an overview. Journal of Volcanology and Geothermal Research 117, 96–100.
Garside, J., 1977. Kinetics of crystallization from solutions. In: Kaldis, E., Scheel, H.J. (Eds), Crystal growth and materials. North Holland, Amsterdam, pp. 484–513.
Genty, D., Blamart, D., Ouadhi, R., Gilmour, M., Baker, A., Jouzel, J., Van Exer, S., 2003. Precise dating of Dansgaard-Oeschger climate oscillations in western Europe from stalagmite data. Nature 421, 833–838.
Genty, D., Quinif, Y., 1996. Annually laminated sequences in the internal structure of some Belgian stalagmites – importance for paleoclimatology. Journal of Sedimentary Research 66, 275–288.
Givarzigov, E.J., 1978. Growth of whiskers by the vapor-liquid-solid mechanism. In: Kaldis, E. (Ed.), Current Topics in Materials Science, Vol. 1. North Holland, Amsterdam, pp. 79–145.
González, L.A., Carpenter, S.J., Lohmann, K.C., 1992. Inorganic calcite morphology: roles of fluid chemistry and fluid flow. Journal of Sedimentary Petrology 62, 382–399.
González, L.A., Lohmann, K.C., 1987. Controls on mineralogy and composition of spelean carbonates: Carlsbad Cavern, New Mexico. In: James, N.P., Choquette, P.W. (Eds), Paleokarst. Springer, New York, pp. 81–101.
Granger, D.E., Stock, G.M., 2004. Using cave deposits as geologic tiltmeters: application to postglacial rebound of the Sierra Nevada, California. Geophysical Research Letters 31, L22501.
Henderson, G., 2006. Caving into new chronologies. Science 313, 620–622.
Hendy, C.H., 1971. The isotopic geochemistry of speleothems. 1. The calculation of the effects of different modes of formation on the isotopic composition of speleothems and their applicability as palaeoclimatic indicators. Geochimica et Cosmochimica Acta 35, 801–824.
Horvatinčić, N., Čalić, R., Geyh, M.A., 2000. Interglacial growth of tufa in Croatia. Quaternary Research 53, 185–195.

Horvatinčić, N., Karajcar Bronić, I., Obelić, B., 2003. Differences in the ^{14}C age, $\delta^{13}C$ and $\delta^{18}O$ of Holocene tufa and speleothem in the Dinaric Karst. Palaeogeography, Palaeoclimatology, Palaeoecology 193, 139–157.
Hill, C.A., Forti, P., 1997. Cave Minerals of the World. National Speleological Society, Huntsville, AL, 484 pp.
Hou, W., Feng, Q., 2005. A simple method to control the polymorphs of calcium carbonate in CO_2-diffusion precipitation. Journal of Crystal Growth 282, 214–219.
Huang, Y., Fairchild, I.J., Borsato, A., Frisia, S., Cassidy, N.J., McDermott, F., Hawkesworth, C.J., 2001. Seasonal variations in Sr, Mg and P in modern speleothems (Grotta di Ernesto, Italy). Chemical Geology 175, 448–499.
Kano, A., Matsuoka, J., Kojo, T., Fujii, H., 2003. Origin of annual laminations in tufa deposits, southwest Japan. Palaeogeography, Palaeoclimatology, Palaeoecology 191, 243–262.
Kaufmann, G., 2002. Karst landscape evolution. In: Gabvrosek, F. (Ed.), Evolution of Karst: From Prekarst to Cessation. Karst Research Institute, Postojna, pp. 243–258.
Kaufmann, G., 2003. Stalagmite growth and palaeo-climate: a numerical perspective. Earth and Planetary Science Letters 214, 251–266.
Kempe, S., 1982. Long-term records of CO_2 pressure fluctuations in fresh waters. In: Degens, E.T., Kempe, S., Soliman, H., and Herrerapp, R. (Eds), Transport of Carbon and Minerals in Major World Rivers. SCOPE-UNEP, Hamburg, pp. 91–323.
Kempe, S., 2004. Natural speleothem damage in Postojnska Jama (Slovenia), caused by glacial cave ice?. Acta Karstologica 18, 265–289.
Kendall, A.C., Broughton, P.L., 1978. Origin of fabric in speleothems of columnar calcite crystals. Journal of Sedimentary Petrology 48, 550–552.
Kim, S.-T., Hillaire-Marcel, C., Mucci, A., 2006. Mechanisms of equilibrium and kinetic oxygen isotope effects in synthetic aragonite at 25°C. Geochimica et Cosmochimica Acta 70, 5790–5801.
Krauskopf, K.B., 1967. Introduction to Geochemistry. McGraw-Hill, New York , 721 pp.
Lacelle, D., Lauriol, B., Clark, I.D., 2004. Seasonal isotopic imprint in moonmilk from Caverne de l'Ours (Quebec, Canada): implications for climatic reconstruction. Canadian Journal of Earth Sciences 41, 1411–1423.
Latham, A.G., Schwarcz, H.P., Ford, D.F., Pearce, G.W., 1982. The paleomagnetism and U-Th dating of three Canadian speleothems: evidence for the westward drift 5.4–2.1 ka BP. Canadian Journal of Earth Sciences 19, 1985–1995.
Lu, G., Zheng, C., Donahoe, R.J., Lyons, W.B., 2000. Controlling processes in a $CaCO_3$ precipitating stream in Huanglong Natural Science District, Sichuan, China. Journal of Hydrology 230, 34–54.
Mallick, R., Frank, N., 2002. A new technique for precise U-series dating of travertine microsamples. Geochimica et Cosmochimica Acta 66, 4272–4361.
Mason, H.E., Frisia, S., Tang, Y., Reeder, R.J., Phillips, B.L., 2007. Phosphorous speciation in calcite speleothems determined from solid-state NMR spectroscopy. Earth and Planetary Science Letters 254, 313–322.
Mattey, D., Lowry, D., Duffet, J., Fisher, R., Hodge, E., Frisia, S., 2008. A 53 year seasonally resolved oxygen and carbon isotope record from a modern Gibraltar speleothem: reconstructed drip water and relationship to local precipitation. Earth and Planetary Science Letters 269, 80–95.
McDermott, F., 2004. Palaeo-climate reconstruction from stable isotope variations in speleothems: a review. Quaternary Science Reviews 23, 901–918.
McDermott, F., Schwarcz, H., Rowe, P.J., 2006. Isotopes in speleothems. In: Leng, M.J. (Ed.), Isotopes in Palaeoenvironmental Research. Springer, Dordrecht, pp. 185–226.

Mickler, P.J., Stern, L.A., Banner, J.L., 2006. Large kinetic isotope effects in modern speleothems. Geological Society of America Bulletin 118, 65–81.

Navarro, C., Carrión, J.S., Munuera, M., Prieto, A.R., 2001. Cave surface pollen and the palynological potential of karstic cave sediments in palaeoecology. Review of Palaeobotany and Palynology 117, 245–265.

Onac, B.P., 1997. Crystallography of speleothems. In: Hill, C., Forti, P. (Eds), Cave Minerals of the World. National Speleological Society, Huntsville, AL, pp. 230–235.

Onac, B.P., Ghergari, L., 1993. Moonmilk mineralogy in some Romanian and Norwegian caves. Cave Science 20, 107–111.

Partin, J.W., Cobb, K.M., Adkins, J.F., Clark, B., Fernandez, D.P., 2007. Millennial-scale trends in the west Pacific warm pool hydrology since the Last Glacial Maximum. Nature 449, 452–455.

Pentecost, A., 1993. British travertines: a review. Proceedings of the Geologists' Association 104, 29–39.

Pentecost, A., 1995. The Quaternary travertine deposits of Europe and Asia Minor. Quaternary Science Reviews 14, 1005–1028.

Perkins, A., 1996. Observations under electron microscopy of magnetic minerals extracted from speleothems. Earth and Planetary Science Letters 139, 281–289.

Perrette, Y., Delannoy, J.-J., Desmet, M., Lignier, V., Destombes, J.-L., 2005. Speleothem organic matter content imaging. The use of Fluorescence Index to characterize the maximum emission wavelength. Chemical Geology 214, 193–208.

Postpischl, D., Agostini, S., Forti, P., Quinif, Y., 1991. Paleoseismicity from karst sediments: "Grotta del Cervo" cave, Central Italy. Tectonophysics 193, 33–44.

Railsback, L.B., Brook, G.A., Chen, J., Kalin, R., Fleischer, C.J., 1994. Environmental control on the petrology of a late Holocene speleothem from Botswana with annual layer of aragonite and calcite. Journal of Sedimentary Research 64, 147–155.

Ramseyer, K., Miano, T.M., D'Orazio, V., Wildberger, A., Wagner, T., Geister, J., 1997. Nature and origin of organic matter in carbonates from speleothems, marine cements and oral skeletons. Organic Geochemistry 26, 361–378.

Reardon, E.J., Allison, G.B., Fritz, P., 1979. Seasonal chemical and isotopic variations of soil CO_2 at Trout Ceek, Ontario. Journal of Hydrology 43, 355–371.

Robledo Ardila, P.A., Durán, J.J., Pomar, L., 2004. Paleocollapse structures as geological record for reconstruction of past karst processes during the Upper Miocene of Mallorca Island. International Journal of Speleology 33, 81–95.

Saether, O.M., De Caritat, P., 1997. Geochemical Processes, Weathering and Groundwater Recharge in Catchments. A.A. Balkema, Rotterdam, 400 pp.

Soligo, M., Tuccimei, P., Barberi, R., Delitala, M.C., Miccadei, E., Taddeucci, A., 2002. U/Th dating of freshwater travertine from Middle Velino Valley (Central Italy): paleoclimatic and geological implication. Palaeogeography, Palaeoclimatology, Palaeoecology 184, 147–161.

Spötl, C., Mangini, A., Frank, N., Eichstädter, R., Burns, S.J., 2002. Start of the last interglacial period at ca. 135 ka: evidence from a high Alpine speleothem. Geology 30, 815–818.

Stepanov, V.I, 1997. Notes on mineral growth. Proceedings University of Bristol Speleological Society 21, 25–42.

Szramek, K., Walter, L.M., 2004. Impact of carbonate precipitation on riverine inorganic carbon mass transport from a mid-continent forested watershed. Aquatic Geochemistry 10, 99–137.

Tinner, W., Theurillat, J.P., 2003. Uppermost limit, extent, and fluctuations of the timberline and treeline ecocline in the Swiss central Alps during the past 11,500 years. Arctic and Antarctic Alpine Research 35, 158–169.

Tucker, M.E., 2008. Sedimentary Petrology (3rd Edition). Blackwell Publishing, Oxford, 262 pp.
Turner, C., Hannon, G.E., 1988. Vegetational evidence for late Quaternary climatic changes in southwest Europe in relation to the influence of the North Atlantic Ocean. Philosophical Transactions of the Royal Society of London 318, 45–485.
Van der Weijden, R.D., Van der Heijden, A.E., Witkamp, G.J., Van Rosmalen, G.M., 1997. The influence of total calcium and total carbonate on the growth rate of calcite. Journal of Crystal Growth 171, 190–196.
Verrecchia, E.P., Verrecchia, K.E., 1994. Needle-fiber calcite: a critical review and a proposed classification. Journal of Sedimentary Research A 64, 650–664.
Viles, H.A., Pentecost, A., 2007. Tufa and travertine. In: Nash, D.J., McLaren, S. (Eds), Geochemical Sediments and Landscapes. Blackwell Publications, Oxford, pp. 172–199.
Wiedner, E., Scholz, D., Mangini, A., Polag, D., Mühlinghaus, C., Segl, M., 2008. Investigation of the stable isotope fractionation in speleothems with laboratory experiments. Quaternary International 187, 15–24.
Williams, I., 2001. Environmental Chemistry. Wiley, Chichester, 388 pp.
Williams, P.W., 2008. The role of the epikarst in karst and cave hydrogeology: a review. International Journal of Speleology 37, 1–10.
Woodhead, J., Hellstrom, J., Maas, R., Drysdale, R., Zanchetta, G., Devine, P., Taylor, E., 2006. U-Pb geochronology of speleothems by MC-ICPMS. Quaternary Geochronology 1, 208–221.
Xie, S., Yi, Y., Huang, J., Hu, C., Cai, Y., Collins, M., Baker, A., 2003. Lipid distribution in a subtropical southern China stalagmite as a record of soil ecosystem response to paleoclimate change. Quaternary Research 60, 340–347.
Žak, K., Ložek, V., Kadlecb, J., Hladıkova, J., Cılek, V., 2002. Climate-induced changes in Holocene calcareous tufa formations, Bohemian Karst, Czech Republic. Quaternary International 91, 137–152.

CHAPTER 7

Invertebrate and Vertebrate Trace Fossils from Continental Carbonates

J.F. Genise[1,*], R.N. Melchor[2], E.S. Bellosi[3] *and* M. Verde[4]

Contents

1. Introduction 320
2. Ichnology of Carbonate-Rich Paleosols 321
   2.1. Paleozoic 321
   2.2. Mesozoic 328
   2.3. Cenozoic 328
3. Ichnology of Lacustrine Carbonates 335
   3.1. Carbonate-dominated lacustrine basins 347
   3.2. Mixed carbonate/siliciclastic lacustrine basins 349
   3.3. Evaporite lacustrine basins 352
4. Ichnology of Carbonate Eolianites, Tufas, and Travertines 353
5. The Ichnofacies Model 354
   5.1. Terrestrial ichnofacies 354
   5.2. Carbonate lacustrine ichnofacies 358
6. Conclusions 359
Acknowledgements 360
References 360

*Corresponding author.
E-mail address: jgenise@mef.org.ar

[1] CONICET-Museo Paleontológico Egidio Feruglio, Av. Fontana 140, 9100 Trelew, Chubut, Argentina
[2] CONICET-Facultad de Ciencias Naturales, Universidad Nacional de La Pampa, Av. Uruguay 154, 6300 Santa Rosa, La Pampa, Argentina
[3] CONICET-División Icnología, Museo Argentino de Ciencias Naturales, Av. Angel Gallardo 470, 1405 Buenos Aires, Argentina
[4] SNI, ANII, Facultad de Ciencias, Universidad de la República, Iguá 4225, CP 11400, Montevideo, Uruguay

Developments in Sedimentology, Volume 61
ISSN 0070-4571, DOI 10.1016/S0070-4571(09)06107-X

Table 1 Summary of key aspects of the case studies of trace-fossil assemblages on carbonate-rich paleosols discussed in the text.

Age	Formation	Country	Trace-fossil-bearing facies	Trace-fossil assemblage	Associated body fossils	Paleoenvironment and paleoecology	Source
Late Pleistocene, Early Holocene	Unnamed	Spain (Canary Islands)	Fluvial sediments and calcretes	*Celliforma*, *Rebuffoichnus*	Land snails	Semi-arid climate with periods of increased rainfall	Alonso-Zarza and Silva (2002), Genise and Edwards (2003)
Late Pleistocene	Homeb Silt	Namibia	Immature calcic paleosols, flash flood deposits	*Termitichnus* and nests?, meniscate burrows	Aquatic (*Bulimus tropicus*, *Biomphaleria pfeifferi*) and terrestrial (*Xerocerastus* sp., *Sculptaria sculptuici*) gastropods; rodent, ostrich and gemsbok remains	Floodplains, flash flood events, semi-arid conditions	Smith et al. (1993)
Pleistocene	Bridgewater	Australia	Lowermost horizon of a colluvial fossil soil, pale yellow clayey calcareous aeolianite	*Palmiraichnus bedfordi*	Vertebrate remains	Colluvial soils with calcretes	Zeuner and Manning, (1976), Houston (1987)
Early Pleistocene	Ensenada	Argentina (Entre Ríos, Santa Fe)	Paleosols with nodular and laminar calcretes	*Coprinisphaera*	Vertebrate remains	Semi-arid to arid climate; paleosols developed during wetter conditions with grass cover	Frenguelli (1938)
Late Pliocene–Early Pleistocene	San Andrés	Argentina (Buenos Aires)	Paleosols with nodular and laminar calcretes	*Coprinisphaera*, termite (*Termes*) and ant nests	Vertebrate remains	Alluvial plains and eolian deposits with intense pedogenesis, arid to semi-arid open areas, steppe or prairie environments	Laza (1995), Zárate and Fasano (1989)

Late Pliocene	Chapadmalal	Argentina (Buenos Aires)	Paleosols with nodular and laminar calcretes	*Coprinisphaera*, termite (*Procornitermes*) and ant nests	Vertebrate remains	Warm temperate savannas, xeric environments, alluvial plains, and loess-like deposits with secondary water reworking	Laza (1995), Zárate and Fasano (1989)
Late Pliocene	Barranca de los Lobos	Argentina (Buenos Aires)	Paleosols with nodular and laminar calcretes	*Tacuruichnus farinai*	Vertebrate remains	Warm and wet climate, savannas and loess deposits	Genise (1997), Zárate and Fasano (1989)
Late Miocene	Paso de las Carretas	Argentina (San Luis)	Paleosols with nodular and laminar calcretes	*Coprinisphaera*, *Celliforma*	Vertebrate remains	Fluvial deposits, calcareous paleosols, arid to semi-arid conditions	Pascual and Bondesio (1981)
Miocene (Aragonian to Vallesian)	"Gypsum and gray shales"	Spain (Calatayud-Teruel Basin)	Green sandy marlstones with millimeter-thick parallel lamination, mudcracks and gypsum nodules, associated with reddish shale and nodular gypsum	*Celliforma* isp. A, *Celliforma* isp. B, *Celliforma* (*Cellicalichnus?*) aff. *habari*, *Rosellichnus* cf. *arabicus*, *Labyrintichnus terrerensis*, *Taenidium barretti*, *Polykladichnus aragonensis*, "branched shafts"	Vertebrate remains	Episodic desiccation and flooding in a shallow-lacustrine setting; interpreted as representing the *Termitichnus* and *Scoyenia* ichnofacies	Uchman and Álvaro (2000)
Early Miocene	Hiwegi (Kibanga Mb.)	Kenya (Rusinga Island)	Marmar Paleosols: pale-red, well indurate, weakly developed paleosols in fluvially reworked deposits of carbonatitic volcanic ash	*Celliforma habari*, "calcareous root traces, burrows and cocoons"	Land snails	Halictinae nest in a volcaniclastic succession	Thackray (1994)
Early Miocene	Hiwegi (Fossil Bed Mb.)	Kenya (Rusinga Island)	Paleosols from a red hard limestone with root traces and relict bedding	Root traces, "ellipsoidal cocoons similar to those of chrysomelid beetles, large flies and small moths"	Diverse mammal remains, terrestrial snails (*Maizania*, *Cerasuta*, and *Ligatella*), millipedes, and endocarps of *Celtis rusingensis*	Well-drained and flooded plains; dry or open dry forest	Retallack et al. (1995)

these authors suggested a rather dry, well-drained, and sparsely vegetated environment.

Other associations dominated by similar trace fossils, mostly Celliformidae and *Rebuffoichnus*, have been described from paleosols bearing nodular or laminar calcretes in terrestrial settings probably with a lower content of carbonate than those mentioned previously (Houston, 1987; Alonso-Zarza and Silva, 2002; Genise and Edwards, 2003; Hembree and Hasiotis, 2007).

2.3.2. Three case studies of the incipient *Celliforma* ichnofacies

2.3.2.1. Calizas del Queguay. The Calizas del Queguay is characterized by quartz sandstones that, according to Veroslavsky and Martínez (1996), were cemented by several generations of calcite locally replacing the clastic fraction. It crops out mainly in the north-western region of Uruguay, at Paysandú, Río Negro and Soriano Counties, and to a lesser extent in the southern region, near Montevideo in Canelones County (Figure 1A). The clastic fraction is similar to the underlying Late Cretaceous Mercedes Formation. These rocks have been known in the literature since the nineteenth century, when Charles Darwin and Priest Dámaso Antonio Larrañaga mentioned them in their diaries. The first detailed mention of these rocks comes from Frenguelli (1930), who recorded the presence of the gastropods *Bulimulus* and *Planorbis* at Estación Quebracho (Paysandú County), and discussed the fossil content, including "solitary vespid" cells, among other materials. Frenguelli (1930) correlated these deposits with those of similar lithology located more than 300 km south in Canelones County. Some years later, Lambert (1940) described these rocks and gave them the informal name "Calizas del Queguay." The current information suggests that the Calizas del Queguay ranges in age from Late Cretaceous to Eocene (Goso and Bossi, 1966; Bossi et al., 1975; Sprechmann et al., 1981; Goso, 1999; Daners and Guerstein, 2004; Martínez and Veroslavsky, 2004), and that it formed under prevailing arid conditions in a lacustrine environment, according to earlier authors (Lambert, 1939; Serra, 1945; Jones, 1956), and in terrestrial settings according to most recent ones (Sprechmann et al., 1981; Bossi and Navarro, 1991; Veroslavsky and Martínez, 1996; Martínez et al., 1997, 2001; Martínez and Veroslavsky, 2004; Veroslavsky and De Santa Ana, 2004). No conclusive data about the age and paleoenvironmental interpretation of the formation is available to date, although it is currently under study (Alonso-Zarza, personal communication).

The presence of fossiliferous paleosols bearing petrocalcic horizons is well known from the Calizas del Queguay. These paleosols are characterized by their tabular geometry, meter-scale thickness, and continuous lateral extent. Two fossil assemblages have been recognized at different localities of the Calizas de Queguay (Martínez et al., 2001).

Figure 1 Calizas de Queguay. All scale bars: 1 cm. (A) Sauce Solo Quarry. (B) *Biomphalaria* sp. (C) *Eoborus charruanus*. (D) Endocarp of *Celtis* sp. (E) *Celliforma germanica*. (F) *Celliforma rosellii*. (G) Cluster of bee cells. (H) *Rosellichnus* isp. (I) *Rebuffoichnus sciuttoi*.

The assemblage recorded from the northern locality, called the *Biomphalaria waltheri* Assemblage after the most common gastropod (Figure 1B), shows a mixture of freshwater and terrestrial fossils, namely gastropods, ostracods (crustaceans), characean oogonia (calcified spiral cells, or gyrogonites, of characean algae), bee cells (clavate or subcylindrical chambers constructed by adult bees to raise their offspring), and pupation chambers (ovoid or elongated structures constructed by many insect larvae to spend the pupal stage) (Table 1). The *Eoborus charruanus* Assemblage (Figure 1C) in the south has only terrestrial fossils (gastropods, bee cells, hackberry endocarps, and rhizoliths; Table 1).

Trace fossils are an important part of these fossil assemblages, with bee cells as the dominant ichnofossil in particular. The solitary club-shaped bee cell *Celliforma germanica* (Figure 1E) and the more stout neck-less subcylindrical *Celliforma rosellii* (Figure 1F) are the most abundant ichnotaxa in the Calizas del Queguay. Other bee nests were found, although they are very scarce. These include the solitary club-shaped or subcylindrical bee cell having a spiral closure cap *Celliforma spirifer*, the cell clusters *Rosellichnus* isp. (Figure 1H), and a new form referable to the Caenohalictini (Apidae, Halictinae) (Figure 1G). *R. sciuttoi*, representing wasp cocoons (a silky envelope woven by wasp larvae to spend the pupal stage) (Genise et al., 2007), is a very common trace fossil in these paleosols (Figure 1I). This trace fossil can be identified by its oval shape, one opening frequently at one end, and surface texture (Genise et al., 2007). The ichnofossil content of this unit is currently under study (Verde, personal communication).

2.3.2.2. Gran Salitral Formation. The Early Eocene Gran Salitral Formation is exposed on the south-western margin of the Gran Salitral depression (La Pampa province, Argentina) (Figure 2A) (Linares et al., 1980; Melchor and Casadío, 2000; Melchor, 2002; Melchor et al., 2002). Dominant lithologies of the Gran Salitral Formation are marls with common pedogenic modification, limestones, red laminated mudstones, fine-grained sandstones, and vitric tuffs. The lower part of the unit is composed of limestones interbedded with sandstones and minor siltstones. The intermediate part includes three carbonate-rich paleosols developed on tuffaceous marl, and also massive limestones, and discrete vitric tuff beds (Table 1). The succession is capped by a laminated lacustrine mudstone with minor interbedded tuff, limestone, and sandstone. Melchor et al. (2002) proposed that this succession was deposited in a periodically exposed saline/alkaline playa-lake setting under palustrine conditions. Carbonate-rich paleosols are poorly developed (Inceptisols) and formed under a semi-arid, seasonal and warm climate, as suggested by stable-oxygen isotope data. The carbonate content, varying from 11 to 36% in volume, consists mostly of micrite and minor sparry calcite.

Figure 2 Gran Salitral Formation. All scale bars: 1 cm. (A) General aspect of the outcrop, showing the three paleosols. (B) *Bostryx* sp. (C) *Bulimulus* sp. (D) *Plagiodontes* sp. 1. (E) *Plagiodontes* sp. 2. (F) *Pomacea* sp. (G) *Teisseirei barattinia*. (H) *Celliforma roselli*. (I) *Rosellichnus* sp. (J) Possible *Rebuffoichnus sciuttoi*. (K) Burrow with knobby surface texture.

Figure 3 The Rosado Member of the Sarmiento Formation. Scale bars of B–D: 0.05 cm; E–F: 1 cm. (A) General aspect of the outcrop at Gran Barranca locality. (B) *Lilloiconcha* sp. (C) *Stephadiscus* sp. (D) *Rotadiscus* sp. (E) *Celliforma rosellii*. (F) *Pallichnus dakotensis*.

Table 2 Examples of trace-fossil assemblages from lacustrine carbonates organized by type of lacustrine basin (carbonate-dominated, mixed, and evaporite) and age.

Age	Formation	Country	Trace-fossil-bearing facies	Trace-fossil assemblage	Associated body fossils	Paleoenvironment and paleoecology	Source
Carbonate-dominated lacustrine basins: low-energy bench, steep gradient							
Late Berriasian–Early Valanginian	La Pedrera de Rubíes	Spain (El Montsec, La Cabrúa locality)	Rhythmically laminated (graded bioclastic and micrite couplets), carbonate turbidites, slump structures	*Cochlichnus anguineus*, *Gordia arcuata*, *Steinsfjordichnus brutoni*, *Undichna britannica*	Tetrapod bones, fish remains (Konservat-lagerstätte)	Anoxic bottom in carbonate-rich lake waters; proximal setting; opportunistic colonization	Gibert et al. (1999, 2000)
Late Berriasian–Early Valanginian	La Pedrera de Rubíes	Spain (El Montsec, La Pedrera locality)	Laminated biomicrites	*Hamipes didactylus*, *Undichna britannica*	Tetrapod bones, fish, plant, insect, and crustacean remains (Konservat-lagerstätte)	Anoxic bottom in carbonate-rich lake waters; basin center	De Gibert (1995), Gibert et al. (1999, 2000)
Late Barremian	La Huérguina Limestone	Spain (Las Hoyas)	Dark laminated rhythmites (graded bioclastic and micrite couplets); sharp bases, hydroplastic deformation, current structures	*Cruziana problematica*, *Helminthoidichnites tenuis*, *Lockeia* isp., *Palaeophycus tubularis*, *Treptichnus pollardi*, *Pteraichnus*?, *Undichna unisulca*, crocodilian trackway	Charophyte, plant, ostracod, crustacean, insect, fish, and bone remains (Konservat-lagerstätte)	Basinal dilute microturbidites in a carbonate-rich stratified lake; oxygen deficient to anoxic bottom; stressed trace fossil assemblage	Fregenal-Martínez et al. (1995), Lockley et al. (1995), Moratalla et al. (1995), De Gibert et al. (1999), Buatois et al. (2000)
Carbonate-dominated lacustrine basins: wave influenced, ramp margin							
Late Eocene–Oligocene	Coatzingo (formerly Pie de Vaca)	Mexico (Puebla)	Massive limestone, mudcracks, load casts, brecciation; oolitic grainstone to packstone; both facies with pisoids, peloids and nodular gypsum	Tracks of camelids, large felids, small artiodactyls, small ungulates, birds (charadriforms, flamingo) and elephant; rhizoconcretions	Ostracod remains	Vegetated and agitated littoral zone of alkaline lake; transition to evaporite facies	Rodríguez de la Rosa et al. (2004), Beraldi-Campesi et al. (2006)

Table 2. (*Continued*)

Age	Formation	Country	Trace-fossil-bearing facies	Trace-fossil assemblage	Associated body fossils	Paleoenvironment and paleoecology	Source
Maastrichtian to ? Danian	El Molino	Bolivia (Cal Orcko)	Alternation of thickly bedded oolitic sandy limestones and calcareous claystones; ripple marks, cross-bedding and mudcracks	Tracks attributed to: wide-gauge sauropods, large and small theropods, ankylosaurs, and small ornithopods	Fish, characeans, ostracods, gastropods, plant remains, stromatolites	Periodically exposed marginal facies of alkaline perennial lakes	Meyer et al. (2001), Lockley et al. (2002)
Late Jurassic	Morrison	USA (Colorado, Purgatoire valley)	Thickly bedded fossiliferous micrite, oomicrite, intramicrorudite, oosparite, intrasparrudite and algal biolithite; associated with shales and sandstones	*Parabrontopodus mcintoshi*, theropod tracks	Unionid bivalves, ostracods, conchostracans, fish remains, dinosaur bone, blue-green algae, charophytes, plant impressions	Nearshore shoals and mudflat in a wave-influenced large lake	Lockley et al. (1986), Prince and Lockley (1989)
Carbonate-dominated lacustrine basins (wave influenced, ramp margin): microbial-caddisfly bioherms							
Oligocene–Miocene	Indusial Limestone Formation	France (Massif Central)	Spheroidal to columnar stromatolitic limestones, ooids and pisoliths; large (20 m thick and 2 m wide) microbial-caddisfly bioherms showing caddisfly-stromatolite couplets	Aggregate of tubes, which are covered by mollusk, ostracod or ooids bound by a calcareous crust	Rare trichopteran pupae inside tubes	Littoral lacustrine setting; annual alternation of caddisfly cases attached to plants and laminated microbial carbonates	Hugueney et al. (1990), Bertrand-Sarfati et al. (1994)

Middle Eocene	Green River (Wilkins Peak Mb)	USA (Wyoming)	Laminated and massive micrite, siltstone, thin tuff beds, and caddisfly-rich stromatolite bioherms; large bioherms (up to 9 m tall and 40 m wide) of calcareous tufa, stromatolite and travertine encased in structureless micrite; microbial-caddisfly couplets	Caddisfly pupal cases composed by small carbonate particles or sand-sized siliciclastic grains; arrays of parallel cases oriented perpendicular and parallel to the substrate	Stromatolites, ostracods, logs, vertebrate bones	Nearshore lacustrine setting during lake transgression; couplets represent annual pupation and microbialite encrustation	Leggitt and Cushman (2001)
Early Eocene	Green River (Tipton Shale Mb)	USA (Wyoming)	Ooid and ostracod packstone with small (dm thick and wide) caddisfly-dominated microbial carbonate mounds (microbial-caddisfly couplets); associated with packstone and grainstone	Caddisfly cases composed of carbonate-coated ostracod carapaces, micritic particles and rare avian eggshell fragments	Avian eggshell fragments, bird and fish bones, gastropods	Platform/bench lake margin with variable wave energy; lowstands of hydrologically closed lake	Leggitt and Loewen (2002), Leggitt et al. (2007)
Early Eocene	Green River (Fossil Butte Mb)	USA (Wyoming)	Massive lime mudstone with tufa-coated tree branches, stromatolites and caddisfly cases	Unoriented caddisfly cases covered with rare carbonate fragments and quartz grains; carbonate coating	Ostracod remains	Subaqueous nearshore lacustrine facies; caddisfly cases attached to subaqueous vegetation or log remains	Loewen et al. (1999)

Table 2. (*Continued*)

Age	Formation	Country	Trace-fossil-bearing facies	Trace-fossil assemblage	Associated body fossils	Paleoenvironment and paleoecology	Source
Late Oligocene to Miocene	Various	Spain	Laminated dolomicrite, mudcracks; reddish siltstone with rhizoliths	"Networks of small rhizoliths"	None	Ephemeral vegetated carbonate ponds in dry mudflats	Rodríguez-Aranda and Calvo (1998)
Late Oligocene to Miocene	Various	Spain	Marl with lenticular gypsum and gypsum nodules; massive lenticular gypsum interbedded with charophyte biomicrite and peloidal micrite; associated with reddish siltstone with rhizoliths; paleokarst	Large rhizoliths, large meniscate burrows, L-shaped burrows, rare networks of small rhizoliths (in marl); tangle-patterned meniscate burrows (in lenticular gypsum)	Charophytes	Transition from siliciclastic mudflat to central saline lake	Rodríguez-Aranda and Calvo (1998)
Late Oligocene to Miocene	Various	Spain	Thick successions of alternating gypsum and marl beds	Tangle-patterned meniscate burrows, rare large meniscate burrows (gypsum); L-shaped burrows (marl)	None	Saline pan deposits	Rodríguez-Aranda and Calvo (1998)
Late Oligocene to Miocene	Various	Spain	Detrital gypsum with cross-lamination associated with green mudstone, primary gypsum, and gypsiferous carbonate	Networks of small rhizoliths, large rhizolits, hyaenid tracks, undeterminated tracks	Charophytes	Vegetated lake margin successions (saline mudflat-saline lake transition)	Rodríguez-Aranda and Calvo (1998)
Late Oligocene to Miocene	Various	Spain	Primary gypsum with karst features and early diagenetic fabrics; associated reddish mudstone	Network of small and large rhizoliths (top of beds); tangle-patterned meniscate burrows, large meniscate burrows	Charophytes	Subaerial exposure of saline pan deposits and interaction with dilute brines	Rodríguez-Aranda and Calvo (1998)

examples in Table 2 are described as a combination of a number of individual trace-fossil assemblages in distinct sedimentary facies. In lacustrine and palustrine carbonates with an abundant bioturbation component, Freytet and Verrecchia (2002) distinguished between lacustrine chalks and marls with a massive micritic fabric (attributed to intense bioturbation) and root traces in palustrine limestones, which can lead to the development of pseudomicrokarst (Plaziat and Freytet, 1978; Freytet and Plaziat, 1982). Both carbonate fabrics commonly lack discrete trace fossils and are not treated in this chapter. The scarcity of Paleozoic ichnofaunas from carbonate lacustrine settings is in agreement with evolutionary and behavioral trends observed in continental ichnofaunas (Buatois et al., 1998; Park and Gierlowski-Kordesch, 2007).

The first undisputed lacustrine ichnofaunas are from Ordovician volcaniclastic rocks (Buatois et al., 1998), although there is no record of lacustrine carbonate ichnofaunas until the Early Devonian. There are only three described examples of Early Paleozoic lacustrine carbonate ichnofaunas, two from the Early Devonian Wood Bay Formation of Spitsbergen and one from the Late Devonian Oneonta Formation of New York (Table 2). Despite the increase in ichnodiversity in Carboniferous and Permian non-marine deposits, only a single example of lacustrine carbonate ichnofaunas has been described to date. The known trace-fossil assemblages in carbonate lacustrine facies can be grouped into three major types of lake basins (Talbot and Allen, 1996): carbonate-dominated basins, mixed carbonate/siliciclastic basins, and evaporite basins.

3.1. Carbonate-dominated lacustrine basins

The examples of trace-fossil assemblages in carbonate-dominated lacustrine basins can be divided (following Platt and Wright, 1991) into (1) low-energy bench, steep-gradient; (2) wave-influenced, ramp-margin; and (3) low-energy, ramp-margin successions. The examples from low-energy bench, steep-gradient successions correspond to the El Montsec and Las Hoyas Konservat-Lagerstätte deposits from the Lower Cretaceous of Spain (Fregenal-Martínez et al., 1995; Moratalla et al., 1995; De Gibert et al., 1999, 2000; Buatois et al., 2000). The typical trace-fossil-bearing facies are rhythmically laminated and graded bioclastic limestone and micrite deposited as turbidites and laminated micrite deposited in a basinal oxygen-deficient setting. The low-diversity trace-fossil assemblages are dominated by simple surface trails and burrows of invertebrates or fishes. These assemblages reflect stress due to poor oxygenation and opportunistic colonization (e.g., Buatois et al., 2000; De Gibert et al., 2000). The carbonate sediments were occasionally emergent at Las Hoyas, as suggested by vertebrate tracks (possible crocodilian and pterodactyl trackways).

Three examples of trace-fossil assemblages are known from wave-influenced, ramp-margin carbonate-dominated lacustrine basins, including stromatolites. These facies are interbedded with fine-grained siliciclastic sedimentary rocks that compose most of the succession. The environmental setting is interpreted as a shallow freshwater or brackish lake (including playa lakes) with fluctuating shoreline and adjacent low-energy, low-gradient carbonate mudflats. The trace-fossil assemblages from this type of basin display a moderate to large ichnodiversity and seem to be dominated by tetrapod tracks. This apparent dominance may be biased because most studies are devoted to the study of this type of trace fossil. In the Cenozoic examples (Moussa, 1968; Breed, 1973; Nations et al., 1981; Antón et al., 2004), the most common trace fossils are mammal and shorebird tracks (Table 2). The trace-fossil assemblages from Mesozoic mixed carbonate/siliciclastic basins (Clemmensen, 1978; Clemmensen et al., 1998; Lockley et al., 1998, 2006; Klein et al., 2006; Lucas and Lerner, 2006; Melchor and De Valais, 2006) are typified by dinosaur (e.g., *Brontopodus*, *Caririchnium*, *Grallator*) and other reptile footprints (e.g., *Brachychirotherium*, *Rigalites*). The single known Permian example (Swanson and Carlson, 2002) contains swimming and wading traces of Early Permian reptiles and amphibians (Table 2). The associated invertebrate traces, irrespective of the age, are simple horizontal trails and burrows (e.g., *Archaeonassa*, *Cochlichnus*, *Palaeophycus*), bivalve resting traces (*Lockeia*), and vertical burrows (e.g., *Skolithos*). In the next paragraph, the trace-fossil assemblage and hosting sedimentary facies of the Ischichuca Formation will be analyzed in detail. Only part of the vertebrate tracks of the unit has been discussed previously (Melchor and De Valais, 2006).

The Ischichuca Formation is part of the lacustrine filling of the Triassic Ischigualasto-Villa Unión Basin, north-west Argentina. The trace-fossil-bearing facies (facies association SL2 of Melchor, 2007) is characterized by meter-scale shallowing upward cycles composed of dark mudstones with paper lamination overlain by mud-cracked dolomitic micrite (Figure 4A). The dolomitic micrite is structureless or finely laminated with planar or undulating laminae, ptygmatic crack infillings, horizontal sheet cracks, current ripples, common mudcracks (Figure 4B), and rare tepee structures. When examined microscopically, the dolomitic micrite exhibits a brecciated or brecciated-nodular microtexture (Armenteros and Daley, 1998). This facies association is interpreted as the product of sedimentation in a moderately deep, perennial playa-lake surrounded by extensive dry mudflats which suffered marked changes in lake level (Melchor, 2007). The trace fossils are restricted to the dolomitic micrite beds. The most common trace fossils are reptile tracks, including trackways assigned to *Rigalites* isp. (Figure 4H) and long trackways of tridactyl bipeds (Figure 4G), large oval-shaped tracks with a raised marginal ridge (Figure 4E), and rare chirotheriid tracks (Figure 4F). Other traces of possible vertebrate origin are two sets of paired spindle-shaped marks with an associated mound of sediment in the external side, showing alternate

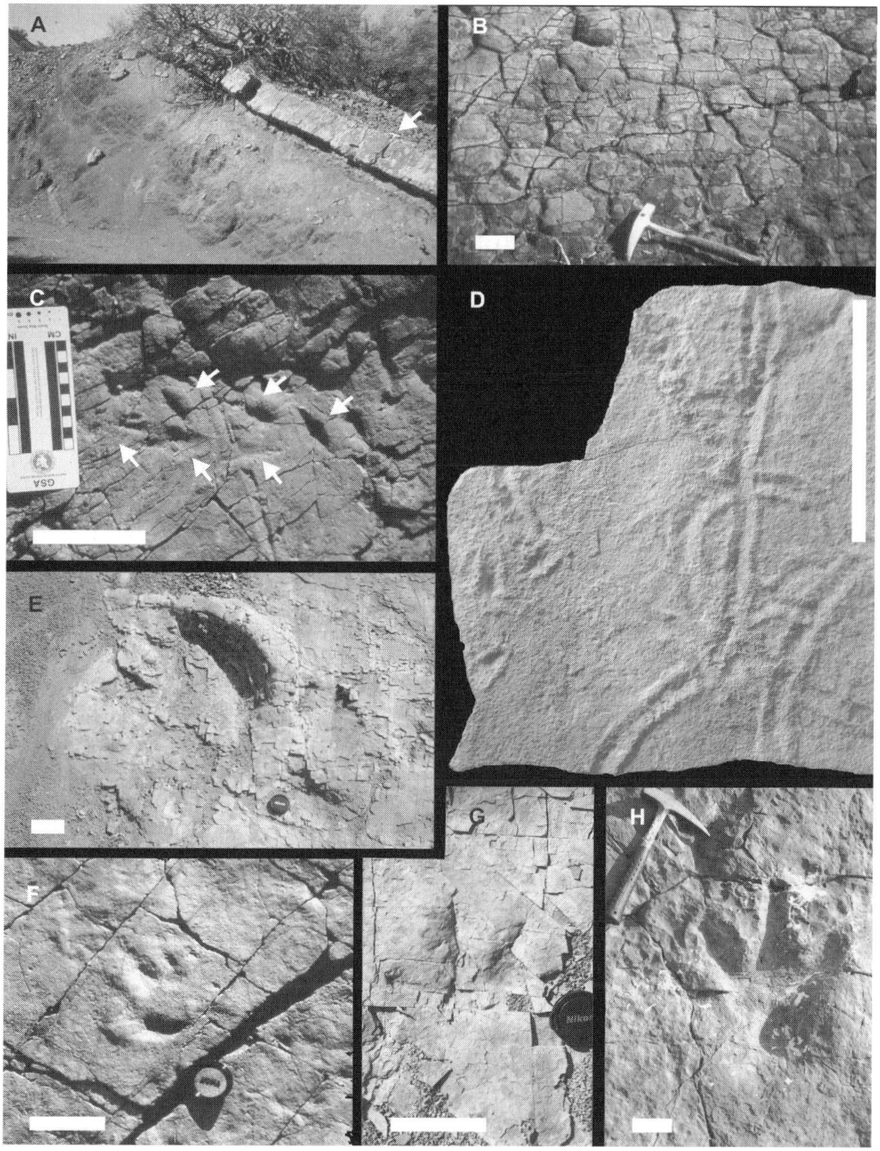

Figure 4 The Triassic Ischichuca Formation, a mixed siliciclastic/carbonate lacustrine basin. The trace fossils illustrated in C–H belong to dolomitic micrite beds. Scale bar is 10 cm, except when indicated. (A) Shallowing upward cycle of the playa lake facies association of the Ischichuca Formation (SL2 facies association of Melchor, 2007), showing dark finely laminated mudstones capped by dolomitic micrite. Hammer (arrowed) is 0.35 m long. (B) Mudcracks on dolomitic micrite. (C) Trackway comparable with *Lunichnium* or *Ichthyoidichnites* (arrowed). (D) *Archaeonassa fossulata*. (E) Large oval footprint with a marginal raised rim. (F) Poorly preserved chiroteriid footprint. (G) Tridactyl footprint showing high digit divarication (divergence). (H) *Rigalites* isp.

symmetry. These traces are provisionally compared with the ichnogenus *Lunichnium*, although the morphology of these traces does not compare favorably with any of the known ichnospecies (e.g., Minter and Braddy, 2006). These traces also are comparable with *Ichthyoidichnites*, although the status of this ichnogenus is uncertain. The invertebrate traces are restricted to trails assigned to *Archaeonassa fossulata* (Figure 4D). As is the case with the Late Triassic Redonda Formation of the south-western USA (Table 2), the trace-fossil assemblage from the Ischichuca Formation includes traces produced in submerged or recently exposed substrates with a high water saturation (*Archaeonassa*, *Lunichnium*?) and tracks produced in fully emergent substrates with a lower water saturation to support the progression of large tetrapods (*Rigalites*, oval-shaped tracks).

3.3. Evaporite lacustrine basins

Sedimentary features of evaporite-lake basins are transitional to some examples of mixed carbonate/siliciclastic lake basins. For the present study, a succession is considered as evaporitic when it contains a substantial amount of evaporite deposits. The nine known examples of trace-fossil assemblages from this type of lacustrine basin are of Cenozoic age (Table 2). The typical trace-fossil-bearing facies are marl, dolomitic micrite, and different forms of gypsum (bedded, primary, detrital, nodular) interbedded with fine-grained siliciclastic sedimentary rocks. The examples are interpreted as deposited in saline mudflats, vegetated ponds in mudflats, central saline pans, and interdune ponds. The trace-fossil assemblages can be subdivided into two groups. The first group is distinguished by a diverse array of mammal (including human) and bird tracks with subordinate invertebrate traces. They have been described from the Early Miocene molasse of the Subcarpathian zone, Romania (e.g., Panin and Avram, 1962; Brustur and Alexandrescu, 1993), the Late Miocene of Hoya de la Sima, Spain (Pérez-Lorente et al., 1999), the Mio-Pliocene Copper Canyon Formation, Death Valley, USA (Scrivner and Bottjer, 1986), and the Holocene Gypsum Lake site, Clare Bay, Australia (Belperio and Fotheringham, 1990). These cases seem to be restricted to saline mudflats. The second group of trace-fossil assemblages is characterized by large and small rhizoliths, tangle-patterned meniscate burrows and larger isolated meniscate burrows, showing rare mammal tracks. The latter were described from several Late Eocene to Early Miocene evaporite successions of Spain by Rodríguez-Aranda and Calvo (1998), who found that highly bioturbated gypsum beds (mostly tangle-patterned meniscate burrows and larger isolated meniscate burrows) are characteristic of saline lakes with brines of intermediate salinity. Other examples of evaporite lacustrine deposits with similar bioturbated gypsum facies (e.g., Salvany et al., 1994;

Ortí et al., 2003) are not treated here because the trace fossils do not occur in limestones.

4. Ichnology of Carbonate Eolianites, Tufas, and Travertines

Carbonate eolianites are rarely recognized in the geological record, although they are widespread in modern coastal tropical settings. These sediments were previously considered as almost devoid of trace fossils (Loope and Abegg, 2001); however, recent descriptions of trace-fossil assemblages from Pleistocene–Holocene carbonate dunes demonstrate that they can hold a moderately diverse assemblage of traces. The case studies include the Late Pleistocene carbonate dunes from Mallorca, Spain (Fornós et al., 2002), Late Pleistocene–Early Holocene eolian oolites from Rhodes, Greece (Milàn et al., 2007), and the Holocene Rice Bay Formation of San Salvador Islands, Bahamas (e.g., White and Curran, 1988; Curran and White, 1991; Curran, 1994; Walker et al., 2003; Martin, 2006). There are also less well-documented cases from other Pleistocene and Holocene eolianites of the Bahamas (Curran, 1994; Martin, 2006). The hosting sedimentary facies are well-sorted cross-bedded grainstones composed of either peloids, ooids, or bioclastic fragments that display sedimentary features typical of eolian successions (the "climbing translatent strata" of Hunter, 1977a,b; grain-fall and grain-flow lamination). The example from the Bahamas also displays interbedded thin micritic crusts. The trace-fossil assemblages from Mallorca and the Bahamas share only two types of traces: root structures and thin, curved, or linear burrows oriented parallel to the stratification with a cemented and pale burrow boundary (named "irregular small burrows" in the Bahamas case, but compared with *Gracilichnus* in the Mallorca assemblage). The example from the Bahamas also contains *Skolithos linearis* (possibly dwelling burrows of arachnids or insects), *Coenobichnus currani* (land hermit-crab trackway), "cluster burrows" (group of vertical shafts radiating upward from a common area of origin, attributed to burrowing wasps), long horizontal rhizoliths associated with micrite crusts, and composite trace fossils of decapod and hymenopteran origin (compared with *Psilonichnus* and *Cellicalichnus* Martin, 2006). Micrite crusts are related to dissolution by rainwater of aragonitic and high-magnesium carbonates and post-depositional precipitation of low-magnesium carbonates. These horizontal rhizoliths are considered as indicative of unvegetated dunes.

The examples from Mallorca and Rhodes contain abundant vertebrate tracks. The first example is remarkable for the abundance of tracks and deformation structures associated with tracks assigned to a single ichnotaxon (*Bifidipes aeolis*), beds with trampled ichnofabric, rodent tracks, and mammal coprolites. *B. aeolis* is attributed to the extinct goat, *Myotragus balearicus*, which is also deemed responsible for the trampled ichnofabric and coprolites (Alcover et al., 1999). The Katavia eolianite from Rhodes includes three eolian dune units with abundant mammal tracks separated by reddish paleosols showing root traces, invertebrate burrows, and shells of land snails. The tracks are mostly preserved in vertical sections and were provisionally assigned to proboscideans and camelids (Milàn et al., 2007).

The cases of trace fossils in tufa (i.e., cool to near ambient freshwater carbonate deposit with common macrophyte and invertebrate remains; Ford and Pedley, 1996), in addition to those mentioned under microbial/caddisfly mounds, are rare. The two known examples include hominid footprints. One is from the Late Pleistocene–Holocene of Coahuilla (Mexico), which corresponds to lithified tufa deposits produced in pools, seeps, and streams in an arid environment (Gonzalez et al., 2007). The remaining example is from a 62+ ka old tufa from the Vârtop cave, Romania (Onac et al., 2005). The Romanian footprints are attributed to *Homo neanderthalensis*. A travertine is considered as a "freshwater" hydrothermal calcium-carbonate deposit that lacks macrophyte or animal remains (Ford and Pedley, 1996). Invertebrate burrows (mainly attributed to chironomids and trichopterans) are considered of minor importance in the modification of the epigean travertine fabric of Pleistocene–Holocene age (Pentecost, 2005). The only known example of pre-Pleistocene bioturbated travertine is from the Paleocene of the Itaboraí Basin, Brazil (Sant'Anna et al., 2004). The identified trace fossils are macroscopic mudstone-filled burrows in a micritic travertine facies, which were deposited in pools located at some distance from springs.

5. The Ichnofacies Model

5.1. Terrestrial ichnofacies

Archetypal or Seilacherian ichnofacies are defined as trace-fossil associations, named after an ichnogenus, recurrent in space and time, which are characteristic of particular environmental conditions (Frey and Pemberton, 1984). Originally, Seilacher (1967), who proposed the model, recognized five marine invertebrate ichnofacies and a continental sixth one, the *Scoyenia* ichnofacies, which was later used erroneously to include all continental trace-fossil associations (Frey et al., 1984). Later, the model of continental ichnofacies began its own development with the recognition of the *Mermia* ichnofacies for lacustrine environments (Buatois and Mángano, 1995), the *Coprinisphaera* ichnofacies for terrestrial environments dominated by herbaceous plants (Genise et al., 2000), which along with the redefined *Scoyenia* ichnofacies (Buatois and Mángano, 1995) and the *Termitichnus* ichnofacies (Smith et al., 1993; Genise et al., 2000) compose the present scenario. The current invertebrate ichnofacies model is still under discussion, although broadly accepted. Criticisms are mostly based on misconceptions about the scale of resolution in which different types of trace-fossil associations may be analyzed and the type of inferences that can be obtained (Goldring, 1995; McIlroy, 2004; Bromley et al., 2007). On the other hand, vertebrate ichnofacies have been at the center of increased attention and debate during the last decade (cf. Hunt and Lucas, 2007; Lockley, 2007). In general, the proposed vertebrate ichnofacies (e.g., Lockley et al., 1994a) does not meet with the requirement of recurrence in time for an archetypal ichnofacies. In most cases, the relationship with invertebrate ichnofacies is not clear, except for the broad inclusion within the *Scoyenia* ichnofacies (Lockley et al., 1994a; Melchor et al., 2006). As a consequence, a comprehensive continental ichnofacies scheme that considers the variety of vertebrate traces and hosting facies is currently lacking.

The proposed model for Seilacherian terrestrial ichnofacies (Genise et al., 2000) was based on the idea of linking terrestrial trace-fossil associations with the main classes of global plant growth (Mueller-Dombois and Ellenberg, 1980). The *Termitichnus* ichnofacies was redefined in a narrow sense to name associations dominated by termite nests, typical of closed forests with plant growth under warm and humid conditions. However, the *Coprinisphaera* ichnofacies was the only one formally defined because it was the only association that showed the necessary recurrent records in time and space to qualify as a Seilacherian ichnofacies (Genise et al., 2000). The *Coprinisphaera* ichnofacies, dominated by insect trace fossils, was related to the occurrence of terrestrial herbaceous communities, which range from steppes to savannas, and which involve a wide range of climates. In accordance, some examples of the *Coprinisphaera* ichnofacies, probably from drier paleoclimates, occur in paleosols bearing nodular and laminar calcretes, such as at Paso de las Carretas, Chapadmalal, San Andrés, and Ensenada, from the Early Miocene to the Early Pliocene of Argentina (Frenguelli, 1938; Pascual and Bondesio, 1981; Laza, 1995) (Table 1).

Genise et al. (2000) called attention to a somewhat recurrent association of insect trace fossils dominated by bee nests in carbonate-rich paleosols, mostly palustrine, bearing freshwater and terrestrial gastropods and hackberry endocarps. Such associations had been recorded three times before, viz. from the Cenozoic of Uruguay, USA, and Kenya, respectively (Retallack, 1984; Thackray, 1994; Martínez et al., 1997). Later, the cases from the Gran Salitral and Sarmiento (Rosado Member) Formations were added to the list (Bellosi et al., 2002; Melchor et al., 2002), along with the cases of the Bembridge Formation and the Fossil Bed Member of the Hiwegi Formation (Retallack et al., 1995; Edwards et al., 1998), which had originally been overlooked. Other possible case examples are those of the Oligocene John Day Formation (Retallack et al., 2000; Retallack, 2004), the lower Oligocene of the Czech Republic (Fejfar and Kaiser, 2005), and

5.2. Carbonate lacustrine ichnofacies

The carbonate lacustrine trace-fossil assemblages are more diverse than terrestrial assemblages, and different case studies are presented to represent distinct ichnofacies. The low-diversity subaqueous trace-fossil assemblages from low-energy bench, steep-gradient successions in carbonate-dominated lacustrine basins (Table 2) are considered as representative of a depleted *Mermia* ichnofacies (Buatois et al., 2000; De Gibert et al., 2000). The main factors controlling the trace-fossil occurrence in these examples are anoxia or oxygen deficiency in lake-bottom waters and the episodic occurrence of oxygenated turbidity currents. The trace-fossil assemblages are considered as reflecting opportunistic colonization (after oxygenation events), critical swimming of fishes in anoxic waters, and possibly doomed pioneers (Buatois et al., 2000, De Gibert et al., 2000).

Cases of carbonate-dominated, wave-influenced, ramp-margin basins from the El Molino and Morrison Formations and an example of a mixed carbonate/siliciclastic basin from the Jindong Formation contain abundant sauropod tracks and have been assigned to the *Brontopodus* ichnofacies (Lockley et al., 1994a,b; Hunt and Lucas, 2007). This proposed vertebrate ichnofacies is typified by the abundance of wide-gauge and narrow-gauge sauropod tracks and associated theropod tracks occurring in carbonate or evaporite facies of tidal flats and alkaline–saline lakes (Lockley et al., 1994b). A number of Cenozoic assemblages from carbonate-dominated wave-influenced successions, mixed carbonate/siliciclastic and evaporate-lacustrine basins are typified by the presence of abundant webbed and/or unwebbed shorebird footprints. These examples include the Coatzingo Formation in Mexico, the Bidahochi, Green River, and Copper Canyon Formations in the USA, the Early Miocene strata of Alava, Spain, and the Early Miocene molasse in Rumania (Table 2). Shorebird tracks occur with mammal footprints, rare invertebrate trails, and root traces. These assemblages have been assigned to the "shorebird ichnofacies" (Lockley et al. 1994a) and to the *Avipeda* ichnocoenosis within the *Grallator* ichnofacies (Hunt and Lucas, 2007). This ichnofacies is still poorly defined, but the large number of examples (besides those of carbonate lacustrine settings discussed in the present review), suggest that there is a good chance of distinguishing a new archetypal ichnofacies or perhaps subdividing the archetypal *Scoyenia* ichnofacies (Melchor et al., 2006).

The *Scoyenia* ichnofacies is characteristic of fluvial or lacustrine substrates that are repeatedly exposed and submerged. Some of the examples of lacustrine trace-fossil assemblages represent this ichnofacies, as indicated by the presence of meniscate burrows and tetrapod footprints (Frey et al., 1984; Buatois and Mángano, 1995). The examples of the *Scoyenia* ichnofacies include low-energy, ramp-margin successions of carbonate-dominated basins (e.g., the Tsondab Sandstone, Namibia) and evaporite-lake basins (e.g., various Cenozoic evaporite basins in Spain). In addition to meniscate burrows (*Taenidium* and *Rutichnus*) and mammal footprints, they commonly contain root traces of different morphology (e.g., Rodríguez-Aranda and Calvo, 1998). The Quaternary cases of carbonate eolianites with trace-fossil assemblages, and those of bioturbation in travertine and tufa, do not represent repetitive assemblages with a significant age range, but the ichnocoenoses from the Bahamas have been assigned to the *Psilonichnus* ichnofacies (Curran and White, 1991).

6. Conclusions

A variety of trace-fossils assemblages have been recorded from continental carbonates. In some cases, they correspond to archetypal ichnofacies or incipient ichnofacies, and in some instances the cases are too few to include the associations in the ichnofacies model. The *Coprinisphaera* ichnofacies, dominated by the eponymous ichnogenus, includes examples that occur in paleosols generally containing a lesser amount of carbonate than the incipient *Celliforma* ichnofacies. The latter has been formerly proposed and is here supported by new study cases for calcretes and mostly palustrine carbonates. This ichnofacies is characterized by Celliformidae, particularly *Celliforma*, *R. sciuttoi*, and subordinate beetle trace fossils, as well as rhizoliths. Also typical of these paleosols is the presence of hackberry endocarps and gastropod shells. The *Celliforma* ichnofacies, where occurring in terrestrial settings, would be indicative of low plant coverage due to an arid climate typical of desert scrubs and dry woodlands. When present in a palustrine environment, it would reflect the presence of bare soil, preferred by wasps and bees to nest, after periodic waterlogging under subhumid to sub-arid climates. In other cases, in palustrine environments, alternating subaqueous and subaerial exposure may involve less time of subaerial exposure than necessary for the development of the well-aerated soil profiles required for insects to nest. In such cases of short subaerial exposure, the *Scoyenia* ichnofacies trace-fossil assemblage may develop. This ichnofacies based on meniscate burrows and vertebrate footprints is present in low-energy, ramp-margin successions of carbonate-dominated basins and evaporitic lake basins. Finally, the subaqueous trace-fossil assemblages from low-energy bench, steep-gradient successions in carbonate-dominated lacustrine basins are considered as representative of a depleted *Mermia* ichnofacies. This ichnofacies is characterized by simple horizontal trails and trackways in permanently subaqueous settings. A single case of carbonate eolianites with trace-fossil assemblages has been assigned to the *Psilonichnus* ichnofacies. There have been some attempts to group vertebrate trace fossils into ichnofacies and to integrate them into the invertebrate-based ones.

The *Brontopodus* ichnofacies is characterized by the abundance of wide-gauge and narrow-gauge sauropod tracks and associated theropod tracks occurring in carbonate or evaporitic facies of tidal flats and alkaline–saline lakes. The "shorebird ichnofacies" and the *Avipeda* ichnocoenosis within the *Grallator* ichnofacies, from carbonate-dominated, wave-influenced successions, mixed carbonate/siliciclastic and evaporitic lacustrine basins, are characterized by shorebird tracks and occur with mammal footprints, rare invertebrate trails, and root traces.

ACKNOWLEDGEMENTS

This research was supported by PICT 13286 from the Agencia Nacional de Promoción Científica y Tecnológica of Argentina. Mariano Verde's research is supported by the Proyecto Fondo Clemente Estable No. 05/010 (DINACYT) FCE 2007/44 (ANII). We thank L. Tanner, D. Loope, and an anonymous reviewer for their helpful comments.

REFERENCES

Alcover, J.A., Pérez-Obiol, R., Yll, E.I., Bover, P., 1999. The diet of *Myotragus balearicus* Bate 1909 (Artiodactyla: Caprinae), an extinct bovid from the Balearic Islands: evidence from coprolites. Biological Journal of the Linnean Society 66, 57–74.

Alonso-Zarza, A.M., 2003. Palaeonvironmental significance of palustrine carbonates and calcretes in the geological record. Earth-Science Reviews 60, 261–298.

Alonso-Zarza, A.M., Silva, P.G., 2002. Quaternary laminar calcretes with bee nests: evidences of small-scale climatic fluctuations, Eastern Canary Islands, Spain. Palaeogeography, Palaeoclimatology, Palaeoecology 178, 119–135.

Ameghino, F., 1906. Les formations sédimentaires du Crétacé superieur et du Tertiaire de Patagonie avec un parallele entre leurs faunes mammalogiques et celles de l'ancien continent. Anales del Museo Nacional de Historia Natural de Buenos Aires 15, 1–568.

Antón, M., López, G., Santamaria, R., 2004. Carnivore trackways from the Miocene site of Salinas de Añana (Alava, Spain). Ichnos 11, 371–384.

Armenteros, I., Daley, R., 1998. Pedogenic modification and structure evolution in palustrine facies as exemplified by the Bembridge Limestone (Late Eocene) of the Isle of Wight, southern England. Sedimentary Geology 119, 275–295.

Batra, S.T.W., 1984. Solitary bees. Scientific American 250, 120–127.

Bellosi, E., in press. Loessic and fluvial sedimentation in Sarmiento Formation pyroclastics of the Middle Cenozoic of Central Patagonia. In: Madden, R., Carlini, A., Vucetich, M.G., Kay, R. (Eds), The Paleontology of Gran Barranca: Evolution and Environmental Change through the Middle Cenozoic of Patagonia. Cambridge University Press, Cambridge.

Bellosi, E., González, M., in press. Paleosols of the Middle Cenozoic Sarmiento Formation, Central Patagonia. In: Madden, R., Carlini, A., Vucetich, M.G., Kay, R. (Eds), The Paleontology of Gran Barranca: Evolution and Environmental Change through the Middle Cenozoic of Patagonia. Cambridge University Press, Cambridge.

Bellosi, E., Miquel, S., Kay, R., Madden, R., 2002. Un paleosuelo mustersense con microgastrópodos terrestres (Charopidae) de la Formación Sarmiento, Eoceno de Patagonia central: significado paleoclimático. Ameghiniana 39, 465–477.

Belperio, A.P., Fotheringham, D.G., 1990. Geological setting of two Quaternary footprint sites, western South Australia. Australian Journal of Earth Sciences 37, 37–42.

Beraldi-Campesi, H., Cevallos-Ferriz, S.R.S., Centeno-García, E., Arenas-Abad, C., Fernández, L.P., 2006. Sedimentology and paleoecology of an Eocene-Oligocene alluvial-lacustrine arid system, Southern Mexico. Sedimentary Geology 191, 227–254.

Bertrand-Sarfati, J., Freytet, P., Plaziat, J.C., 1994. Microstructures in Tertiary nonmarine stromatolites (France). Comparison with Proterozoic. In: Bertrand-Sarfati, J., Monty, C. (Eds), Phanerozoic Stromatolites II. Kluwer Academic Publishers, Netherlands, pp. 155–191.

Blomeier, D., Wisshak, M., Joachimski, M., Freiwald, A., Volohonsky, E., 2003. Calcareous, alluvial and lacustrine deposits in the Old Red Sandstone of central north Spitsbergen (Wood Bay Formation, Early Devonian). Norwegian Journal of Geology 83, 281–298.

Bossi, J., Ferrando, L., Fernández, A., Elizalde, G., Morales, H., Ledesma, J., Carballo, E., Medina, E., Montaña, J., 1975. Carta Geológica del Uruguay a Escala 1:1.000.000. Dirección de Suelos y Fertilizantes, Montevideo, 32 pp.

Bossi, J., Navarro, R., 1991. Geología del Uruguay. Departamento de Publicaciones de la Universidad de la República, Montevideo, 496 pp.

Breed, W.J., 1973. New avian fossils from the Bidahochi Formation (Pliocene), Arizona. In: Fasset, J.E. (Ed.), Cretaceous and Tertiary Rocks of the Colorado Plateau, Four Corners Geological Society Memoir, pp. 144–147.

Bromley, R., Buatois, L., Genise, J.F., Labandeira, C., Mángano, G., Melchor, R., Schlirf, M., Uchman, A., 2007. Comments on the paper "Reconnaissance of Upper Jurassic Morrison Formation ichnofossils, Rocky Mountain Region, USA: paleoenvironmental, stratigraphic, and paleoclimatic significance of terrestrial and freshwater ichnocoenoses" by Stephen T Hasiotis. Sedimentary Geology 200, 141–150.

Brustur, T., Alexandrescu, G., 1993. Paleoichnological potential of the Lower Miocene molasse from Vrancea (East Carpathians). Revue Roumaine de Géologie 37, 77–94.

Buatois, L.A., Mángano, M.G., 1995. The paleonvironmental and paleoecological significance of the lacustrine *Mermia* ichnofacies: an archetypical subaqueous nonmarine trace fossil assemblage. Ichnos 4, 151–161.

Buatois, L.A., Mángano, M.G., 1998. Trace fossil analysis of lacustrine facies and basins. Palaeogeography, Palaeoclimatology, Palaeoecology 140, 367–382.

Buatois, L.A., Mángano, M.G., Fregenal-Martínez, M.A., De Gibert, J.M., 2000. Short-term colonization trace-fossil assemblages in a carbonate lacustrine *Konservat-Lagerstätte* (Las Hoyas Fossil Site, Lower Cretaceous, Cuenca, Central Spain). Facies 43, 145–156.

Buatois, L.A., Mángano, M.G., Genise, J.F., Taylor, T.N., 1998. The ichnologic record of the continental invertebrate invasion: evolutionary trends in environmental expansion, ecospace utilization, and behavioral complexity. Palaios 13, 217–240.

Cabrera, A.L., Willink, A., 1973. Biogeografía de América Latina, Editorial Secretaría General de la Organización de Estados Americanos, Monografías Científicas (Serie Biología), 13, 1–120.

Carlson, K.J., 1968. The skull morphology and estivation burrows of the Permian lungfish *Gnathorhiza serrata*. Journal of Geology 76, 641–663.

Clemmensen, L.B., 1978. Lacustrine facies and stromatolites from the Middle Triassic of East Greenland. Journal of Sedimentary Petrology 48, 1111–1128.

Clemmensen, L.B., Kent, D.V., Jenkins, F.A., Jr., 1998. A Late Triassic lake system in East Greenland: facies, depositional cycles and palaeoclimate. Palaeogeography, Palaeoclimatology, Palaeoecology 140, 135–159.

Curran, H.A., 1994. The paleobiology of ichnocoenoses in Quaternary, Bahamian-style carbonate environments: the modern to fossil transition. In: Donovan, S.K. (Ed.), The Paleobiology of Trace Fossils. Wiley, Chichester, pp. 83–104.

Curran, H.A., White, B., 1991. Trace fossils of shallow subtidal to dunal ichnofacies in Bahamian Quaternary carbonates. Palaios 6, 498–510.

Daners, G., Guerstein, R., 2004. Dinoflagelados del Maastrichtiense-Paleógeno en la Formación Gaviotín, Cuenca de Punta del Este. In: Veroslavsky, G., Ubilla, M., and Martínez, S. (Eds), Cuencas sedimentarias de Uruguay. Geología, Paleontología y Recursos Naturales. Cenozoico. DIRAC, Universidad de la República, Facultad de Ciencias, Montevideo, pp. 37–62.

De Gibert, J.M., 1995. Ichnology of the Lower Cretaceous lithographic limestones of El Montsec (Lleida, Spain). Extended Abstracts II Internacional Symposium on Lithographic Limestones, 85–88.

De Gibert, J.M., Buatois, L.A., Fregenal-Martínez, M.A., Mángano, M.G., Ortega, F., Poyato-Ariza, F.J., Wenz, S., 1999. The fish trace fossil *Undichna* from the Cretaceous of Spain. Palaeontology 42, 409–427.

De Gibert, J.M., Fregenal-Martínez, M.A., Buatois, L.A., Mángano, M.G., 2000. Trace fossils and their palaeoecological significance in Lower Cretaceous lacustrine conservation deposits, El Montsec, Spain. Palaeogeography, Palaeoclimatology, Palaeoecology 156, 89–101.

Demicco, R.V., Bridge, J.S., Cloyd, K.G., 1987. A unique freshwater carbonate from the Upper Devonian Catskill Magnafacies of New York. Journal of Sedimentary Petrology 57, 327–334.

Demko, T., Currie, B., Nicoll, C., 2004. Regional paleoclimatic and stratigraphic implications of paleosols and fluvial/overbank architecture in the Morrison Formation (Upper Jurassic), Western Interior, USA. Sedimentary Geology 167, 115–135.

Doyle, P., Wood, J.L., George, G.T., 2000. The shorebird ichnofacies: an example from the Miocene of southern Spain. Geological Magazine 137, 517–536.

Dunagan, S.P., Driese, S.G., 1999. Control of terrestrial stabilization on Late Devonian palustrine carbonate deposition: Catskill Magnafacies, New York, U.S.A.. Journal of Sedimentary Research 69, 772–783.

Edwards, N., Jarzembowski, E.A., Pain, T., Daley, B., 1998. Cocoon-like trace fossils from the lacustrine-palustrine Bembridge Limestone Formation (Late Eocene), Southern England. Proceedings of the Geologists Association 109, 25–32.

Ekdale, A.A., Brown, F.H., Feibel, C.S., 1989. Nonmarine macroborings in early Pleistocene algal biolithites (stromatolites) of the Turkana Basin, northern Kenya. Palaios 4, 389–396.

Fejfar, O., Kaiser, T.M., 2005. Insect bone-modification and paleoecoloy of Oligocene mammal-bearing sites in the Doupov Mountains, Northwestern Bohemia. Palaeontologia Electronica 8 (Sigue 1; 8[a]), 11p. http://palaeo-electronica.org/2005_1/fejfar8/issue1_05.htm

Feruglio, E., 1949. Descripción Geológica de la Patagonia. II. Dirección General de Yacimientos Petrolíferos Fiscales, Buenos Aires, 349 pp.

Ford, T.D., Pedley, H.M., 1996. A review of tufa and travertine deposits of the world. Earth-Science Reviews 41, 117–175.

Fornós, J.J., Bromley, R.G., Clemmensen, L.B., Rodríguez-Perea, A., 2002. Tracks and trackways of *Myograptus balearicus* Bate (Artiodactyla, Caprinae) in Pleistocene aeolianites from Mallorca (Balearic Islands, Western Mediterranean). Palaeogeography, Palaeoclimatology, Palaeoecology 180, 277–313.

Fregenal-Martínez, M.A., Buatois, L.A., Mángano, M.G., 1995. Invertebrate trace fossils from Las Hoyas fossil site (Serranía de Cuenca, Spain). Paleoenvironmental interpretations. Extended Abstracts, II Internacional Symposium on Lithographic Limestones, 67–70.

Frenguelli, J., 1930. Apuntes de geología uruguaya. Boletín del Instituto de Geología y Perforaciones (Uruguay) 11, 1–47.

Frenguelli, J., 1938. Bolas de escarabeidos y nidos de véspidos fósiles. Physis 12, 348–352.

Frey, R.W., Pemberton, S.G., 1984. Trace fossil facies models. In: Walker, R.G. (Ed.), Facies models. Geoscience Canada Reprint Series 1, pp. 189–207.

Frey, R.W., Pemberton, S.G., Fagerstrom, J.A., 1984. Morphological, ethological, and environmental significance of the ichnogenera *Scoyenia* and *Ancorichnus*. Journal of Paleontology 58, 511–528.

Freytet, P., Plaziat, J.C., 1982. Continental carbonate sedimentation and pedogenesis – late Cretaceous and early Tertiary of Southern France. Contributions to Sedimentology (E. Schweizerbart'sche Verlagbuchhandlung, Stuttgart), 12, 213pp.

Freytet, P., Verrecchia, E.P., 2002. Lacustrine and palustrine carbonate petrography: an overview. Journal of Paleolimnology 27, 221–237.

Genise, J.F., 1997. A fossil termite nest from the Marplatan stage-age (late Pliocene) of Buenos Aires province, Argentina, as paleoclimatic indicator. Palaeogeography, Palaeoclimatology, Palaeoecology 136, 139–144.

Genise, J.F., 2004. Ichnotaxonomy and ichnostratigraphy of chambered trace fossils in palaeosols attributed to coleopterans, termites and ants. In: McIlroy, D. (Ed.), The Application of Ichnology to Palaeoenvironmental and Stratigraphic Analysis, Geological Society of London Special Publications 228, pp. 419–453.

Genise, J.F., Edwards, N., 2003. Ichnotaxonomy, origin, and paleoenvironment of Quaternary insect cells from Fuerteventura, Canary Islands, Spain. Journal of the Kansas Entomological Society 76, 320–327.

Genise, J.F., Mángano, M.G., Buatois, L.A., 2004. Ichnology moving out of the water: a model for terrestrial ichnofacies. Abstract Book First International Congress on Ichnology, 38.

Genise, J.F., Mángano, M.G., Buatois, L.A., Laza, J.H., Verde, M., 2000. Insect trace fossil associations in paleosols: the *Coprinisphaera* ichnofacies. Palaios 15, 49–64.

Genise, J.F., Melchor, R., Bellosi, E., González, M., Krause, M., 2007. New insect pupation chambers (Pupichnia) from the Late Cretaceous of Patagonia (Argentina). Cretaceous Research 28, 545–559.

Goldring, R., 1995. Organisms and the substrate: response and effect. In: Bosence, D.W.J., Allison, P.A. (Eds), Marine Palaeoenvironmental Analysis from Fossils, Geological Society of London Special Publication 83, pp. 151–180.

Gonzalez, A.H.G., Lockley, M., Rojas, C., Espinoza, J.L., Gonzalez, S., 2007. Notes on the re-discovery of a 'lost' hominid footprint site from the Cuatro Ciénagas Basin (Coahuilla), Mexico. In: Lucas, S.G., Spielmann, J.A., Lockley, M.G. (Eds), Cenozoic Vertebrate Tracks and Traces, New Mexico Museum of Natural History and Science Bulletin 42, pp. 11–15.

Goso, C., 1999. Análise estratigráfica do Grupo Paysandú (Cretáceo) na Bacia do Litoral, Uruguai. Tese de Doutoramento IGCE-UNESP, Rio Claro, 184 pp.

Goso, H., Bossi, J., 1966. Cenozoico. In: Bossi, J. (Ed.), Geología del Uruguay. Departamento de Publicaciones de la Universidad de la República, Montevideo, pp. 259–301.

Grimaldi, D.A., Engel, M.S., 2005. Evolution of the Insects. Cambridge University Press, New York, 755 pp.

Hasiotis, S.T., 2004. Reconnaissance of Upper Jurassic Morrison Formation ichnofossils, Rocky Mountain Region, USA: paleoenvironmental, stratigraphic, and paleoclimatic significance of terrestrial and freshwater ichnocoenoses. Sedimentary Geology 167, 177–268.

Hasiotis, S.T., Wellner, R.W., Martin, A.J., Demko, T.M., 2004. Vertebrate burrows from Triassic and Jurassic continental deposits of North America and Antarctica: their paleoenvironmental and paleoecological significance. Ichnos 11, 103–124.

Hembree, D.I., Hasiotis, S.T., 2007. Paleosols and ichnofossils of the White River Formation of Colorado: insight into soil ecosystems of the North American Midcontinent during the Eocene-Oligocene transition. Palaios 22, 123–142.

Hester, P.M., Lucas, S.G., 2001. Lacustrine depositional environments of the Upper Triassic Redonda Formation, east-central New Mexico. In: Lucas, S.G., Ulmer-Scholle, D. (Eds), Geology of the Llano Estacado. New Mexico Geological Society Guidebook, 52nd Field Conference, Socorro, pp. 153–168.

Hope Jahren, A., Gabel, M.L., Amundson, R., 1998. Biomineralization in seeds: developmental trends in isotopic signatures of hackberry. Palaeogeography, Palaeoclimatology, Palaeoecology 138, 259–269.

Houck, K.J., Lockley, M.G., 2006. Life in an active volcanic arc: petrology and sedimentology of dinosaur track beds in the Jindong Formation (Cretaceous), Gyeongsang Basin, South Korea. Cretaceous Research 27, 102–122.

Houston, T.F., 1987. Fossil brood cells of stenotritid bee (Hymenoptera, Apoidea) from the Pleistocene of South Australia. Transactions of the Royal Society of South Australia 3, 93–97.

Hugueney, M., Tachet, H., Escuillé, F., 1990. Caddisfly pupae from the Miocene Indusial Limestone of Saint-Gerand-Le-Puy, France. Palaeontology 33, 495–502.

Hunt, A.P., Lucas, S.G., 2007. Tetrapod ichnofacies: a new paradigm. Ichnos 14, 59–68.

Hunter, R.E., 1977a. Basic types of stratification in small eolian dunes. Sedimentology 24, 361–387.

Hunter, R.E., 1977b. Terminology of cross-stratified sedimentary layers and climbing-ripple structures. Journal of Sedimentary Petrology 47, 697–706.

Jones, G., 1956. Memoria explicativa y mapa geológico de la región oriental del Departamento de Canelones. Boletín del Instituto Geológico del Uruguay 34, 1–193.

Klein, H., Lucas, S.G., Haubold, H., 2006. Tetrapod track assemblage of the Redonda Formation (Upper Triassic, Chinle Group) in east-central New Mexico – re-evaluation of ichnofaunal diversity from studies of new material. New Mexico Museum of Natural History and Science Bulletin 37, 241–250.

Lambert, R., 1939. Memoria explicativa del mapa geológico de los terrenos sedimentarios y las rocas efusivas del Departamento de Durazno. Boletín del Instituto de Geología del Uruguay 25, 1–37.

Lambert, R., 1940. Memoria explicativa de un mapa geológico de reconocimiento del Departamento de Paysandú y los alrededores de Salto. Boletín del Instituto de Geología del Uruguay 27, 1–41.

Laza, J.H., 1995. Signos de actividad de insectos. In: Alberdi, T.M., Leone, G., and Tonni, E.P. (Eds), Evolución Biológica y Climática de la Región Pampeana durante los Últimos Cinco Millones de Años. Consejo de Investigaciones Científicas de España, Madrid, pp. 347–361.

Leggitt, V.L., Biaggi, R.E., Buchheim, H.P., 2007. Palaeoenvironments associated with caddisfly-dominated microbial-carbonate mounds from the Tipton Shale Member of the Green River Formation: Eocene Lake Gosiute. Sedimentology 54, 661–699.

Leggitt, V.L., Cushman, R.A., Jr., 2001. Complex caddisfly dominated bioherms from the Eocene Green River Formation. Sedimentary Geology 145, 377–396.

Leggitt, V.L., Loewen, M.A., 2002. Eocene Green River Formation "Oocardium tufa" reinterpreted as complex arrays of calcified caddisfly (Insecta: Trichoptera) larval cases. Sedimentary Geology 148, 139–146.

Linares, E., Llambias, E.J., Latorre, C.O., 1980. Geología de la provincia de La Pampa, República Argentina y geocronología de sus rocas metamórficas y eruptivas. Revista de la Asociación Geológica Argentina 35, 87–146.

Lockley, M.G., 2007. A tale of two ichnologies: the different goals and potentials of invertebrate and vertebrate (tetrapod) ichnotaxonomy and how they relate to ichnofacies analysis. Ichnos 14, 39–57.

Lockley, M.G., Houck, K.J., Prince, N.K., 1986. North America's largest dinosaur trackway site: implications for Morrison formation paleoecology. Geological Society of America Bulletin 97, 1163–1176.

Lockley, M.G., Houck, K.J., Seong-Young, Y., Matsukawa, M., Lim, S.-K., 2006. Dinosaur-dominated footprint assemblages from the Cretaceous Jindong Formation, Hallyo Haesang National Park area, Goseong County, South Korea: evidence and implications. Cretaceous Research 27, 70–101.

Lockley, M.G., Hunt, A., 1995. Dinosaur Tracks and Other Fossil Footprints of the Western United States. Columbia University Press, New York, 338 pp.

Lockley, M.G., Hunt, A., Meyer, C., 1994a. Vertebrate tracks and the ichnofacies concept: implications for palaeoecology and palichnostratigraphy. In: Donovan, S.K. (Ed.), The Paleobiology of Trace Fossils. Wiley, Chichester, pp. 241–268.

Lockley, M.G., Hunt, A.P., Meyer, C.A., Rainforth, E.C., Schultz, R.J., 1998. A survey of fossil footprint sites at Glenn Canyon National Recreation Area (western USA): a case study in documentation of trace fossil resources at a national preserve. Ichnos 5, 177–211.

Lockley, M.G., Logue, T.J., Moratalla, J.J., Hunt, A.P., Schultz, R.J., Robinson, J.W., 1995. The fossil trackway *Pteraichnus* is pterosaurian, not crocodilian: implications for the global distribution of pterosaur tracks. Ichnos 4, 7–20.

Lockley, M.G., Meyer, C.A., Hunt, A.P., Lucas, S.G., 1994b. The distribution of sauropod tracks and trackmakers. In: Lockley, M.G., Dos Santos, V.F., Meyer, C.A., Hunt, A.P. (Eds), Aspects of Sauropod Paleobiology, Gaia, 10, pp. 233–248.

Lockley, M.G., Schulp, A.S., Meyer, C.A., Leonardi, G., Kerumba Mamani, D., 2002. Titanosaurid trackways from the Upper Cretaceous of Bolivia: evidence for large manus, wide-gauge locomotion and gregarious behaviour. Cretaceous Research 23, 383–400.

Loewen, M.A., Leggitt, V.L., Buchheim, H.P., 1999. Caddisfly (Trichoptera) larval cases from Eocene Fossil Lake, Fossil Butte National Monument. In: Santucci, V.L., McClelland, L. (Eds), National Park Service Paleontological Research, Vol. 4. United States Department of the Interior National Parks Service, Geological Resource Division, Lakewood, CO, pp. 72–77.

Loope, D.B., Abegg, F.E., 2001. Recognition and geologic preservation of ancient carbonate eolianites. In: Abegg, F.E., Harris, P.M., Loope, D.B. (Eds), Modern and Ancient Carbonate Eolianites: Sedimentology, Sequence Stratigraphy, and Diagenesis, SEPM Special Publication 71, pp. 3–16.

Lucas, S.G., Hunt, A.P., Lockley, M.G., 2001. Tetrapod footprint ichnofauna of the Upper Triassic Redonda Formation, Chinle Group, Quay County, New Mexico. In: Lucas, S.G., Ulmer-Scholle, D. (Eds), Geology of the Llano Estacado. New Mexico Geological Society Guidebook, 52nd Field Conference, Socorro, pp. 177–180.

Lucas, S.G., Lerner, A.J., 2006. Invertebrate ichnofossil assemblages of the Upper Triassic Redonda Formation at Mesa Redonda, east-central New Mexico. New Mexico Museum of Natural History and Science Bulletin 37, 122–127.

Mack, G.H., James, W.C., Monger, H.C., 1993. Classification of paleosols. Journal of Geology 102, 360–362.

Martin, A.J., 2006. A composite trace fossil of decapod and hymenopteran origin from the Rice Bay Formation (Holocene), San Salvador, Bahamas. In: Gamble, D., Davis, R.L. (Eds), 12th Symposium of the Geology of the Bahamas. Gerace Research Center, San Salvador, Bahamas, pp. 99–112.

Martínez, S., Veroslavsky, G., 2004. Registros continentales no depositacionales del Terciario temprano. In: Veroslavsky, G., Ubilla, M., and Martínez, S. (Eds), Cuencas Sedimentarias de Uruguay. Geología, Paleontología y Recursos Naturales. Cenozoico. DIRAC, Universidad de la República, Facultad de Ciencias, Montevideo, pp. 63–82.

Martínez, S., Veroslavsky, G., Verde, M., 1997. Primer registro del Paleoceno en el Uruguay: paleosuelos calcáreos en la Cuenca de Santa Lucía. Revista Brasileira de Geociências 27, 295–302.
Martínez, S., Veroslavsky, G., Verde, M., 2001. Paleoecología de los paleosuelos calcáreos fosilíferos ("Calizas del Queguay"-Paleoceno) de las regiones sur y litoral del Uruguay. Actas del XI Congreso Latinoamericano de Geología y III Congreso Uruguayo de Geología, Contribución 219.
McIlroy, D., 2004. Some ichnological concepts, methodologies, applications and frontiers. In: McIlroy, D. (Ed.), The Application of Ichnology to Palaeoenvironmental and Stratigraphic Analysis, Geological Society of London, Special Publication 228, pp. 3–27.
Melchor, R.N., 2002. Depósitos continentales eocenos del sudoeste de La Pampa, Argentina: Edad, estratigrafía y paleoambientes. In: Cabaleri N., Cingolani, C.A., Linares, E., López de Luchi, M.G., Ostera, H.A., Panarello, H.O. (Eds), Actas del XV Congreso Geológico Argentino I, pp. 694–698.
Melchor, R.N., 2007. Changing lake dynamics and sequence stratigraphy of synrift lacustrine strata in a half-graben: an example from the Triassic Ischigualasto-Villa Unión Basin, Argentina. Sedimentology 54, 1417–1446.
Melchor, R.N., Bedatou, E., de Valais, S., Genise, J.F., 2006. Lithofacies distribution of invertebrate and vertebrate trace fossil assemblages in an Early Mesozoic ephemeral fluvio-lacustrine system: implications for the Scoyenia ichnofacies. Palaeogeography, Palaeoclimatology, Palaeoecology 239, 253–285.
Melchor, R.N., Casadío, S.A., 2000. Descripción geológica de la Hoja 3766-III "La Reforma" (1:250.000), provincia de La Pampa. Boletín del Servicio Geológico Minero Argentino 295, 1–56.
Melchor, R.N., De Valais, S., 2006. A review of the Triassic tetrapod track assemblages from Argentina. Palaeontology 49, 355–379.
Melchor, R.N., Genise, J.F., Miquel, S., 2002. Ichnology, sedimentology and paleontology of Eocene calcareous paleosols from a palustrine sequence, south west La Pampa, Central Argentina. Palaios 17, 16–35.
Mérida, E., Athor, J., 2006. Talares Bonaerenses y su Conservación. Fundación de Historia Natural "Félix de Azara", Buenos Aires, 259 pp.
Meyer, C.A., Hippler, D., Lockley, M.G., 2001. The Late Cretaceous vertebrate ichnofacies of Bolivia – facts and implications. Asociación Paleontológica Argentina, Publicación Especial 7, pp. 133–138.
Michener, C.D., 1979. Biogeograhy of the bees. Annals of the Missouri Botanical Garden 66, 277–347.
Milàn, J., Bromley, R.G., Titschack, J., Theodorou, G., 2007. A diverse vertebrate ichnofauna from a Quaternary eolian oolite, Rhodes, Greece. In: Bromley, R.G., Buatois, L.A., Mángano, M.G., Genise, J.F., Melchor, R.N. (Eds), Sediment-Organism Interactions: A Multifaceted Ichnology, SEPM Special Publication 88, pp. 333–343.
Minter, N.J., Braddy, S.J., 2006. The fish and amphibian swimming traces *Undichna* and *Lunichnium*, with examples from the Lower Permian of New Mexico, USA. Palaeontology 49, 1123–1142.
Miquel, S., Bellosi, E., 2007. Microgasterópodos terrestres (Charopidae) del Eoceno medio de Gran Barranca (Patagonia Central, Argentina). Ameghiniana 44, 121–131.
Miquel, S., Bellosi, E., in press. Middle Eocene–Oligocene gastropods of the Sarmiento Formation, Central Patagonia. In: Madden, R., Carlini, A, Vucetich, M.G., Kay, R. (Eds), The Paleontology of Gran Barranca: Evolution and Environmental Change through the Middle Cenozoic of Patagonia. Cambridge University Press, Cambridge.
Moratalla, J.J., Lockley, M.G., Buscalioni, A.D., Fregenal-Martínez, M.A., Meléndez, N., Ortega, F., Pérez-Moreno, B.P., Pérez-Asensio, E., Sanz, J.L., Schultz, R.J., 1995.

A preliminary note on the first tetrapod trackways from the lithographic limestones of Las Hoyas (Lower Cretaceous, Cuenca, Spain). Geobios 28, 777–782.
Moussa, M., 1968. Fossil tracks from the Green River Formation (Eocene) near Soldier Summit, Utah. Journal of Paleontology 42, 1433–1438.
Moussa, M., 1970. Nematode fossil trails from the Green River Formation (Eocene) in the Uinta Basin, Utah. Journal of Paleontology 44, 304–307.
Mueller-Dombois, D., Ellenberg, H., 1980. Aims and methods of vegetation ecology. Wiley, New York, 525 pp.
Nations, J.D., Hevly, R.H., Blinn, D.W., Landye, J.J., 1981. Paleontology, paleoecology, and depositional history of the Miocene-Pliocene Verde Formation, Yavapai County, Arizona. Arizona Geological Society Digest 13, 133–149.
Onac, B.P., Viehmann, I., Lundberg, J., Lauritzen, S.-E., Stringer, C., Popiţă, V., 2005. U-Th ages constraining the Neanderthal footprint at Vârtop Cave, Romania. Quaternary Science Reviews 24, 1151–1157.
Ortí, F., Rosell, L., Anadón, P., 2003. Deep to shallow lacustrine evaporites in the Libros Gypsum (southern Teruel Basin, Miocene, NE Spain): an occurrence of pelletal gypsum rhythmites. Sedimentology 50, 361–386.
Paik, I.S., 2005. The oldest record of microbial-caddisfly bioherms from the Early Cretaceous Jinju Formation, Korea: occurrence and palaeoenvironmental implications. Palaeogeography, Palaeoclimatology, Palaeoecology 218, 301–315.
Panin, N., Avram, E., 1962. Noe urme de vertebrate in Miocenul Subcarpaţilor romineşti. Studii şi Cercetări de Geologie 7, 455–484.
Park, L.E., Gierlowski-Kordesch, E.H., 2007. Paleozoic lake faunas: establishing aquatic life on land. Palaeogeography, Palaeoclimatology, Palaeoecology 249, 160–179.
Parrish, J.T., Peterson, F., Turner, C., 2004. Jurassic 'savannah'-plant taphonomy and climate of the Morrison Formation (Upper Jurassic, Western USA). Sedimentary Geology 137–162.
Pascual, R., Bondesio, P., 1981. Sedimentitas Cenozoicas. Relatorio del VIII Congreso Geológico Argentino, pp. 117–153.
Pentecost, A., 2005. Travertine. Spinger-Verlag, Berlin, 445 pp.
Pérez-Lorente, F., Rodríguez, T., Mancheño, M.A., Aerrano, F., Romero, M., 2004. Pisadas fósiles de mamíferos en el Mioceno Superior de la Hoya de la Sima (Jumilla, Murcia, España). Memorias de Arqueología 12, 15–28.
Pérez-Lorente, F., Serrano, F., Rodríguez, T., Mancheño, M.A., Romero, M., 1999. Pisadas fósiles de mamíferos en el Mioceno Superior de la Hoya de la Sima (Jumilla, Murcia, España). Revista Española de Paleontología 14, 257–267.
Platt, N., Wright, V., 1991. Lacustrine carbonates: facies models, facies distributions and hydrocarbon aspects. In: Anadón, P., Cabrera, L., Kelts, K. (Eds), Lacustrine Facies Analysis. International Association of Sedimentologists, Special Publication 13, Blackwell Scientific Publications, Oxford, pp. 57–74.
Plaziat, J.C., Freytet, P., 1978. Le pseudomicrokarst pédologique: un aspect particulier des paléopédogénèses développées sur les dépôts calcaires lacustres dans le Tertiaire du Languedoc. Comptes Rendus de l'Académie des Sciences Paris D 286, 1661–1664.
Prince, N.K., Lockley, M.G., 1989. The sedimentology of the Purgatoire tracksite region, Morrison Formation of southeastern Colorado. In: Gillette, D.D., Lockley, M.G. (Eds), Dinosaur Tracks and Traces. Cambridge University Press, Cambridge, pp. 155–163.
Rădulescu, C., Ştiucă, E., Brustur, T., Zaharia, S., 1995. Neogene mammalian fauna from the Bend Zone of the East Carpathians. Romanian Journal of Stratigraphy 76 (Suppl. 6), 13–25.
Retallack, G.J., 1983. A paleopedological approach to the interpretation of terrestrial sedimentary rocks: the mid-Tertiary fossil soils of Badlands National Park, South Dakota. Geological Society of America Bulletin 94, 823–840.

Subject Index

accommodation space, 120
accretion rates, 251
acicular fabrics, 298
aggradation, 149
  of the floodplain, 121
aggrading neomorphism, 148
algae, 14
alluvial, 255
  basins, 121, 251
  fan facies, 238
  fans, 250
alpha microfabrics, 232
Altiplano, 8
alveolar structures, 107
Alveolar-septal structures, 233–234, 243
anhydrite, 25
annual cycles, 197
anoxia, 358
anoxic conditions, 155
Antalya, 178
Antarctica, 10
Appalachian Basin, 17
aquatic invertebrates, 328
aquifer systems, 236
Aragonite, 107, 187
Archaeonassa, 350
archetypal ichnofacies, 320
Argentina, 321
Aridisol, 335
As Pontes Basin, 22
associations of facies, 149
asymmetrical basins, 253
Avipeda ichnocoenosis, 358

bacteria, 115, 178, 195, 232
Bahamas, 353
balance-fill basins, 120
Barkly karst, 159
Barrage, 141, 149, 158, 160
barrage tufas, 349
barrage-waterfall deposits, 143
bars, 139
bedrock configuration, 135
bee cells, 332
Bembridge Formation, 329
beta microfabrics, 232
bicarbonate, 271
Bicorb Basin, 17, 41
Bifidipes aeolis, 353
Big Soda Lake, 39
Bioclastic limestones, 140, 143, 146, 155
Biogenic features, 232, 234
bioherms, 19, 348
Biomphalaria waltheri Assemblage, 332
bioturbation, 50
bird footprints, 348
bivalves, 12
Blackleaf Formation, 328
blocky peds, 328
Bohemian Massif, 329
boiling springs, 182
borings, 349
Bostryx, 334
Boundstone
  of bryophytes, 137, 141, 148
  of charophytes, 140, 143
  of hanging stems, 143
  of mosses, 143, 146, 156, 159
  of stems, 141, 143
Brachychirotherium, 350
brecciation, 111
brecciated horizon, 243
brecciated limestone, 106
brecciated-nodular microtexture, 350
brecciation, 59, 241
Brontopodus, 350
Brontopodus ichnofacies, 358
brood cells, 329
Brule Formation, 328, 329
Bulimulus, 334
burrows, 320
bush-like filamentous microbes, 146

$\delta^{13}C$, 162, 283
C and O isotopes, 254
C. *germanica*, 334
C_3 to C_4 vegetation, 253
caddisfly cases, 348
Cal Orcko, 348
Calatayud Basin, 25, 150
calcareous paleosol, 329
calcareous tufa, 276
Calcaric Cambisols, 329
calcic argillisols, 328
calcic soils, 239
calcified
  filaments, 118, 233, 241
  root cells, 234, 235

hygrophytes, 149
hymenopterous cocoons, 329
hyperthermal, 183

ichnofacies model, 321
ichnofauna, 349
ichnofossils, 320
ichnological record, 321
ichnology, 320
Ichthyoidichnites, 352
illuvial concentrations, 228
impact craters, 7
Inceptisols, 332
incision, 149
indurated limestones, 117
inorganic carbonate precipitation, 231
Insect, 19, 320
  larvae, 332
  pupation chamber, 334
  trace fossils, 355
  tufa/travertine, 202
interdune pond, 349
Intraclastic, 138
  limestones, 109
intraclasts, 107, 118
invertebrate, 328
  trails, 358
iron mobilisation, 115
iron oxide, 39
Ischichuca Formation, 321, 350
Isotopic fractionation, 167

John Day Formation, 328
jumps, 141, 152
Juniata Formation, 321

K horizon, 228
Kelly Lake, 57
Kenya Rift Valley, 178
kerogen, 48
K-fabric, 228
Knowlton Lake, 15
Konservat-Lagerstätte, 347
Konya Basin, 10

Lac d'Annecy, 69
lacustrine, 137, 255
  muds, 106
Lagoa Feia Formation, 67
Lake Awassa, 69
Lake Baikal, 9
Lake Balaton, 12, 13
Lake Biel, 13
Lake Bogoria, 56
Lake Elsinore, 69
Lake Eyre, 7
Lake Geneva, 18
Lake Juusa, 15
Lake Kassjön, 35
Lake Kopais, 20
Lake Lahontan, 56
lake level, 108
lake margins, 112
Lake Ohrid, 9
Lake Ontario, 15
Lake Rõuge, 35
lake shores, 137, 139, 141
Lake Superior, 35
Lake Tanganyika, 67
Lake Turkana, 39
Lake Zürich, 13
lakes, 103
laminae couplets, 197
Laminar calcretes, 111, 242–244
laminar horizon, 242–243
laminated deposits, 147
laminated marls, 140
land, 320
land microgastropods, 335
Las Hoyas, 15
Las Tablas de Daimiel wetlands, 105
Late Archean, 320
lateral migration, 149
Lauca Basin, 43
Lilloiconcha, 335
Limagne Graben, 19
limestones, 138–139
liquefaction, 59
Lisan Formation, 35
Lithoclastic-bioclastic tufa and travertine, 205
Littlefield Lake, 57
Lockeia, 350
low-gradient, 155, 157
low-gradient systems, 156
low-magnesian calcite, 107
low-sinuosity, 155
low-sinuosity multichannel system, 157
lowstands, 252
Lunichnium, 352

maar lakes, 35
Macroflora, 229
macrophyte, 155
  stems, 147
Madrid Basin, 25
Mae Sot Basin, 69
Mae Tip Basin, 69
magnesite, 4
Malham Tarn, 50
Mallorca, 353
mammal
  coprolites, 353
  footprints, 358
  tracks, 352
marl, 352

marsh, 112
meandering, 155
Mechanisms of accumulation, 231
Medicine Lake, 22
Medicine Lodge, 65
meniscate burrows, 352
Mequinenza Formation, 41
Mercia Mudstone Group, 57
Mermia, 320
  ichnofacies, 354
mesothermal, 183
meteogene tufa, 280
meteoric diagenesis, 227
micrite, 107, 301
  laminae, 243
micritic coatings, 111
Micritic tufa/travertine, 205
micritization, 24
microbes, 14, 232
microbial remains, 146
Microbial/caddisfly mounds, 348
microbialite, 349
Microbioclastic, 139
Microcodium, 235
microcrystalline fabric, 295
microflora, 229
micromorphology, 227
  of Calcretes, 232
micro-organisms, 109, 147
millipedes, 321
mixed carbonate/siliciclastic basins, 347
Molasse Basin, 44
Molluscs, 19
Mono Lake, 8
moonmilk, 276
morphological stages, 239
Morrison Formation, 328
mosaic fabrics, 301
moss buildups, 155
moss mounds, 152
mosses, 148
mottled limestone, 106–107
mottling, 334
mudflat environments, 250
mudflats, 350
mud-size sediment, 107
Mudstones, 140, 142

Nar Gölü, 39
needle-fibre calcite, 112, 233, 234
Newark Supergroup, 6
nodular horizon, 227, 241–242
nodular limestone, 106, 108
Nodules, 233
Nonacho Basin, 7
Non-biogenic features, 232
non-crystallographic dendrites, 190
non-pedogenic carbonates, 236
non-stepped conditions, 157
non-stepped fluvial and fluvio-lacustrine, 155
Nördlinger Ries, 56
North America, 320

$\delta^{18}O$, 162, 283
oil shale, 48
Old Red Sandstone, 321
oncoids, 112, 117, 146–147, 152, 203
oncolites, 143
Oncolitic, 139
onditions, 155
ooids, 56, 203
oolites, 56
oolitic grainstones, 348
open lake, 123
Orcadian Basin, 50
Organic
  filaments, 111
  molecules in sparitic speleothems, 309
  rich marlstone, 112
  rich muds, 141
  rich sediments, 156
ornamented burrow fills, 334
Ostracodes, 18, 107, 332
Overbank deposits, 139
overbank processes, 152
overfilled basins, 120
ovoid structures, 334
oxygen deficiency, 358
Oxygen isotope analysis, 23

Packstones and sands, 139
palaeoclimatic data, 230
Palaeophycus, 350
palaeosols, 121, 320
paleosurface, 335
Pallichnus, 320
Pallichnus dakotensis, 328
palustrine, 154, 255
  areas, 149
  calcrete facies associations, 114
  carbonates, 103, 254
  facies, 106
  limestone, 104
  microenvironment, 114
  settings, 141
Peat, 15
peat beds, 117
pedofacies, 251
pedogenesis, 114, 245
pedogenic calcretes, 226–227
Pedogenic Profile, 239
Peloidal limestone, 111
peloids, 57, 107
per ascensum, 230
per descensum, 230

perched deposits, 159
petrocalcic, 321
  Alfisol, 328
pH, 271
phreatic, 108
  calcretes, 236
phytoclast, 143, 149
Phytoclastic, 138
Phytoherm limestones, 140, 141
Piceance Basin, 39
pisoids, 203
pisolith, 241
Pisolithic horizons, 243
Plagiodontes spp., 334
plankton, 11
Planolites, 349
Planorbis, 330
Plant and microbial tufa/travertine, 202
plant formations, 356
plant roots, 232
plants, 178, 229
plastic deformation, 59
Platy horizon, 243
playa lakes, 250, 349
Plitvice Natural Park, 159
Pomacea sp., 334
Ponded areas, 140–141
ponds, 103, 113
porosity, 148
precession cycles, 122
primary
  dolomite, 13
  lacustrine deposit, 104
  structures, 241
Prior calcite precipitation (PCP), 304
prismatic horizon, 241
prismatic structures, 109
proboscideans, 354
profiles, 227
profundal zone, 20
progradation, 149
protodolomite, 107
pseudomicrokarst, 109, 111, 347
Psilonichnus, 353
  ichnofacies, 359
pupation chambers, 332
Purgatoire Valley, 348
Pyramid Lake, 56

Quaternary, 135, 150
Quaternary calcretes, 248

R. sciuttoi, 328
Radiocarbon dating, 248
Radiodiscus, 335
Raft tufa/travertine, 205
rainfall, 229
Ranu Lamongan, 69
Rays, 297
Rebuffoichnus, 320
recent fresh-water lakes, 107
recrystallization, 24, 209
reefs, 67
residence time of water, 162
retrogradation, 149
reworked calcrete, 252
reworking, 114, 247
rhizoconcretions, 349
rhizogenic calcretes, 244
Rhizoliths, 235, 241, 352
rhizosphere, 109
Rhodes, 353
rhodoids, 203
rhythmites, 26
rich-organic layers, 143
rift, 253
Rigalites, 350
river banks, 137
rodent tracks, 353
root
  cavities, 108
  cracks, 111
  cretes, 244
  hairs, 234
  systems, 243
  traces, 347
Rosado Member of, 321
Rosellichnus, 320
Rotadiscus, 335
Rubielos de Mora Basin, 59
Rudstone
  of coated plants, 137
  of oncoids, 137, 139, 143
  of phytoclasts, 138, 143
  of stems, 153
Ruidera pool system, 150, 154, 159
Rutichnus, 349

Sahel, 18
saline mudflats, 352
sandstones, 142
sapropel, 154
Sarliève lake, 13
Sarmiento Formation, 321
saturation index, 136
sauropod, 348
  tracks, 358
scandulitic dendrites, 207
Scots Bay Formation, 53
Scoyenia, 320
  beerboweri, 321
  ichnofacies, 354
seasonal cycles, 147
seasonal wetlands, 116
seasonality, 119

sediment input, 120
sediment supply, 119
sedimentary facies models, 135, 157, 160
Sedimentary Facies, 137
Sedimentation rates, 136, 148
sedimentation, 250
Seilacherian ichnofacies, 354
sequence
  boundaries, 256
  of facies, 149
  stratigraphy, 120, 252
Serranía de Cuenca Basin, 37
shallow phreatic zone, 226
shelfstones, 310
shorebird
  footprints, 358
  ichnofacies, 358
  tracks, 350
shrinkage cracks, 233
Shrub tufa/travertine, 203
Shulu Sag, 43
Shuttle Meadow Formation, 43
siliciclastic sedimentation, 157
Silicification, 24
skeletal crystals, 190
Skolithos helicoidalis, 328
Skolithos linearis, 334
sloped cascades, 152
small jumps, 137
snails, 354
soda straws, 279
soil classification, 228
soil productivity, 22
sources of the carbonate, 123
sparitic speleothems, 276
sparmicritization, 148
Speleothem, 142
Spherulites, 235
spray areas, 141
spray zones, 159
springs, 116, 155, 177
$^{87}Sr/^{86}Sr$, 230
Sr isotopes, 9
Stable carbon isotopes, 165–166
Stages I–III, 239
stages of calcrete development, 245
stagnant, 141
Stalactite tufa and travertine, 204
stalactites, 152, 276
stalagmites, 276
Stephadiscus, 335
Stephanoda, 335
stepped fluvial conditions, 155
stromatolite facies, 137
stromatolites, 56, 139, 143, 147, 244, 349
strontium, 24
subaerial exposure, 105
Sublacustrine Calcareous Spring Deposits, 210
subsidence, 121, 250
Sucker Lake, 57
supply of carbonate, 230
surface texture, 334
sweat bee nest, 328
symmetrical basins, 253
synchronous crystallization, 290
synchrotron radiation, 307
systems tract, 252

Taenidium, 349
tectonic activity, 250
tectonics, 135, 168
tectonism, 120
Teisseirei, 320
Teisseirei barattinia, 334
termite nests, 355
Termitichnus, 329
Termitichnus ichnofacies, 354
terrestrial settings, 330
terrestrial vertebrates, 328
Teruel Basin, 329
thermal springs, 182, 183
theropod tracks, 348, 358
topographic breaks, 158
topography, 136, 168
trace-fossil, 320
trace-fossil associations, 356
trampled ichnofabric, 353
travertines, 113, 185, 280, 354
trichopterans, 354
trigonal dendrite crystals, 207
tufa, 113, 134, 185, 354
tuffaceous marl, 332
turbidites, 49, 347
turbidity currents, 358

unconformities, 256
underfilled basins, 120
Upper Freeport Limestone, 37
upstream dammed zones, 153
Uruguay, 321, 330
U–Th series, 249

vadose, 108
vadose silt, 106
vapour–liquid–solid (VLS), 284
Vârtop cave, 354
varves, 26
vegetal debris, 141
vegetation cover, 107, 229
vertebrate ichnofacies, 355
vertebrate tracks, 347
Vertical waterfalls, 141
vitric tuffs, 332